T0256973

MARKET DRIVEN ENTERPRISE

MARKET DRIVEN ENTERPRISE

Product Development, Supply Chains, and Manufacturing

Amiya K. Chakravarty

Tulane University

A. B. Freeman School of Business

New Orleans, LA

JOHN WILEY & SONS, INC.

New York • Chichester • Weinheim • Brisbane • Singapore • Toronto

To Indira—
a serendipitous meeting of a lifetime. You color my world with sunshine.

To Deepak and Nisha—
thinking, striving, learning, and conquering challenges:
our pride and joy.

To Maa—
sculptress par excellence of value and dignity.

This publication is designed to provide accurate and authoritative information in regard to the subject matter covered. It is sold with the understanding that the publisher is not engaged in rendering professional services. If professional advice or other expert assistance is required, the services of a competent professional person should be sought.

Library of Congress Cataloging-in-Publication Data:

Chakravarty, Amiya K.
    Market driven enterprise : products, supply chains, and manufacturing / Amiya K. Chakravarty.
        p.   cm.
    Includes bibliographical references and index.
    ISBN 0-471-24492-9 (cloth : alk. paper)
        1. Business logistics.   2. Product management—Technological innovations.
    3. Design, Industrial—Management.   4. Manufacturing industries—Technological
    innovations—Management.   5. Marketing—Management.   I. Title.

HD 38.5 .C38 2001
658.5—dc21   00-043694

10   9   8   7   6   5   4   3   2   1

CONTENTS

Foreword   viii
Preface   xi
Acknowledgments   xiv

Part I: Interfaces and Decisions in an Enterprise

Chapter 1:   Domain and Process Views of an Enterprise   3
    1.1   Introduction   3
    1.2   Domain View   4
    1.3   Process View   15
    1.4   Decisions and Information Flow   23
    1.5   Dynamic Enterprise   27
    References   30

Chapter 2:   Manufacturing Marketing Interface   33
    2.1   Product Definition and Process Selection   33
    2.2   Inter-Domain Linkages   36
    2.3   Interface Sets   45
    2.4   Distributed Coordination   48
    2.5   Information Flows for Coordination   61
    References   63

Chapter 3:   Knowledge Organization for Domain Decisions   64
    3.1   Information for Decisions   64
    3.2   Domain Knowledge   66
    3.3   Linking Nodes in a Chain   71
    3.4   Semi-structured Decisions   75
    3.5   Relationship Chain Construction   81
    3.6   Knowledge Modularization   85
    3.7   Enterprise Resource Planning   88
    3.8   Implementation   92
    References   94

Part II: Product Design and Time-to-Market

Chapter 4:   Marketing Approaches to Product Design   99
    4.1   Product Positioning   100
    4.2   Discrete Choice Models   114
    4.3   Conjoint Analysis   121
    4.4   Deterministic Models   127

v

4.5   Product Ranking   129
4.6   Mass Customization   132
References   135

Chapter 5:   **Design Engineering**   **136**
5.1   The Design Process   136
5.2   Conceptual Design   137
5.3   Form Design   145
5.4   Designing Tolerances   150
5.5   Design Sequence   157
5.6   Cost of Design   162
References   169

Chapter 6:   **Concurrent Mapping of Product Features**   **170**
6.1   Relevance of Mapping   170
6.2   Quality Function Deployment   171
6.3   Mathematical Modeling of Quality Function
      Deployment   176
6.4   Modeling Design Attribute Interactions   182
6.5   Pattern Recognition Model of Mapping   186
6.6   Managing the Mapping Process   188
6.7   Concurrency in Mapping   197
References   200

Chapter 7:   **Product Platform and Variety**   **201**
7.1   Scope of Variety   201
7.2   Role of Product Variety   203
7.3   Aggregate Design Benchmark   205
7.4   Product Platform Concepts   208
7.5   Platform Model for Product Breadth   221
7.6   Technology Driven Products   230
7.7   Manufacturing Cost of Variety   235
7.8   A Strategy for Managing Variety   239
References   241

Chapter 8:   **Product Realization**   **243**
8.1   Product Realization Factors   243
8.2   Product Definition   245
8.3   Product Launch Date   254
8.4   Technical and Market Uncertainties   263
8.5   Overlapping Development Tasks   271
8.6   Design Factory   286
References   295
Appendix   298

Part III: Supply Chains and Responsive Manufacturing

Chapter 9:   **The Extended Enterprise:**
             A Supply Chain Perspective   **303**
9.1   Leveraging the Value Chain   303
9.2   Product and Supply Chain Architectures   306

9.3 Product and Supply Chain Business Models   308
9.4 Supply Chain Dynamics   315
9.5 Capabilities in the Chain   320
9.6 Ownership and Coordination   333
9.7 Supply Chain Strategies   340
References   343

Chapter 10: Electronic Chains of Suppliers and Customers   345
10.1 Dimensions of Differentiation   345
10.2 Digital Connectivity   346
10.3 New Business Models   350
10.4 Digital Value   355
10.5 Reengineering Ownership in the Value Chain   361
10.6 Electronic Business Communities   375
10.7 Virtual Enterprise   384
10.8 Intelligent Agents   391
References   396
Appendix   398

Chapter 11: Supply Chain Models   402
11.1 Scope of Quantitative Modeling   402
11.2 Supply Chain Configuration   403
11.3 Supply Chain Platform   410
11.4 Coordination in a Supply Chain   417
11.5 Impact of Uncertainty Reduction   422
11.6 Sharing Inventory in a Network   429
11.7 Supply Contract   436
11.8 Supply Chain Accounting   445
References   448

Chapter 12: Responsive Manufacturing   451
12.1 Paradigm Shift   451
12.2 Product-Process Interactions   455
12.3 Acquisition of Advanced Manufacturing
      Technology   460
12.4 Manufacturing Capability and Flexibility   469
12.5 Real-Time Flexibility   476
12.6 Product Customization Through Process Design   485
12.7 New Product Phase-In   489
12.8 Proactive Response to Product Promotions   493
12.9 Production on Demand   498
References   505

Subject Index   509

Author Index   513

FOREWORD

Professor Amiya K. Chakravarty of the A.B. Freeman School of Business at Tulane University has written a remarkable book. I describe this book as remarkable for at least two reasons. The first is currency. The book's focus is on today's news. It is not, as is so often the case with scholarly work, related to yesterday's problems. The second reason is that in addition to being timely, or perhaps because of it, the approach is courageous. To explain both of these points, let me note that the overall goal of the book is to provide insights into reorganizing the way business is conducted in response to the new e-forces that are driving change. Insights are not prescriptions. Fortunately, the author is willing to share with us his ideas about how to model the dynamic forces that are evolving.

Dr. Chakravarty sandwiches marketing (which is the driver) between organizational knowledge and information systems on the one hand, and supply chain management on the other. *Market Driven Enterprise* is not a name chosen carelessly. New markets demand goods and services in new ways using electronic means to communicate these demands. This is an e-shift that forces firms to change their character. Organizations that cannot learn to cope with the need for speed (order processing and delivery, new product development, etc.), and multitudinous demands for products and services (from anywhere at anytime) probably will not survive. They surely will not thrive.

Part I of this book starts with traditional structures of organizations and then modifies their design for survival in the new economy. Three chapters attend to major shifts that are required in how information is coded, sent, received, sorted, combined and evaluated, and then acted upon. Part I develops the fundamental concepts of "domain and process architectures" of enterprises. Important attention is paid to ideas about global and local knowledge; about social (or explicit) and tacit knowledge; and about how information must flow to linked domains. Emphasis on knowledge management capabilities for e-business is correct and appropriate. There is no doubt that it will play an increasingly important role in reducing response time for problem-solving and decision-making. Professor Chakravarty presents his ideas about how to treat organizational knowledge in a way that permits scholars to debate the pros and cons of various approaches.

Part II, "Product Design and Time-to-Market," ranges far and wide, touching on marketplace phenomena and market research methods for quickly ascertaining what customers want. The reader encounters solid discussions of conjoint analysis and mapping product features. Both of these are nominally market research tools. Quality from the production point of view is related to the marketing viewpoint using quality function deployment (QFD—widely used in Japan) and the House of Quality (widely used in the U.S.). These mixed-function (marketing-production) tools permit broad systems analyses to discern and react to *the voice of the customer.* Chapter 5, "Design Engineering," relates product design and production tolerances. The issue of tolerances is too often overlooked in dealing with the marketing perception of optimal product design. *Market Driven Enterprise* reveals how eclectic an approach is required to achieve an integrated, competitive system. Tolerances are more readily associated with an educated engineering point of view rather than with marketing. This book will appeal to engineers, but it will require them to learn a great deal about issues that have never been part of the traditional field of industrial engineering. If this text was traditional, it would not contribute to survivability rules in the new economy.

Part III consists of 4 chapters that get into the business of emerging digital supply chains. The operational and managerial issues in evolving electronic business communities are the heart of successfully navigating the e-shift. Chapter 9, "The Extended Enterprise: A Supply Chain Perspective," starts the ball rolling. Business to business (B2B) commerce is no longer an aggregate of isolated input-output entities. Each firm is a member of a chain of interdependent players that stretches around the globe 24/7/52. That is, a system that is never turned off. Leveraging the value chain of suppliers at one end and customers at the other requires understanding the properties of supply chains as well as the dynamic forms of electronic communications used to integrate them. Dr. Chakravarty's treatment of the supply chain is stimulating. It probes the many aspects of this complex system of interactions using qualitative and quantitative modeling as befits the subject.

This book highlights the proper issues to be considered. It provides answers that provoke thoughtful review. Chapter by chapter it offers insights concerning opportunities for research. This book suggests structures that will keep Ph.D. candidates busy for a long time to come. Perhaps that is the greatest tribute that can be paid to a book.

Market Driven Enterprise is a remarkable compilation of materials that impinge on the competitive capabilities of rapid-response organizations of the millennium. There are numerous anecdotes and facts about actual companies in real time. The reference list is a pleasure to behold.

Martin K. Starr
Professor Emeritus, Columbia University, New York
Professor, Rollins College, Orlando, Florida

PREFACE

A market-driven enterprise must react quickly to changes in customer preferences, technology, and competition. It can reposition itself in its value chain or reengineer those of its business processes that may impact flows of goods, information, and funds. These options require the company to be agile and nimble, like a fruit fly. For long-term viability, however, it must also be capable of mobilizing the mass of a dinosaur. It is clear that a company can leverage digital access to its suppliers and customers to substantially increase its agility. At the same time, by positioning itself appropriately in a business community, it can bring the mass of the entire community to bear upon a predator. It can achieve this with a low financial burden, if ownership of the value chain is shared with many partners. The company can thus restructure its business processes quickly and modify its collaborative arrangements with its partners so as to minimize handoff inefficiency.

There are many interface issues in a value chain linking product development to supply chains and manufacturing. For example, an attractive product feature in automobiles such as rapid acceleration may increase sales, but it may also come with a big increase in manufacturing cost. The cost of increasing market share, however, can be prohibitively high, as preferences for product features may vary among customers. Other examples of synchronization would be aligning product promotions with the production schedule, moving products from supplier to supplier as components are assembled, and delaying product differentiation. It is therefore clear that a realistic discussion of interfacing and/or alignment must include issues in product development, supply chain management, and manufacturing. Many companies have begun addressing product development and supply chain issues, albeit in a piecemeal fashion. These approaches tend to be static and specific to a company's business environment. Companies are finding that even with such quick-and-dirty approaches, they can generate substantial savings, and this has led to a rapid growth in interest in supply chain models. Issues such as reacting to market forces through agility in supply chains, however, have largely remained unaddressed. Additionally, the roles of product development and manufacturing in enhancing this agility are not well understood. While business-to-business (B2B) e-commerce is flourishing, there have not been consistent attempts to incorporate B2B in the business strategy of an extended enterprise.

I intend to fill this gap by providing state-of-the-art concepts, models, and frameworks that help in understanding how an adequate response to market forces can be shaped. The discussions have been motivated by industry practices, and in particular by the digital supply chains of a few leading companies. Real-world examples and cases, where appropriate, have been included. To motivate research in some new areas, I have included discussion of a few select mathematical models. These models include product platform, product launch, supply chain coordination, and coordination with manufacturing.

I have organized the 12 chapters of the book into three sections, covering interfaces and decisions in an enterprise, product development, and responsive supply chains. Part I examines how operational decisions are made in an enterprise. This starts, in Chapter 1, from an exploration of the domain and process architectures of enterprise in the context of global and local knowledge, enterprise architecture, and virtual organization. In Chapter 2 I study domain interfaces in greater depth in terms of inter-domain alignment and business processes pertaining to them, and I develop a framework to facilitate measurement of alignment in distributed coordination. In Chapter 3, I elaborate how domains and processes can be linked by information flows and organize the flows for building modular knowledge appropriate for modular supply chains.

Part II is where I discuss both marketing and engineering approaches for product design and time to market. In Chapter 4 I survey existing approaches in the marketing domain for product positioning. Conjoint analysis and mass customization are discussed in depth. The engineering approach to product design, involving concept and form design, tolerances in product specification, and design task sequencing, are elaborated in Chapter 5. Concurrent mapping of product features, which builds upon the discussion in Chapter 2, is examined in Chapter 6. Mathematical models of quality function deployment, including the pattern recognition model of mapping, are emphasized. Chapter 7 discusses the importance of product variety and design benchmarks in different industry settings.

Mathematical models for designing product breadth and depth for both consumer-driven and technology-driven products are developed. The strategic dimensions of product variety in relation to product, customer, and system economics are discussed. In Chapter 8, I discuss how a product can be realized and brought to market quickly. The trade-off among product performance, early freezing of design, and launch date uncertainty are modeled analytically. Overlapping of development tasks in different scenarios is discussed, and models for a better understanding of design productivity, design reviews, and resource utilization are explored.

In Part III I discuss digital supply chains and responsive manufacturing. In Chapter 9 I elucidate the concept of extended enterprise from a supply chain perspective, using different business models. The implications for coordination issues of fragmentation of ownership are explained. In Chapter 10 I extend these concepts to digitally connected supply chains. How digital connectivity may add value is explained using industry examples. Value chain reengineering from perspectives of companies such as Cisco Systems and Dell Computer are discussed. Managerial issues in emerging electronic business communities and their governance are outlined. In Chapter 11 I discuss several quantitative supply chain models from both industry and academic perspectives. Supply chain configuration and supply chain platform are discussed in terms of the entire network, comprising suppliers, manufacturers, distributors, retailers, and customers. The discussion of supply chain coordination is focused at reducing handoff inefficiencies through resource sharing, incentives, and contracts, in presence of uncertainty. Changes in manufacturing operations brought about by new business models (digital connectivity, and supply chain) are analyzed in Chapter 12 through product-process interactions. Impact of market forces on technology choice, new products, process design, and manufacturing costs is modeled. Models of real-time flexibility, setup cost reduction, capacity mix planning, product promotion, and on-demand production are studied.

ACKNOWLEDGMENTS

A book of this nature is not possible without the support of a large number of individuals. Foremost among them are my research colleagues from several universities and companies, who helped shape many of my ideas in their early stages. This is a long list but some stand out in my mind: Sherwin Dahn, Abraham Seidmann, Martin Starr, Sanjoy Ghose, Ricardo Ernst, Panos Kouvelis, and Nagraj Balakrishnan. I am also indebted to the company executives from General Motors, Ford Motor Company, Lucent Technologies, Allen-Bradley (Rockwell group), Kerney & Trecker, and Laitram Corporation, where I have made presentations and/or have participated in disucssions on many topics included in this book. Professional societies including INFORMS and POMS deserve special mention for providing the forum for panel discussions, special sessions, and tutorials on these topics where my thoughts were crystallized further.

I am thankful for support from the A. B. Freeman School of Business at Tulane University through the Seinsheimer endowment, the Goldring Institute, and through sabbatical leave. To get it off the ground, however, it needed a boost from an industry host who provided valuable "laboratory" for experimenting with ideas.

I am fortunate to have worked closely with former students, Bijay Naik, Jon Baum, Mikko Tarkkala, Noah Flom, and Ravi Sonnad, who contributed greatly with valuable insights. Current students, Jarkko Nurminen and Marketa Sykarova, deserve special mention for editorial work, and for sounding out ideas. I remain grateful to Rhonda Butler for important administrative support over the years that has allowed me to devote more time to this book.

The John Wiley team was superb. Bob Argentieri believed in this project from the start. Maury Botton kept the project on schedule. Stacey Rympa was the savior whenever I needed help. Sue Warga improved the text tremendously. Paul DiNovo did a superb job with cover design. Anne Tull Gudovitz brought the book to market.

I could not, of course, have written this book without the help, support, patience and understanding of my family. They are the ones who lived through the difficult months in the "trenches." I remain indebted forever.

I

Interfaces and Decisions in an Enterprise

Structuring the value-adding activities that make an organization competitive in the dimensions of innovation, rapid-response, scope, and quality. Understanding the roles of information and material-flow in creating interfaces among functions in terms of decentralized coordination.

1

Domain and Process Views of an Enterprise

1.1 INTRODUCTION

How a business unit is structured and organized has major implications for its competitiveness. Competitive advantage primarily comprises two elements: the values a company can create for its customers (in excess of the cost of creating the value) and innovations. While value to the customer creates an immediate competitive advantage to the firm, innovations provide sustenance to this advantage. Value created for a customer is a multidimensional entity comprising price, variety, quality, and response time, among other factors.

To add value to a product, it must be put through a series of activities (process), from acquisition of raw material to distribution of the product to retail stores. The sequence of activities has variously been called a value chain (Porter 1985), commercial chain (Hayes and Wheelwright 1984), and supply chain. While the models differ in detail, all of the above chains advocate clusters of activities, usually called functions, joined together by a linking mechanism. We call this the domain view of the enterprise, where each function is an individual domain. Proponents of business process reengineering (BPR) strongly argue for a process view of the enterprise, where the emphasis is on individual processes rather than on functions (Davenport 1993). This is also known as the horizontal view.

1.2 DOMAIN VIEW

The domain view, where activities are organized into functions, is the more traditional view. This view, in all likelihood, has evolved from the need to differentiate one function from another because of their inherent dissimilarities. In the marketing and manufacturing domains the factors causing differentiation would be expectations (profit and sales vs. cost and timely production), complexities (capacity utilization and sales forecast vs. cost of capacity), managerial orientation (customer focus vs. plant focus), and organizational culture (impressionist vs. craftsman). Thus the nature of tasks in any two functions may require totally different sets of skills, resources, and organization culture. As the company grows, these functions evolve to assume their own distinct identities. Managers in the manufacturing domain, for example, may know nothing about brand management, advertising, promotion, sales, channels, and pricing. Marketing managers likewise may be ignorant of manufacturing technology, process plans, machine capabilities, maintenance, and in-process inventory.

Such differences may become a major impediment to the firm in achieving competitive advantage. If, for example, customers demand greater product variety and/or more frequent delivery, the marketing domain would need manufacturing to agree to broaden product mix and/or to implement a just-in-time (JIT) type of production schedule. In the absence of meaningful linkages among functions, such problems become major sources of conflict. Reducing the number of domains would be one way of minimizing interdomain conflict.

Malone and Rockart (1991) report how Frito-Lay reduced organization layers by using technology to increase coordination. With a sales force of 10,000, a hierarchical organization structure would have required many layers to enable communication and coordination between the salespeople and the management. Instead, each salesperson was given a handheld computer to record sales data on 200 grocery products. The sales data is transmitted daily to a central computer, which sends back information on changes in product pricing and promotion to the salespeople. A similar application of computer technology (laptops) was reported by Hewlett-Packard, which used it to provide information to salespeople during customer meetings (Berger, Angiolillo, and Mason 1987). Time spent in meetings decreased by 46%, travel time was cut by 13%, and sales rose by 10%.

The linkages between any two functions are meaningful if they lead to a maximization of the surplus for the two units together. Consider again marketing and manufacturing functions. Let the total revenue from the current sales of products be denoted by R_0, and the corresponding manufacturing cost by M_0, so that the surplus is $R_0 - M_0$. Assume now that the revenue can be increased by R units by providing a greater product variety to the customers, but that the corresponding increase in manufacturing cost would be M. Let M and R vary with product variety, as shown in Figure 1.1.

It is obvious that the marketing function would like to increase the value of R up to R_C, corresponding to the product variety denoted by point C in Figure 1.1, and the manufacturing function would like to hold the value of M at M_0, at point A. A meaningful relationship, on the other hand, would assess the shapes of the M and R curves and establish that the maximum surplus of $R_0 + R_B - (M_0 + M_B)$ is obtained at point B. Assessment

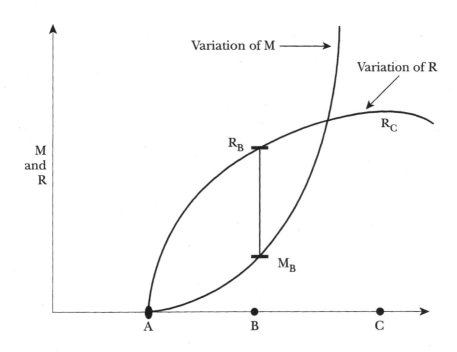

FIGURE 1.1: VARIATION OF REVENUE AND COST

of the shapes of R and M curves becomes a harder proposition when other dimensions of competitive advantage such as quality, delivery lead time, and life cycle are included in this consideration.

It is clear that although a linkage between marketing and manufacturing functions maximizes the firm's surplus (at point B), both marketing and manufacturing incur costs. The cost to manufacturing is $M_B - M_0$, and the cost to marketing is $R_C - R_B$. Porter (1985) discusses several types of costs in creating relationships between different business units. As pointed out above, similar costs are also incurred by different functions within a business unit. These costs, as in Porter, can be summarized as cost of compromise, cost of coordination, and cost of inflexibility.

Cost of compromise results when neither function in a linkage is allowed to achieve its own optimum. In Figure 1.1 we observed how both marketing and manufacturing functions incur cost when required to compromise on product variety. Consider a second example, delivery lead time. Increasing the frequency of product delivery to the customer decreases the customer's inventory-related costs. The customer would therefore be willing to pay more for the product and/or buy more annually. The potential of increased revenue would motivate the marketing function to increase delivery frequency until the marginal marketing surplus becomes negative (similar to point C in Figure 1.1). The increase in delivery frequency, however, would increase the annual cost of machine setups, or the manufacturer would have to invest heavily in new processes and technology to implement a JIT-type production schedule. Through an appropriate form of linkage, a compromise on frequency of delivery can be reached (similar to point B in Figure 1.1) that would have cost implications for marketing (as point B is lower than point C) and for manufacturing (as point B is higher than point A). Another example of compromise would be in planning product promotions and production scheduling. Marketing functions may carry out special promotions of a few select products that may be under threat from competitors. Most such promotions are time bound projects in which sales increase significantly but with only a moderate increase in net revenues. After the promotion ends, sales are scaled back but (typically) not as far back as before the promotion, creating a longer-term advantage in revenue. The sudden short-term surge in sales, before reaching a new steady state, requires manufacturing to build up inventory of the product or to contract for additional manufacturing capacity. This may be very costly to the firm. There will therefore be a compromise that would constrain marketing in terms of the frequency of promotions and require manufacturing to incur additional cost to satisfy the periodic surges in sales.

Cost of coordination results from the need to have the parts of the linked system perform in unison. Consider the case of product modification. From time to time marketing may come up with requests for product modifications, based on feedback from customers. Such a modified product (since it is initiated by the customers) has a high demand and hence a high revenue potential. A modified product differs from the original product in only a few aspects. Many of the components between the modified and old products may be common, but there may also exist a significant number of components that are not common. An instant and total switch to the modified product would cause major upheavals for manufacturing for several reasons. First, a product is usually committed to customers for several periods in advance, and they may not always agree to switch to the modified product. Second, since commitments are made for component production and/or purchase several periods in advance of assembly of the product, an instant switch to the modified product would make the noncommon components obsolete, or a heavy penalty would have to be paid to the suppliers for contract violation. To minimize such coordination-related costs, the manufacturing function may prefer to switch to the modified product in stages, gradually reducing the quantity of the old product and increasing the quantity of the modified product. These quantities need to be phased in carefully by reassigning the common components to the modified product and minimizing the obsolescence of the noncommon components.

Boeing faced a similar problem in the early eighties (Garvin 1991) when it planned to modify its 767 aircraft design from a three-person cockpit to a two-person cockpit, preferred by some airlines. For the 30 aircraft that would be affected, Boeing estimated that it would require 2 million hours of additional labor. The redesign also required over 12,000 modifications related to seating arrangement, carpet color, and wiring and part changes. Such modifications needed to be phased in because of their technological dependencies. Though the costs of coordination and new material were significant, the potential for long-term revenue increase was also very high.

Finally, cost of inflexibility results from the fact that the cost of making changes in one function will be very high if that function is coupled tightly with several others. The impact of changes in one function ripple through other functions and may become amplified in the process. Toyota's supply chain is based on small "mom-and-pop" suppliers several layers down the chain (supplier's supplier's supplier, etc.). Even a small change in the final product can cause havoc for these tiny suppliers, who may depend on Toyota's business for almost 100% of their operations.

It should be clear from the foregoing discussion that well-designed linkages help interface one domain (manufacturing) with another (marketing). It is also obvious that not all activities in marketing need interfacing with manufacturing. For example, shelf management at a retail store (for the firm's products) may not have a significant effect on how the products are manufactured. Similarly, whether or not computer-aided design (CAD) is used for engineering design may not have a significant effect on the way distribution channels are managed by marketing. There is, however, a set of activities that significantly impact both manufacturing and marketing. We examine some of them to understand the nature of interface required. Shapiro (1977) and Montgomery and Hausman (1986) have also outlined the nature of this conflict.

Activities at the Interface

The activities that appear to have most impact across the interface between marketing and manufacturing are product design, quality assurance, demand and capacity management, inventory holding, supply chain, production scheduling, and costing.

Product Design

Because of the increasing importance of two factors—decreasing product life cycles and shortening time to market—product design is assuming strategic importance. It is a given that marketing people would want products to mean all things to all people (Shapiro 1977). However, since customer preferences and customer ability to pay vary widely, and since customers appreciate variety, with emphasis on new products, it may not be cost-effective to try to satisfy all customers in all market segments at all times. Increasing product variety requires a larger number of unique components to be designed and manufactured, increasing cost. Modular product design increases component sharing, but it may also reduce product variety to a less-than-desirable level for customers. Flexible manufacturing equipment can produce a wide range of components on the same machine, but the cost of designing and implementing a flexible manufacturing system (including personnel training) may be very high. Similarly, to reduce the time to market, product development time must be reduced. One way of doing that is to overlap (in parallel) development activities such as prototyping and testing, which would otherwise be done sequentially. Overlapping such activities can be very risky, however, as a design error found in one test may require all other tests, done in parallel, to be repeated.

An application of product design to reduce time to market of single-use 35 mm cameras is reported by Kodak, who used a well-structured database and a computer-aided procedure to frequently exchange design drawings among different functions (Davenport 1993). This transformed a sequential design process into one where components could be designed in parallel.

Quality Assurance

There are several dimensions of product quality (Garvin 1984), but the two major dimensions are conformance quality and performance quality. Market share can be increased (albeit in different segments) by improving conformance quality (practiced by Japanese companies) or by strengthening performance quality (practiced by German companies such as BMW). To enhance performance, it is crucial that emphasis be placed on technology innovation and its incorporation in product design. This requires product designs to be modified as and when new technology appears. In the case of BMW (Pisano 1996) this has meant that product designs could not be frozen even at advanced prototyping stages, and so expensive modifications in manufacturing processes were required even during production ramp-up. This ensures the latest technology in products, but the time to market may become long and uncertain, and the cost of product development could be high. Japanese companies, on the other hand, aim at conformance quality. They meticulously practice freezing designs at a certain point, so any new technology innovations beyond those time fences are left to be incorporated in a future modification. During prototyping they lay emphasis on process simplification, appropriate material use, and so on, so that conformance quality can be improved. The strategy (performance or conformance) to be followed obviously depends on the market segments the products are targeted to.

Linking product warranty with product quality is a vexing issue. Product warranty may be perceived as a substitute for product quality by many customers. The trade-off between the cost of servicing a product warranty and the cost of improving product quality must therefore be incorporated in product design and coordinated with the design specifications of the warranty (such as length of warranty for each component or assembly covered, group warranty, and encouraging customer maintenance of products).

Demand and Production Capacity

Uneven sales from one period to another are a fact of life in marketing. If production quantities were synchronized exactly with sales, manufacturing would have a severe capacity management problem on its hands. Capacity cannot be increased at short notice, as construction of new buildings, acquisition and commissioning of complex machinery, and training of personnel take a considerable amount of time.

The type of capacity to be added and its location will have a profound impact on competitiveness. The facility can be focused or it can be flexible. It can be automated or it can be manual. Chakravarty (1987) describes possible interactions of facility types with market dimensions (related to customer service) and control dimensions (related to software control) and suggests possible paths of upgrading the facility as market conditions change. An automated mass-production assembly line is an extreme case of facility focus. This would be appropriate for cost-based competition in a mature industry such as textiles or paper. For manufacturers in industries such as electronics and auto, a flexible manufacturing system (FMS) would be very desirable. Allis Chalmers Company of Milwaukee was perhaps the first to implement a flexible manufacturing system, in the 1970s. However, the farm equipment industry at that time was stagnant and could not sustain product innovations. Allis Chalmers was unable to use its flexibility for market advantage, and the company folded in the mid 1980s, as the investment in its FMS was very high. Allis Chalmers' market analysis, obviously, was far from satisfactory. Many companies in the United States have rushed headlong to invest in FMS but have ended up using it for dedicated mass production with disastrous results (Jaikumar 1986). In Japan, on the other hand, companies have carefully acquired FMS and have exploited its flexibility to maximum advantage. Naik and Chakravarty (1992) discuss a framework for linking the strategic priorities of the company (low cost, product differentiation, mass customization, etc.) to the choice of a manufacturing system. They use an adaptation of a technique called quality function deployment (Hauser and Clausing 1988), to create multidimensional linkages layer by layer (for several layers), and use a qualitative model for choosing an appropriate set of linkages.

Demand management, where customer demands are shifted from one time period to another, using financial or other incentives is another way of reducing cost of capacity. Service industries such as electrical utilities, telephone companies, airlines, and doctors' clinics are prime examples. This approach can be tried when providing flexibility, in a manufacturing or service capacity, is either not feasible or is prohibitively expensive. It will obviously be optimal to try to use a mix of capacity flexibility and demand management.

Outsourcing manufacturing or service is a third approach that has become popular, as it frees the company from the headaches of managing a production facility (especially for those companies that do not possess core competency in the production function). The third-party service, provided by firms that combine orders from different companies, can better exploit economies of scale. While the cost advantage of outsourcing is not in doubt, the optimal level of outsourcing may not be apparent. Hayes and Wheelwright (1984) address the issue by asking where on the commercial chain a company should position itself. A simplistic response would obviously be to retain core competencies and outsource the rest. But what about developing new competencies in view of market opportunities (Hamel and Prahalad 1989)? The point is that outsourcing policy should be carefully examined in terms of the company's core competencies and emerging market opportunities.

Inventory

Inventories build up in a system for two main reasons: fluctuations in customer demand and cost of setups in the system. Substantial labor and material costs are involved in changing over a machine to produce a different product. A setup cost is also incurred in preparing and processing a purchase order and inspecting the goods delivered.

If setup costs in the system are high, the number of times (per week) a machine is set up for a product change and/or the number of purchase orders (per week) will have to be kept low. The production quantity per setup and/or the quantity per purchase order will have to be large to cover demand for the longer duration between setups. This will increase the average inventory (cycle inventory) carried in the system. Without sufficient inventory in the system, customers would experience shortages toward the end of a production cycle. Reduction in setup time, therefore, is an important issue in reducing the conflict between manufacturing and marketing and propelling the system toward JIT.

To safeguard against fluctuations in demand, most firms like to carry a safety stock over and above the cycle inventory as described above. The size of the safety stock will, of course, depend upon the extent of uncertainty in demand and the degree of customer service planned. Note that if production and/or purchases are done just in time, both cycle and safety inventories will be driven to zero simultaneously. This may not be achievable, however, for several reasons. If demand fluctuations are high, the cost of capacity fluctuation induced by JIT may be too high. To avoid up-and-down capacity variations, most companies that have implemented JIT allow excess capacities. This, in effect, substitutes the cost of excess capacity for the cost of safety inventory. Before implementing JIT, therefore, it is necessary to reduce setup time and cost by completely reengineering the setup process (Monden 1983), which may also require investment in new technology used in the setup process, production process, and product design. Such investments can be substantial, as experienced by companies such as Harley-Davidson and Johnson Controls (Mishina 1993) in their switch to JIT from MRP (materials requirement planning). Investing in a flexible manufacturing system is an alternative to reengineering the setup process, as in a flexible system manufacturing can be switched from one product to another without a substantial setup cost. Irrespective of whether the setup process is reengineered or flexible manufacturing is used, it may never be possible to drive the setup cost to absolute zero. Hence inventory carried in the system may not be totally eliminated, as evidenced by the number of kanban cards (>1) circulating in any JIT system. For this reason, Japanese companies such as Toyota try to maintain level production but use a mix of model types, carefully determined, to match variations in demand (Miltenburg 1989).

Product family is a concept that can be used to reduce setups in the system. Products that are similar in terms of setup needs such as tool requirements, unit costs, and demand rates are grouped together. The entire group of products is manufactured together without requiring a major setup of the machine (Goyal 1974; Chakravarty 1984a). Chakravarty (1984b) shows how such groups can be used to obtain price discounts based on the total value of a purchase. For products with demand fluctuations, however, determination of such groups becomes very complex.

Another issue is the trade-off between carrying cost and lead time. When sales are uneven, sales personnel would like to carry an inventory of finished goods to maximize the probability of making a delivery to a customer on time. Because of the value-added effect, which can be substantial in a high-tech or complex product, manufacturers like to hold inventories of raw material and components rather than finished goods. While this reduces inventory-carrying costs, customer service may suffer, as the length of manufacturing lead time, from component manufacture to final assembly, may be substantial. A mixed strategy of holding inventory at different levels of the bill of material (BOM) and using some manufacturing flexibility may be optimum.

Supply Chain

If customers require rapid response for a wide variety of products, the effectiveness of a supply chain from plant to customer becomes very important. Major issues are configuration of the supply chain and structure of the supply contract. While a centralized manufacturing plant requires the least investment, it may also be the most unreliable from the point of delivering goods on time to different locations, and the cost of distribution can be extremely high. How many plants to have and where to locate them, obviously, would depend upon the structure of the supply contract. In designing such a contract, the trade-offs among delivery-time window, unit price, plant proximity to customer, and inventory holding policy must be exploited. Penalties for cancellation of orders by a customer, order revision, and the supplier's tardiness must also be weighed in. While marketing would like to minimize costs and risks involved in the distribution (supply) chain by having plants in all major locations, manufacturing would attempt to minimize the cost of total investment in plants. Design of an efficient supply chain may be totally overlooked in this process of negotiating a compromise between the marketing and manufacturing domains.

It is not cost-effective to design a quick-response supply chain for a commodity product such as toothpaste. On the other hand, for products with high value added and/or short life cycles, such as cars and electronic goods (including PCs), supply chains must be designed to be responsive and fast (Fisher 1997). Federal Express, in an industry with short delivery cycle requirements, has developed an elaborate computer-based supply chain that has received high acclaim (Blackmon 1996). Tracking products by time and location is an important element of control in a supply chain of this kind. To respond to customer queries about package status, Federal Express uses a method in which each package is scanned several times while in transport. Customers can now tap into the company's central computer in Memphis to build Web sites to promote their own products. It would also allow online transactions.

To avoid bottlenecks and to eliminate less promising drugs early, Johnson and Johnson has developed a database to track the progress of drugs through research and development cycles (Davenport 1993).

Production Scheduling

To determine daily or weekly production quantities, most manufacturers use control systems that can be some form of MRP or JIT. While MRP maximizes capacity utilization, JIT is geared to satisfying individual customer orders. The three-way trade-off among capacity utilization, inventory holding, and customer service determines the appropriate control system. MRP is associated with a high cost of inventory and schedule inflexibility, and JIT can incur high capacity cost (due to unused capacity). The marketing function would naturally want to use JIT to maximize customer service, but the manufacturing function would like to use a hybrid of JIT and MRP to minimize the total cost.

At Bethlehem Steel, for example, the process of scheduling a customer order into the production schedule was extremely chaotic and inefficient. It led to customer dissatisfaction and sales force frustration, at one end, and high manufacturing cost, at the other. Bethlehem had to completely reengineer its scheduling process (Davenport 1993).

To maximize throughput and/or to minimize congestion on the shop floor, dispatching rules that prioritize jobs to be processed on a machine are used. Such priority rules are based on a combination of factors such as processing time of the job, its due date, order of job arrival at the machine, and number of remaining processing steps. Rules such as priority by shortest processing time or priority by earliest due date will obviously vary in their impact on the system. To optimize the system, dispatching rules need to be selected dynamically based on the shop floor status. This requires an intelligence component to be built into shop floor scheduling. Artificial-intelligence approaches have been suggested that use decision trees (Shaw, Park, and Raman 1992; Naik 1995), data envelopment analysis (Chakravarty 1997), or neural nets (Cho and Wisk 1993) to make a selection for a given shop floor status from a given set of dispatching rules.

Costing

The unit cost of a product is (usually) used as the base for determining unit price of the product. How overheads are allocated to products impacts the unit cost. Activity-based-costing (ABC) allocates overheads to products based on the extent of resources utilized by the product. While it is a big improvement from allocating overheads based on direct labor hours, it suffers from not being able to incorporate market-related concerns. Depending on the shapes of the price-demand curves of individual products, it may be possible to increase total profit if overhead allocations, different from that suggested by ABC, were permitted. The other issue is that ABC cannot be adapted easily to a flexible system, necessary in a market-responsive company. The problem lies in the fact that ABC does not permit multiple-allocation schemes for overhead. Consider, for example, a flexible system, where parts can be routed to more than one machine at any processing step. The total overhead allocated to a part will now be a function of its routing, which in turn will be a function of shop floor status (congestion, tool availability, machine breakdowns, etc.). Since the profit margin is determined on a product type and not on individual copies of the product, some kind of averaging of overhead allocation for different routings will have to be done.

1.3 PROCESS VIEW

A process is an activity or a set of activities with clearly defined inputs, outputs, structure (rules and conditions) for action, and resource use. Outputs link a process to its successor processes, and inputs link it to processes of which it is a successor. Structure of the process defines how it is to be performed and what conditions need to be satisfied. For example, in an engineering heat treatment process it may stipulate that the component is to be heated in an oven up to a certain temperature and then immediately quenched in an oil bath. Resource use defines the resources, such as equipment, tools, skills, and so on, required to complete the process.

The whole enterprise can thus be conceived of as a linked system of processes that can be used by an organization to synchronize its efforts to the customer's needs. In a domain view of the enterprise, the handoffs between functions may not be coordinated. Since performance of a function is often measured by the rate of output or capacity utilization, any difference between production rate at manufacturing and sales rate at marketing may lead to either an inventory buildup or dissatisfied customers. In a process view this is less likely to happen, as the output rate of a process must match the desired input rate at a successor process. An assembly line, if it could respond to changes in customer demand by varying its throughput, would be a good example of a process view. Control systems such as JIT have used a process view to design a synchronization mechanism that links the output and input rates at processes with kanban. A process can be defined as broadly as "order management" or as narrowly as "recording a transaction" or "moving material." A large enterprise can thus be structured as a set of a few broadly defined processes (usually 15 to 20 broad processes for an enterprise [Davenport 1993]) with several narrowly defined subprocesses for every broad process. In Table 1.1, the typical sets of broad processes employed by IBM and Xerox are shown.

Process Modeling

To construct a horizontal process view (in contrast to a vertical domain view), the processes must be interrelated in terms of inputs and outputs. A modeling construct called $IDEF_0$, developed by the U.S. Air Force, can be used as a graphic tool. An $IDEF_0$ representation of broadly defined manufacturing processes—perform process planning, perform MRP function, perform shop floor control, perform workstation activities, and move material—is shown in Figure 1.2. $IDEF_0$ models of the subprocesses corresponding to two of the broad processes (MRP function and shop floor control) are also shown in Figures 1.3 and 1.4.

IBM	XEROX
Market information capture	Customer engagement
Market selection requirement	Inventory management and logistics
Development of hardware	Product maintenance
Development of software	Product design and engineering
Development of services	Technology management
Production	Production and operations management
Customer fulfillment	Market management
Customer relationship	Supplier manager
Customer feedback	Information management
Human resources	Financial management
Marketing	Human resource management
Financial analysis	Legal
Accounting	Capital asset management
IT infrastructure	

Table 1.1: Broadly Defined Processes

(Source: Davenport 1993. Adapted by permission of Harvard Business School Press, © Ernst and Young 1993.)

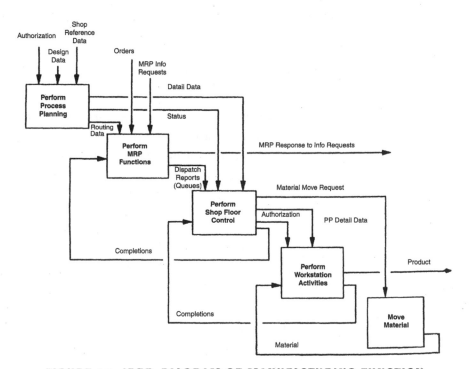

FIGURE 1.2: IDEF$_0$ DIAGRAM OF MANUFACTURING FUNCTION

(Source: Hsu 1994. Reprinted by permission of John Wiley & Sons.)

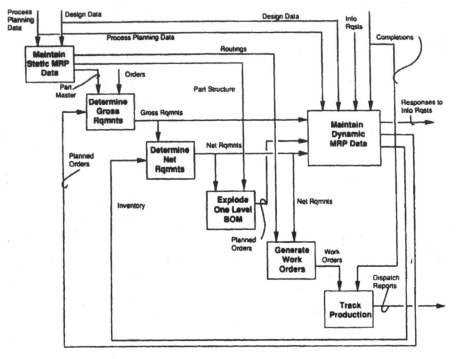

FIGURE 1.3: IDEF$_0$ DIAGRAM OF MRP

(Source: Hsu 1994. Reprinted by permission of John Wiley & Sons.)

While IDEF$_0$ provides a very useful horizontal process view, it lacks the decision-making perspective of coordination, as the metrics of performance and parametric relationships are not defined. Second, as seen in Figures 1.2 to 1.4, it may not be possible to keep the relationship totally horizontal. That is, a certain degree of parallelism cannot be avoided in a single diagram. Various logical-relationship diagrams, such as the data flow diagram (DFD), which links the processes with information flows, have been suggested. Systems algebra and mathematical functional relationships have also been used. Corresponding to IDEF$_0$, we may define a mathematical relationship linking the output to other factors as

$$\tilde{O} = f(\tilde{T}_i, \tilde{S}_i, \tilde{R}_i) \qquad (1)$$

where
$\tilde{O} = f(O_{i1}, O_{i2} \ldots O_{ik})$ is a vector of outputs from process i to all other processes

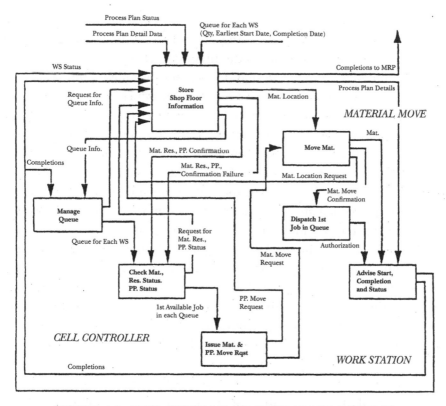

FIGURE 1.4: IDEF$_0$ DIAGRAM OF SHOP FLOOR CONTROL

(Source: Hsu 1994. Reprinted by permission of John Wiley & Sons.)

$\tilde{I}_i = (I_{1i}, I_{2i} \ldots I_{ih})$ is a vector of inputs at process i from all other processes

$\tilde{S}_i = (S_{i1}, S_{i2} \ldots S_{im})$ is a vector of possible structure criteria at process i

M ($m \in M$) is the set of criteria

$\tilde{R}_i = (R_{i1}, R_{i2} \ldots R_{in})$ is a vector of resources available at process i

N ($n \in N$) is a set of resources.

The above relationship suggests that outputs from a process vary according to the type of resource availability and controls on process decisions, and the nature and quality of inputs at that process. Such relationships can be empirically determined for a few select scenarios, which can then be used to predict the behavior of the function in other scenarios. Chakravarty and Ghose (1993) have shown how a functional relationship of inputs and outputs can be used to determine a minimum number of required processes (such as advertising and warranty) that would increase quality perception of a product to the highest level. They use a mathematical programming formulation that determines the above (minimum) set of processes in response to queries related to product improvement. In a different study Chakravarty (1997) has used data envelopment analysis (DEA) to link the choice of dispatching rules (outputs) to shop floor status (inputs). An artificial neural net model is another form of functional relationship of inputs and outputs. Balakrishnan, Chakravarty, and Ghose (1997) have successfully modeled the link between customer-desired and engineering-designed attributes to predict design specifications (outputs) of cars, targeted to specific market segments (inputs).

Barua, Lee, and Whiston (1996) argue that some of the variables in a relationship such as in equation 1.1 are complementary instead of being independent. For example, payoff from investment in process reengineering is not significant unless there is a corresponding investment in information technology. Using supermodular functions as defined by Topkis (1994), they establish how complementarity affects payoff, and how complementary variables in the process view of an enterprise should be varied together to maximize payoff. In their model they suggest using variables such as level of intraprocess sharing, level of access to resources, functionality of interprocess interface, functionality of decision aids, level of performance monitoring, transaction simplicity, level of process integration, size of customer base, and unit operating cost. The problem is that most such variables are hard to quantify, and the authors do not provide any information on how to obtain them from empirical data.

Implications of Process View

A process view, in theory, is an action-oriented plan where the interrelationship between any pair of processes is clearly defined. In Figure 1.5 we provide a contrast between a domain and a process view.

With a domain view of the enterprise, a product development project would lead to the so-called over-the-wall design (Clark and Wheelwright 1993). That is, marketing intelligence on new products is passed on to engineering only periodically (usually in a batch mode), engineering completes the design without manufacturing input and then passes it on

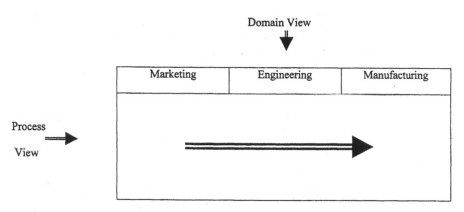

FIGURE 1.5: DOMAIN AND PROCESS VIEWS

for prototype tests, and so on. In the process view, on the other hand, a heavyweight cross-functional team would be assembled that would be responsible for pushing the product's development from marketing to manufacturing and beyond. The heavyweight team is empowered to obtain resources from engineering, manufacturing, and other departments as and when required.

Modicon Inc. is a good example of what a process view can do in product development. It reduced the time to market of six products in automation control by more than 70% by using a cross-functional team for product design (Byrne 1993). In the past, manufacturing did not get involved in the design process until the design was brought to the factory, at which stage the cost of any design changes would be exorbitant. Now the team of 15 managers from engineering, marketing, and manufacturing routinely works together on the design process.

Some proponents of process reengineering advocate a total process view with no domains. While such a structure would be very efficient in terms of responding quickly to the market changes, it is by no means certain that it would be effective from a cost perspective or from a strategic perspective. Consider two separate product development projects, A and B. Since the projects would have parallel horizontal views, there would be no resource sharing in engineering or manufacturing between these two products. In an enterprise where hundreds of such projects may be in progress, the cost of resources would soon get out of control. It is not clear how one would provide interproject coordination and overall control of projects without a higher-level control system (metasystem). But the very existence of such a metasystem would be in conflict with the notion of parallel and horizontal process views. Companies may use a combination of broadly defined processes, as in Table 1.1, and narrowly defined processes or subprocesses to overcome some of the above difficulties. A broadly defined process, since it is a collection of subprocesses that are mostly unique to it, can look a lot like a domain.

From a strategic point of view, the role of innovations in the enterprise is crucial. Davenport (1993) suggests a series of questions, answers to which should form the building blocks of innovation. These are: (1) How could we do things differently? (2) How will it work? (3) How well will it work? (4) What things have to go right? (5) Why might they not go right? Consider again the two projects for products A and B and try to imagine how the design engineer for project A will answer question (1) (how things could be done differently). Unless there is a diffusion of knowledge from other projects to project A, engineer A's search for alternatives would be limited at best. We can easily extend this argument to each of the other four questions to see that A's search for alternatives would not go very far. Therefore, without a functional structure (domain view), the expertise related to a function would be hard to evolve. Since there is no ownership of processes, the innovation potential of the enterprise will be questionable, although the individual processes could do very well in performing assigned tasks. Business consulting firms such as Ernst and Young (Chard 1997) and KPMG (Alavi 1997) have developed special organizational structures to capture consultants' experience (knowledge) from individual consulting assignments so as to make it available to others in the company on demand.

Finally, since processes may change quite often, an organization structure based on process view is not likely to be very stable. Virtual organizations, where at the start of a new project its process view is added to the system and on completion it is deleted from the system, have been proposed. While this may be feasible in certain industries such as consulting, oil exploration, and construction, it would not be cost-effective in a manufacturing company. A structure in which a process view is embedded within each function may be more appropriate for such firms.

1.4 DECISIONS AND INFORMATION FLOW

Whether an enterprise is structured as a domain view, a process view, or a hybrid, the means of coordination (or linkages) between domains or between processes need to be structured carefully. Market economists suggest a decentralized structure where the functions (or processes) are coordinated by transfer price (Alles and Datar 1998), incentives (Porteus and Whang 1991), price discounts (Chakravarty and Martin 1991; Dada and Srikant 1987; Banerjee 1986), and negotiations (Nash 1950). Other suggested approaches are information flow (Ives and Learmunth 1984), agency theory (Baiman 1982), ownership rights (Jensen and Meckling 1992), and intelligent agents (Wooldridge and Jennings 1995). Information system planners suggest a more integrated (centralized) structure.

Coordination Mechanisms

There are a number of coordinating mechanisms a company can choose from, depending on how it is structured. Two important determinants of organization structure are the distribution of decision-making authority and access to relevant information at any decision-making point (Gurbaxani and Whang 1991; Hill and Jones 1995). The two coordinating mechanisms commonly discussed in the literature are centralization and decentralization of decision making. Another mechanism is called transnational in an international context (Bartlett and Ghoshal 1991) and fully distributed in a domestic setting (Anand and Mendelson 1997).

The three coordination mechanisms—centralized, decentralized, and fully distributed—represent different trade-offs between information availability and the quality of decisions. For domains such as manufacturing and marketing, information categories used for decision making are different. In marketing, such information categories describe price, customer preference, market share, market segment, promotions, and distribution. In manufacturing, they describe processing rates of machines, process plans, machine setups, in-process inventory, operations scheduling, and capacity availability. Most of these information types are measurable, and they are transferable between domains. We call this social knowledge, as it is available to all. In any domain there exists another form of knowledge, called tacit knowledge. For example, whereas the processing rate of a machine is measurable (social knowledge), the way to tweak it so that its output gets closer to specification is not (tacit knowledge). Similarly, while the price charged for a product in a market can be quantified, personal relationships with the customers that help increase sales cannot be. Anand and Mendelson (1997) show how the three different coordination mechanisms can be differentiated by the mix of tacit and social knowledge used for decision making in each, and by how profit expressions are obtained for each coordination mechanism.

While profit (based on unit price, unit cost, and sales) is one measure of system effectiveness, there are many other less tangible factors that are equally important. These include responsiveness to customers, innovations, and coordination of production with customer needs. From our earlier discussion, it is clear that in a decentralized system, innovations and responsiveness to customers are likely to improve. The manufacturing cost, on the other hand, is likely to increase, since manufacturing would not be able to take advantage of component and other forms of commonalities. In centralized coordination, production will be planned according to a demand pattern that would be conveyed to the central decision maker by the marketing domain in the form of social knowledge. However, since the tacit knowledge about relationships with customers and the idiosyncrasies of the customers cannot be conveyed to the central planners, production plans will not be as synchronized with the market as in decentralized coordination. Observe that in fully distributed coordination, decisions would be made in each domain individually (as in decentralized coordination), but the social knowledge from all domains along with the tacit knowledge of the individual domain will be made available for decision making in that domain.

To understand the trade-off between information processing and information flow, consider the following:

number of social knowledge entities in domain $i = s_i$

number of tacit knowledge entities in domain $i = t_i$

number of domains $= n$

(a)   Centralized Coordination

$$\text{cost of information transfer} = \lambda \sum_{i=1}^{n} s_i$$

$$\text{cost of information processing} = \mu \sum_{i=1}^{n} s_i$$

where λ and μ are constants.

(b)   Decentralized Coordination

$$\text{cost of information transfer} = 0$$

$$\text{cost of information processing} = \mu \sum_{i=1}^{n} (s_i + t_i)$$

(c)   Fully Distributed Coordination

$$\text{cost of information transfer} = \lambda (n-1) \sum_{i=1}^{n} s_i$$

$$\text{cost of information processed} = \mu \left\{ n \sum_{i=1}^{n} s_i + \sum_{i=1}^{n} t_i \right\}$$

It is clear that the costs of information transfer and processing in fully distributed coordination will exceed their corresponding values in centralized and decentralized coordination mechanisms, but the quality of the decision made would be far superior, as the decision maker in each domain will have access to *all* social knowledge in the system and that domain's own tacit knowledge.

To compare total benefits with total costs, let W and Π denote the total cost (information and process) and benefits, respectively, and C, D, and F denote the three coordination mechanisms, respectively. It is clear from examining the above expressions that for $n = 2$ (i.e., two domains such as manufacturing and marketing), we can write

$$W_F = W_C + W_D$$

However, for $n \geq 3$,

$$W_F > W_C + W_D$$

That is, information-related cost in fully distributed coordination increases faster than the combined costs of centralized and decentralized coordination when the number of domains to be coordinated increase.

Therefore, for the fully distributed coordination to be desirable, Π_F must also increase much faster than Π_C or Π_D with the number of domains. If that is not the case, distributed coordination should be applied between pairs of domain groups (instead of individual domains), and the domains within each group will have within-group distributed coordination.

Intelligent Agents for Coordination

An approach for simplifying information retrieval and transfer is the use of intelligent software agents that may reside at the interface of domains. Such agents are being developed rapidly in the context of the Internet (Sycara and Zeng 1996; Wooldridge and Jennings 1993). Such agents would be useful when the number of information entities being transferred across an interface is large and information content is either ambiguous or changing constantly due to innovations within individual domains. The idea of intelligent agents is an extension of the ideas used in agency theory. That is, an agent is assigned a specific set of tasks with well-defined expectations. Unlike real agents, they do not receive any compensation, however. The agents are endowed with decision-making characteristics to optimize their performance. The agents are empowered to retrieve data from databases in one domain, process them, derive inferences, and deliver them in the form specified by the users in another domain, as shown in Figure 1.6. Applets, used in the Java programming language, have some of the above characteristics. We shall see in more detail in Chapters 2 and 10 how such agents can enhance decision quality at the interface between marketing and manufacturing. Mathematicians such as Goertzel (Petzinger 1998) have begun creating software agents that would roam across networks, seeking patterns in numbers, words, and the behavior of other agents. They claim that such networks could eventually recognize context, nuance, concepts, and the rules of grammar. Ex-

FIGURE 1.6: INTERFACE AGENT

periments are under way by a Wall Street firm to assess how well such a network, designed as a company's intranet, performs the company's market analysis.

1.5 DYNAMIC ENTERPRISE

The combination of structure (domain and process views) and coordination mechanism (centralized, decentralized, and fully distributed) generate six possible scenarios, shown in Table 1.2, where the salient characteristics of each scenario are shown. The competitive advantage of an enterprise, as discussed earlier, is born out of innovations, price, variety, quality, and response time. Using the information in Table 1.2, the two views of an enterprise can be positioned as in Table 1.3.

It is clear that for an enterprise to compete in all four dimensions shown in Table 1.3, it should ideally be able to switch back and forth between domain and process views. This is a form of agility that requires that the enterprise be structured to be able to project any of the two views as

	Centralized	**Decentralized**	**Fully Distributed**
Domain View	• Total system social knowledge is used • Tacit knowledge is unused	• Domain social and tacit knowledge are used individually within each domain	• Total system social knowledge is used by all domains • Tacit knowledge is used within individual domains
Process View	• Rules of each process are determined centrally • Information flow from adjacent processes is used for updating input and control operations	Chaos	• Intelligent interface agents between pairs of processes • Individual processes decide which other processes to interface with

**Table 1.2: Interactions of Domain and Process Views
and Coordination Mechanism**

	Domain View	Process View
Innovation	• Ideal for using shared knowledge and knowledge diffusion • Pride and ownership of knowledge, not products • Can use tacit knowledge for breakthroughs • Decentralized or fully distributed coordination	• Fragmented knowledge • Too focused • Fully distributed coordination with intelligent agents may be used
Quick Response	• Nonexistent ownership of products • Batch mode of operations is common • "Over-the-wall" syndrome • Centralized coordination would be necessary	• Geared to tracking of products • Tight control on the process inputs and outputs • Can create exception to push through tardy jobs • Centralized coordination
Variety	• Can run multiple product development projects concurrently • Decentralized or fully distributed coordination	• Structured for horizontal parallel views • Not cost-effective for multiple products • Fully distributed coordination required
Quality	• Manufacturing craftsmanship and quality in decentralized coordination • Product quality requires customer input and fully distributed coordination for product development	• Extremely effective for controlling quality of individual processes with SPC • Good for process improvement

Table 1.3: Competitive Advantage and Enterprise Views

and when required—domain view primarily for innovation and variety, and process view primarily for quick response and process quality.

The enterprise may be structured as a domain view, where the domains are interfaced with one another by intelligent software agents. These agents can then be used to create a process view, on demand, by pulling together relevant information from different domains, corresponding to the processes of the product or service in question. This would represent an online implementation of permanent cross-functional teams (Hill and Jones 1995). The cross-functional teams shown in Figure

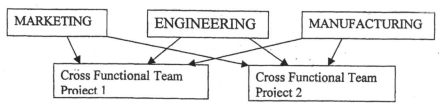

FIGURE 1.7: CROSS-FUNCTIONAL TEAMS IN A DOMAIN VIEW

1.7 are not physical entities separated from the domains of marketing, engineering, and manufacturing. They are created online by software agents that track individual projects and identify the resources associated with a project in each domain.

In such a structure ownership will have to be divided: process ownership to the domains and product ownership to the cross-functional heavyweight teams. How resources are allocated between innovations and value creation will obviously depend on the competitive posture of the enterprise. Competition based on quick response, time to market, focused production, and focused market segments will require the process view to get priority resource allocation. To compete with innovative products, variety, and total quality, the domain view would get prominence.

To get the best out of the manufacturing-marketing interface, it must be subjected to two types of forces: the drive from marketing for increased variety, quality, and quicker response, and the force of moderation from manufacturing in terms of commonalities, focus, substitutes, and reengineering. For competitive advantage, these two forces should be balanced, and they can be balanced in a productive way only if both manufacturing and marketing domains are proactive. For a domain such as manufacturing to be proactive, several things need to happen. First, it should be continuously innovating new processes, materials, and equipment. Second, it should be able to access databases in other domains such as marketing to be able to anticipate what customers may be looking for and how the marketing domain plans to respond. Third, the intelligent agents that scan the inter-domain and external environments must be fine-tuned and protocols must be perfected. Finally, a mechanism should be in place to quickly assess the impact of alternative proposals on the manufacturing domain.

Observe that if both manufacturing and marketing functions are proactive, it becomes easier to implement a fully distributed coordination that can find synergy between the drive of the marketing domain and the moderation of the manufacturing domain.

REFERENCES

Alavi, M. (1997), "KPMG Peat Marwick U.S.: One Giant Brain," Harvard Business School Case 9-397-108.

Alles, M., and S. Datar (1998), "Strategic Transfer Pricing," *Management Science,* vol. 44, no. 4, pp. 451–461.

Anand, K., and H. Mendelson (1997), "Information and Organization for Horizontal Multimarket Coordination," *Management Science,* vol. 43, no. 12, pp. 1609–1627.

Baiman, S. (1982), "Agency Research in Managerial Accounting: A Survey," *Journal of Accounting Literature,* vol. 1, pp. 154–213.

Balakrishnan, N., A. Chakravarty, and S. Ghose (1997), "Role of Design Philosophy in Interfacing Manufacturing with Marketing," *European Journal of Operational Research,* vol. 103, pp. 453–469.

Banerjee, A. (1986), "A Joint Economic Lot Size Model For Purchaser and Vendor," *Decision Sciences,* vol. 17, pp. 292–311.

Bartlett, C., and S. Ghoshal (1991), *Managing Across Borders: The Transnational Solution,* Harvard Business School Press, Cambridge, MA.

Barua, A., C. Lee, and A. Whiston (1996), "The Calculus of Reeingineering," *Information Systems Research,* vol. 7, no. 4, pp. 409–428.

Berger, J., P. Angiolillo, and T. Mason (1987), "Office Automation: Making it Pay Off," *Business Week,* October 21, pp. 134–146.

Blackmon, D. (1996), "Fed Ex Plans to Establish Marketplace in Cyberspace," *The Wall Street Journal,* October 9.

Byrne, J. (1993), "The Horizontal Corporation: It's About Managing Across, Not Up and Down," *Business Week,* October 20, pp. 76–81.

Chakravarty, A. (1984a), "Deterministic Lot Sizing For Coordinated Families of Production/Inventory Items," *European Journal of Operational Research,* vol. 17, no. 2.

Chakravarty, A. (1984b), "Joint Inventory Replenishment with Discount Based on Invoice Value," *Management Science,* vol. 30, no. 9.

Chakravarty, A. (1987), "Dimensions of Manufacturing Automation," *International Journal of Production Research,* vol. 25, no. 9.

Chakravarty, A. (1997), "A Model For Switching Dispatching Rules in Real Time in a Flexible Manufacturing Cell," *Production and Operations Management,* vol. 6, no. 4, pp. 398–418.

Chakravarty, A., and S. Ghose (1993), "Tracking Product-Process Interactions:

A Research Paradigm," *Production and Operations Management,* vol. 2, no. 2, pp. 72–93.

Chakravarty, A., and G. Martin (1991), "Operational Economies of a Process Positioning Determinant," *Computers and Operations Research,* vol. 18, no. 6, pp. 515–530.

Chard, A. (1997), "Knowledge Management at Ernst and Young," Harvard Business School Case 9-397-108.

Cho, H., and R. Wysk (1993), "A Robust Adaptive Scheduler for an Intelligent Work Station Controller," *International Journal of Production Research,* vol. 31, no. 4, pp. 771–789.

Clark, K., and S. Wheelwright (1993), *Managing New Product and Process Development,* The Free Press, New York.

Dada, M., and K. Srikant (1987), "Pricing Policies for Quantity Discounts," *Management Science,* vol. 33, pp. 1247–1252.

Davenport, T. (1993), *Process Innovation: Reengineering Through Information Technology,* Harvard Business School Press, Boston.

Fisher, M. (1997), "What Is the Right Supply Chain for Your Product," *Harvard Business Review,* vol. 75, pp. 105–116.

Garvin, D. (1984), "What Does Product Quality Really Mean?," *Sloan Managemen Review,* vol. 26, pp. 25–43.

Garvin, D. (1991), "The Boeing 767: From Concept to Production," Harvard Business School Case 9-688-040.

Goyal, S. (1974), "Determination of Optimum Packaging Frequency of Items Jointly Replenished," *Management Science,* vol. 21, pp. 436–443.

Gurbaxani, V., and S. Whang (1991), "The Impact of Information Systems on Organizations and Markets," *Communications of the Association of Computing Machines,* vol. 34, no. 1, pp. 59–73.

Hamel, G., and C. Prahalad (1989), "Strategic Intent," *Harvard Business Review,* vol. 67, pp. 63–76.

Hauser, J., and D. Clausing (1988), "The House of Quality," *Harvard Business Review,* vol. 66, pp. 63–73.

Hayes, R., and S. Wheelwright (1984), *Restoring Competitive Edge: Competing Through Manufacturing,* John Wiley, New York.

Hill, C., and G. Jones (1995), *Strategic Management: An Integrated Approach,* Houghton Mifflin, Boston.

Hsu, C., (1994), "Manufacturing Information Systems," in *Handbook of Design, Manufacturing, and Automation,* Dorf and Kusiak (eds.), John Wiley and Sons, New York.

Ives, B., and G. Learmunth (1984), "The Information System as a Competitive Weapon," *Communications of the Association of Computing Machines,* vol. 27, pp. 1193–1201.

Jaikumar, R. (1986), "Postindustrial Manufacturing," *Harvard Business Review,* vol. 64, pp. 69–76.

Jensen, M., and W. Meckling (1992), "Specific and General Knowledge, and Organizational Structure," in *Contract Economics*, Werin and Hajkander (eds.), Basil Blackwell, Cambridge, Mass.

Malone, T., and J. Rockart (1991), "Computers, Networks, and the Corporation," *Scientific American*, September, pp. 128–136.

Miltenburg, J. (1989), "Level Schedules for Mixed Model Assembly Lines in Just-In-Time Production Systems," *Management Science*, vol. 35, no. 2, pp. 192–207.

Mishina, K. (1993), "Johnson Controls, Automotive Systems Group: The Georgetown Kentucky Plant," Harvard Business School Case 9-693-086.

Monden, Y. (1983), *Toyota Production System: Practical Approach to Production Management*, Industrial Engineering and Management Press, Atlanta.

Montgomery, D., and W. Hausman (1986), "Managing the Marketing-Manufacturing Interface," *Gestion 2000: Management and Perspective*, vol. 5, pp. 69–85.

Naik, B. (1995), "A Knowledge-Based Approach to the Management and Control of Flexible Manufacturing Systems," Ph.D. dissertation, University of Wisconsin–Milwaukee.

Naik, B., and A. Chakravarty (1992), "Strategic Acquisition of New Manufacturing Technology: A Framework for Research," *International Journal of Production Research*, vol. 30, no. 7, pp. 1575–1601.

Nash, J. (1950), "The Bargaining Problem," *Econometrika*, vol. 18, pp. 155–162.

Petzinger, T. (1998), "Mathematician Sees Mind as a Model for Company Intranets," *The Wall Street Journal*, May 22, p. 31.

Pisano, Gary (1996), "BMW: The T-Series Project," Harvard Business School Case 9-692-083.

Porter, M. (1985), *Competitive Advantage: Creating and Sustaining Superior Performance*, Free Press, New York.

Porteus, E., and S. Whang (1991), "On Manufacturing/Marketing Incentives," *Management Science*, vol. 37, no. 9, pp. 1166–1181.

Shapiro, B. (1977), "Can Marketing and Manufacturing Coexist?," *Harvard Business Review*, vol. 55, pp. 104–114.

Shaw, M., S. Park, and N. Raman (1992), "Intelligent Scheduling with Machine Learning Capabilities: The Induction of Scheduling Knowledge," *IIE Transactions*, vol. 24, no. 2, pp. 156–168.

Sycara, K., and D. Zeng (1996), "Coordination of Multiple Intelligent Software Agents," *International Journal of Cooperative Information Systems*, vol. 5, no. 2, pp. 1–31.

Topkis, D. (1994), "Modern Manufacturing Revisited," Technical Report, University of California-Davis, Davis, Calif.

Wooldridge, M., and N. Jennings (1995), "Intelligent Agents: Theory and Practice," *Knowledge Engineering Review*, vol. 34, no. 1, pp. 59–73.

2

Manufacturing Marketing Interface

The activities on the value chain of a company, from acquisition of raw materials to delivery of products to customers, belong to two distinct sets (Porter 1985). The first set of activities relates to product definition and market intelligence and generally falls in the marketing domain. The second set of activities defines processes related to manufacturing, product development, and technology innovation (including some R&D), and generally falls in the manufacturing domain. To appreciate the nature of the interface required between the two domains, we first need to understand the nature of activities in individual domains. Activities in a domain transform a set of inputs to a set of outputs, but this transformation may be realized in different ways for different activities. These differences are manifestations of the interactions among inputs and outputs (technology-centered or customer-centered), resources used, controls, and time frame.

2.1 PRODUCT DEFINITION AND PROCESS SELECTION

As shown in Figure 2.1, a product is defined in the marketing domain as a bundle of attributes chosen on the basis of market benefits and marketing costs. Customer profiles and market intelligence on customer preferences help determine the optimal bundles, which may not always mirror customer preferences. The two major reasons for deviations are the range of customer preferences and the cost of manufacturing and delivering the

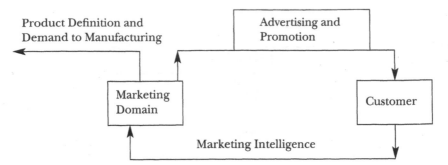

FIGURE 2.1 MARKETING DOMAIN: INPUTS AND OUTPUTS

products. Marketing may decide to concentrate on a few market segments, and it may attempt to alter customers' perception of the company's products. Marketing instruments that help in this purpose are advertising, product promotion, pricing, discounts, quick response to customer's needs, and removing delays in communicating orders to manufacturing. Technological capabilities determine the limits of manufacturing processes, as shown in Figure 2.2. These limits may make it impossible to manufacture products exactly as defined by the marketing function. To elevate process limits, manufacturers need to innovate with technological flexibility, component commonality, production synchronization, inventory reduction, and plant-to-market proximity, among others.

Issues that would be of interest in the marketing domain are patterns

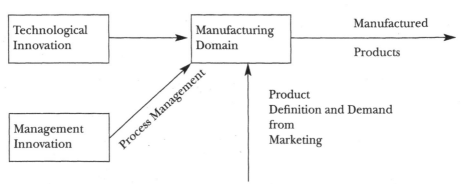

FIGURE 2.2: MANUFACTURING DOMAIN: INPUTS AND OUPUTS

in the choices people make in their buying decisions and customer-related variables that best explain this pattern (income, age, race, education, neighborhood of residence, etc). They would also be interested in how customers are drawn to a specific retail store and which attributes customers find attractive in a product (shelf display, labeling, packaging, internal attributes, etc.) (Kotler 1988). Conditions under which customers prefer product promotions to everyday low prices and the best way to work with different distribution channels are some other issues of interest to marketing managers. The concerns in the manufacturing domain revolve around questions such as how to make processes simple so that operator errors are minimized, how to make processes flexible so that the same process can be shared by several components, how to maximize use of standard components in assemblies, how to coordinate processes so as to minimize in-process inventory buildup, and how to design processing sequences (process plans) so as to avoid bottlenecks in the system (Chang and Wysk 1985).

From the above description of typical tasks in marketing and manufacturing it is clear that there is very little that is common between the two domains. Issues of interest in the marketing domain are mostly customer-centered, and those in the manufacturing domain are technology-centered. Since each domain must be proactive and excel by itself (so as to compete with competitors' marketing and manufacturing functions), the two domains' goals may become discordant, especially where decentralized coordination is in vogue. To ensure that the marketing drive for maximizing customer satisfaction does not become too obsessive and that manufacturing cost minimization does not become too restrictive, a mechanism is needed to provide a meaningful linkage between domains. It is obvious that it is not necessary to link all marketing activities with all manufacturing activities; for example, manufacturing process simplification may have very little to do with what brings a customer to a specific retail store. What, then, should be the constituents of such a linkage between manufacturing and marketing? The objective clearly is to link issues or variables in the marketing domain to issues or variables in the manufacturing domain. While it may be sufficient to link a subset of manufacturing issues with a subset of marketing issues, we do not know how to pick such subsets in any domain. Neither do we know what would be the sufficient set of links constituting such a linkage.

2.2 INTER-DOMAIN LINKAGES

The two dominant forms of coordination, centralized and decentralized, were discussed in Chapter 1. That is, for production and sales decisions we can let the manufacturing domain decide what it wants to produce and let the marketing domain decide what it should sell (decentralized), or we can let a central planner make decisions for both domains (centralized). A better way, as we discussed, is to let the two domains make decisions in a decentralized mode but coordinate them in a distributed coordination mode. We also mentioned briefly the nature and extent of the information transfer required between domains in order to enable distributed coordination. We use the term *interface* to characterize inter-domain coordination. We say two domains or two activities are interfaced when these domains or activities can influence each other's decision-making process without sacrificing individual creativity. Thus, if a product that manufacturing is good at producing does not sell well, it should not cause manufacturing to abandon its core competencies. Such a strategy would lead to high manufacturing costs in the long term. Instead, manufacturing (and marketing) should find a creative way of modifying the product so that it has good sales potential as well as low cost expectations. Manufacturing, of course, cannot determine sales potential by itself. It needs marketing advice on different variations of its product. Similarly, marketing would need advice on cost expectations associated with a modified product. A sound structure for inter-domain advice giving (and advice taking) is a central tenet of domain interfacing, and we will elaborate on this shortly.

It is clear that the building blocks of inter-domain interface are the links between two sets of activities, one from manufacturing and one from marketing. Interfacing two activities may be desirable for several reasons, such as functional interdependence and resource interdependence. The activity of satisfying car buyers with a need for high horsepower, for example, is functionally dependent on designing and building a car with a larger number of cylinders. Fuel economy, on the other hand, requires fewer cylinders. Thus both horsepower and fuel economy preferences should be interfaced with the number of cylinders in a car. Note that the dominant interdependence in inter-domain interface would be between functions, while in intra-domain interface it would be resources. The terms *interface* and *linkage* (as used by Porter) are not too different in the context of coordination, and so we use these two terms interchangeably.

What is interfaced (or linked) between two domains are activities, issues, concepts, and decisions. We use the terms *entity* and *variable* interchangeably to refer to them.

To determine which entities should be linked, interdependence of entities provides an important clue. Thus one can ask the question "What manufacturing entities, such as components, equipment, skills, suppliers, and knowledge, are required to provide a product attribute such as fuel economy of 50 miles per gallon?" Functional interdependence would provide a list of manufacturing/engineering entities as the answer; these can then be defined as interfaced entities for fuel economy. All such links will be necessary for establishing the cost expectation and sales potential of a product. All of the links, however, may not be equally important in all scenarios of creating competitive advantage. In cost-based competition, similarly, links will be established among entities such as process improvement, process efficiency, and inventory cost. In quick-response-based competition, entities of interest would be order cycle, production cycle, and product development cycle. To understand this better, we examine in some detail what links among entities would be relevant for each of the competitive advantage dimensions, such as innovation, quick response, variety, quality, and cost.

Innovation Linkage

Innovation in technology in our context can be of two types: product innovation and process innovation. Let us look at innovation in product technology first.

The impact of product technology has perhaps been greatest in the computer industry. Consider the design of a product such as a terminal, which facilitates interaction of users with a computer. Such terminals can be categorized as dumb, intelligent, and portable. Some of the components that make up a terminal are storage, microprocessor (16-bit or 32-bit), integrated circuits, CRT display, keyboard, and printer. In Table 2.1, the technology requirements for the three product variants are shown.

Assume that the product market shares are S_D, S_I, and S_P, for dumb, intelligent, and portable terminals, respectively. If S_I is high compared to S_D and S_P, the company should emphasize development of all technology types except the printer (see Table 2.1). If, on the other hand, S_D is high,

	Technology					
Product	Storage	Microprocessor	Integrated Circuits	CRT Display	Keyboard	Printer
Dumb Terminal	No	No	Yes	Yes	Yes	No
Intelligent Terminal	Yes	Yes	Yes	Yes	Yes	No
Portable Terminal	No	No	Yes	No	Yes	Yes

Table 2.1: Technology-Product Matrix

only three technology types—integrated circuits, CRT display, and keyboard—will be emphasized. Note also that irrespective of the relative market shares, two technology types—integrated circuits and keyboard—can be developed without any risk, as they are common to all three products. The above conclusions must, however, be moderated by the development-related costs of each technology type. If the cost of the microprocessor (developed in-house or purchased) is very high, an intelligent terminal may not be a viable product for the company.

The above simple example helps to portray the relationship of product technology to technology functionality, technology capability, acquisition/manufacturing cost, and market share. Observe that such a relationship cannot be expressed as a mathematical function, continuous or discrete. The linkage among the entities, on the other hand, can be expressed as a general relationship, shown below.

$$Technology\ choice = R_1 \begin{pmatrix} functionality,\ capability,\ customer \\ preference,\ acquisition\ cost, \\ manufacturing\ cost \end{pmatrix}$$

In many instances, the relationship $R_1(\cdot)$ may be expressed as a set of logical statements, as we shall see in Chapter 3, and can be used to derive important conclusions about the nature of the interface required.

Process technology comprises elemental processes such as forging, heat treatment, welding, electrolysis, and etching printed circuits, and equipment used for processes, such as automated machines, flexible manufacturing devices, and robotics. Innovations in process technology may create new processes and equipment such as laser welding, moldless foundry, and magnetic resonance imaging, or they may enhance the capabilities of existing processes. New or enhanced processes may be required for new products, for increasing product variety, for increasing efficiency, and for enabling quotation of shorter due dates. Thus the linkage can be expressed as

$$process\ Technology = R_2 \begin{pmatrix} product\ variety,\ manufacturing \\ cost,\ process\ flexibility,\ product \\ due\ date \end{pmatrix}$$

Product Variety Linkage

Both manufacturing and marketing domains are responsible for managing product variety. Variety is perceived by customers as differentiation of one product from another in terms of what it does, how much it costs, and how convenient it is to use. If the product is a car, its attributes would include fuel economy, acceleration, interior comfort, dealer's price, and ease of handling. Observe that these attributes are usually defined by the customer, and manufacturing technology is not much of a factor. What the marketing function does is to group these attributes into bundles and identify each bundle of attributes as a product. For example, if two levels are specified for attribute 1, three levels for attribute 2, and five levels for attribute 3, the maximum number of variations possible with these attributes would be 30. In the manufacturing domain, on the other hand, variety is associated with the physical composition of the product that enables it to perform its expected functions. The components of a car that enable a braking distance of 50 feet may be substantially different from those that enable braking in 20 feet. Thus product variety can also be conceived of in terms of components and manufacturing processes used to make that product. The larger the number of unique components and unique processes used, the larger the variety a manufacturing system has to cope with. Competitive advantage is created by *maximizing* variety in the marketing domain and *minimizing* variety in the manufacturing domain.

ITT (even after its acquisition in 2000) is a major supplier of anti-lock braking systems (ABS) for cars, trucks, and vans. The major components of ABS are a pump, a motor, and a valve block. Several variants are possible for each component. ITT chose to have only two variants of the pump, two variants of the valve block, and two variants of the motor, creating 12 varieties of ABS (Pisano 1994). According to the ITT's estimates, the 12 variants satisfied the needs of most but not all of their customers. The variety level of 6 that ITT chose was obviously a compromise between the variety the customers (and most likely the marketing domain) wanted and the variety that the manufacturing domain could provide economically.

To make a decision, as ITT did, to limit variety to a certain number of options, one needs to assess the loss of business from such limitation and the cost of providing additional variety. To be able to do that, a link must be provided between the customer desired attribute set, and the corresponding set of components and manufacturing processes.

Let S_R define a set of customer-desired product attributes, with each attribute specified at a certain level. Let S_C be a set of components constituting the product, and let S_P be the set of manufacturing processes required by this product. We can express the variety linkage as

$$S_R = R_1 (S_C, S_P)$$

The nature of the relationship R_1 is what the engineering designers attempt to estimate from engineering theory and empirical observation. Since the set S_R may not satisfy all customers, the satisfaction level will be defined as

$$\text{Satisfaction} = R_2 (S_R)$$

The nature of the relationship R_2 is what the marketing domain must establish from the "distance" between S_R and customers' ideal products. The satisfaction level or the product's utility will be further impacted by customer profiles, advertising, and the presence or absence of competition.

The cost of engineering and manufacturing can be expressed as

$$\text{Cost} = R_3 (S_C, S_P)$$

The nature of the relationship R_3 is what the manufacturing domain must establish by including material, labor, and equipment costs to complete the linkage.

Observe that the functions R_1, R_2, and R_3 may not be well-defined mathematical functions. Instead they may be logical expressions or a collection of qualitative descriptions, as mentioned earlier.

In section 2.4 we will explore further why managing the variety linkage in manufacturing and marketing domains can be complex, and how appropriate incentive schemes can help.

Quick-Response Linkage

According to Wheelwright and Clark (1992), if a product is introduced six months before the competitor's product, its life cycle profits can increase by as much as three times over what it would be in the case of simultaneous introduction. Factors that contribute to a delay in product introduction are product development time, manufacturing cycle time, order cycle, and delivery cycle time. It has been demonstrated (Christensen 1992) that a process view of the company, with a heavyweight cross-functional team responsible for the development project, can substantially reduce product development time. One way of reducing development time is to at least partially overlap the product's design task and build task (Krishnan, Eppinger, and Whitney 1997; Loch and Terwiesch 1998). A cross-functional team can minimize the risk of having to repeat overlapped tasks in case of redesign of upstream design tasks. Manufacturing cycle time (or lead time) can be cut by reducing the amount of time jobs wait on the shop floor. Dynamic scheduling approaches to minimize system bottlenecks are required for optimization (Chakravarty 1997; Chang, Sueyoshi, and Sullivan 1996). Finally, to reduce delivery time, one needs to rationalize the supply chain (up to the customer) in terms of market mix selection, customer classification, and product classification (Lilien and Kotler 1983).

Stalk and Hout (1990) describe the problems at a customized industrial equipment manufacturer where the promised delivery dates varied from 4 to 35 weeks. The ordering cycle varied from 1 to 6 weeks, and the production cycle varied from 3 to 30 weeks. The manufacturer had implemented a design-and-make-to-order policy. That is, product design was driven by salespeople who passed on information concerning product specifications desired by customers. A high-priority order from the salespeople would interrupt ongoing product design and manufacturing, thereby causing loss of control in design and production cycles. In addition, different manufacturing departments were permitted to plan their production quantities independently of downstream departments. Stalk and Hout emphasize how important it was for the company to map the interactions among departments by tracking both customer orders and semi-finished products as they moved through the system. For this company, the response time was related to the inflexible linkage between order processing and design and to the lack of linkages among manufacturing departments.

Murakoshi (1994) describes how Matsushita Electric has linked order entry to shipping, in real time, for bicycle manufacturing. As shown in Figure 2.3, Matsushita has introduced several new technologies to help produce individual bicycles. These include a personal-computer network for production control, a CAD system linked directly to order entry, and a BOM-oriented system to prevent mismatches between materials and bicycle requirements. Since all information is available electronically, production can start immediately after order entry. Matsushita has reduced its order-to-delivery cycle for a bicycle to 14 days. Observe in this case, however, that the price of a reduced cycle time is a higher manufacturing cost and, more important, relegation of manufacturing to a pure reactive role.

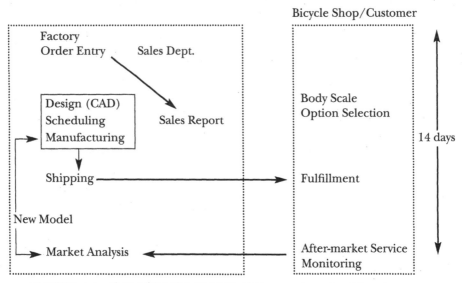

FIGURE 2.3: MATSHUSITA ELECTRIC'S BICYCLE MANUFACTURING SYSTEM

(Source: Murakoshi 1984. Reprinted with permission from Elsevier Science.)

Everything is planned centrally, all the way down to the details of work schedules and process instructions for every workstation.

The quick-response linkage can thus be expressed in terms of time to market as

$$T = R_1 \, (T_P, \, T_M, \, T_D, \, T_O)$$

where T_P is the product development time, T_M the manufacturing cycle time, T_D the delivery cycle time, and T_O the order cycle time. A reduction in T can be obtained by decreasing one or more of T_P, T_M, T_D, and T_O. The question is, by how much must each of the three variables decrease? Therefore, one needs to stipulate relationships of the kind

$$T_P = R_2 \, (T_M, \, T_D, \, T_O)$$

$$T_M = R_3 \, (T_P, \, T_D, \, T_O)$$

$$T_D = R_4 \, (T_P, \, T_M, \, T_O)$$

$$T_O = R_5 \, (T_P, \, T_M, \, T_D)$$

In view of the foregoing discussion, we may also express T_P, T_M, T_D, and T_O as

$$T_P = R_6 \text{ (overlap, rapid prototyping, design review)}$$

$$T_M = R_7 \text{ (schedules, resources, flexibility, waiting time, control)}$$

$$T_D = R_8 \text{ (channels, customers, products, fleet size)}$$

$$T_O = R_9 \text{ (order processing, sales engineering, production capacity)}$$

We shall see later how the quick-response linkage can be operationalized at the interface of the manufacturing and marketing domains.

Quality Linkage

According to Garvin (1984), there are eight dimensions of quality—performance, features, reliability, conformance, durability, serviceability, aesthetics, and perceived quality. We have discussed performance and features under product variety linkage. Reliability is measured by MTBF (mean time between failures,) and durability by the mean time to first failure. Conformance is a measure of how close the product features are to specifications, and serviceability is a measure of how quickly the product can be put back in operation after a failure. Aesthetics relate to the shape, feel, and appearance of the product, and perceived quality relates to the customer's perception of quality, which can be altered by an appropriate form of communication such as advertising and product warranty. Note that a majority of the above quality measures are impacted by how the product is designed, how complex it is to manufacture, what quality of material is used, how well the manufacturing processes are designed and controlled, and how skillful the production operators are.

Thus we can define the quality linkage as

Reliability $= R_1$ (product design, process design, material)
Durability $= R_2$ (product design, process design, material)
Conformance $= R_3$ (specification planning, product design, process design, material, skill level, SQC)
Serviceability $= R_4$ (product design, repair sequence, repair parts)
Aesthetics $= R_5$ (product design, material)
Perception $= R_6$ (advertising, warranty, after-sale service, performance, features, reliability, durability, and others)

To increase the customer's perceived quality, the marketing domain will be responsible for advertising, warranty, and after-sale service. The manufacturing domain will be responsible for the rest. There would therefore be direct linkages, such as

Warranty $= R_7$ (performance, reliability, serviceability, conformance)
Advertising $= R_8$ (quality desired by customers, quality built in the product, aesthetics)
After-sale service $= R_9$ (reliability, serviceability)

It is clear that to compete on quality, the company will have a large number of relationships to manage, not all of which are independent of one another. Techniques such as house of quality can help manipulate these relationships in a systematic way. As we shall see in Chapter 6, although such techniques help, there remain many unresolved issues.

Cost Linkage

Components of manufacturing cost are material cost, labor cost, equipment cost, and overhead cost. Manufacturing cost can be reduced in two different ways: (1) by making the system more efficient, and (2) by exploiting scale economies, including the advantage of learning. A manufacturing system can be made more efficient by reducing material waste, increasing worker training and motivation, and maintaining production equipment so that it does not break down. To take advantage of scale economies, on the other hand, one needs to produce a large quantity of a product without frequent changeovers. This also contributes to workers' learning, as they get better by repeatedly doing a task. Observe that the cost-saving options related to manufacturing system efficiency are all internal to the manufacturing domain. Scale economy, on the other hand, has a crossover impact in the marketing domain as well. It is clear that scale economy will preclude the company from producing too many varieties of a product. Thus a competitive strategy based on low cost will work only if customers are happy with relatively few varieties of a product. Modern technologies such as flexible manufacturing, which enables quick changeover from one product to another, can alleviate this situation. Loss of worker learning will persist, however, as only small volumes of a product will be produced. Thus we may express cost linkage as

$$Cost = R \begin{pmatrix} \textit{manufacturing efficiency, changeover cost,} \\ \textit{product variety, technology flexibility,} \\ \textit{learning} \end{pmatrix}$$

2.3 INTERFACE SET

It is clear that inter-domain linkages by themselves are not sufficient to manage coordination of manufacturing with marketing. We need to position inter-domain linkages relative to intra-domain linkages (explained shortly) to understand how the two domains can be motivated to excel and how to keep the flow of necessary innovations manageable. We first define a few constructs.

We call the set of entities or variables that appear in an inter-domain linkage the interface set. It is clear from the foregoing discussion that there will be multiple views of the interface set, depending on how the company decides to compete. Let us consider the competitive advantage of product variety in greater detail.

The interface set will consist of three entities: the customer's preference for product attributes, the set of components and modules used to assemble the product, and the set of manufacturing processes required to make the components. In Figure 2.4, the interface set for product variety is shown in the context of the manufacturing and marketing domains.

Observe that the manufacturing and marketing domains may comprise several other entities in addition to the three in the interface set. In a domain there is thus an interdependency between the set of entities in the interface set and those outside it. For example, if the component and process sets within the interface set are known, the manufacturing domain can determine a feasible configuration (there may be more than one) of process sequences, production equipment, suppliers, and schedules that would be compatible with the above component and process sets. Determination of an optimal configuration may also require a modification of component and process choices in the interface set. In the marketing domain, similarly, for a given set of product attributes, feasible choices for distribution channels, advertising media, promotion frequencies, and sales force can be arrived at. Thus three types of linkages, L_I, L_F, and L_K, would be involved, as shown in Figure 2.5. L_I is the interface linkage, dis-

Marketing Domain			Manufacturing Domain
	Interface Set		
• Channel management • Advertising • Promotion • Selling	• Product 　attributes	• Component set • Process set	• Process plan • MRP/JIT • Production capacity • Supplier management

FIGURE 2.4: INTERFACE SET FOR PRODUCT VARIETY

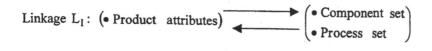

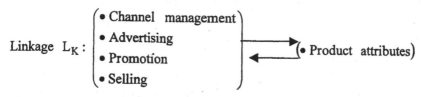

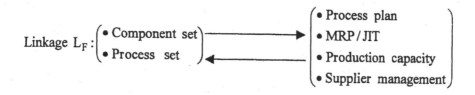

FIGURE 2.5: LINKAGES FOR PRODUCT VARIETY

cussed earlier; L_F is the linkage within manufacturing domain; and L_K is the linkage within marketing domain. The last two are called intra-domain linkage.

Constructing an interface set, as above, helps us avoid having to explicitly analyze all indirect linkages between channel management, advertising, promotion, and selling, on one hand, and process plan, MRP/JIT scheduling, production capacity, and supplier management, on the other. Thus an interface can be conceived of as a "gatekeeper" of links between the two domains. All links between interface and non-interface entities in a domain terminate at the interface "gate," and they are reorganized and conveyed, through the smaller number of established links in the interface linkage L_I, as in Figure 2.4, to the other domain. It is obvious that links between domains, within the interface set, must be kept to a minimum. This can be done by minimizing the number of entities (variables) in the interface set from either domain. Care must be taken, however, to ensure that the variables chosen are sufficient for linkage L_I to propagate all necessary information between the domains.

Chakravarty and Ghose (1992) provide two simple tests to ensure that the interface set of entities is a necessary as well as sufficient set. These tests are described next.

Test 1a: If the competitive advantage (say, product variety) increases or decreases and the value of an entity (variable) in the interface set does not, the entity does not belong in the interface set. (Example: Production capacity need not necessarily increase if product variety increases.)

Test 1b: If the competitive advantage does not increase or decrease and the value of an entity in the interface set does, the entity does not belong in the interface set. (Example: Increasing product promotion frequency need not necessarily increase product variety.)

Test 2: If the increase (or decrease) in competitive advantage cannot be explained by variations in values of any of the entities in the interface set, the interface set is not sufficient. (Example: The increase in variety in terms of packages cannot be explained if the activity called packaging is not included as an entity in the interface set.)

The strength of the domain interface lies in enabling the three linkages L_I, L_F, and L_K to be operationalized *independently* of one another. Thus if an increase in product variety is sought as a competitive advantage, linkage L_I will "fire" first to determine the compatibility of component and process sets with the customer's notion of increased variety. Next, both L_F and L_K would fire in parallel to determine the variations necessary in other domain variables corresponding to the changes in the interface set variables. If cost estimates are excessive or revenue estimates are not sufficient, feedback will be provided to the interface set and L_I will fire again, repeating the sequence of firing L_F and L_K, and so on.

2.4 DISTRIBUTED COORDINATION

Distributed coordination, while aimed at creating synergy among domains, must allow full control to the decision-making processes in individual domains. This means that there should be no constraints on innovations, decision styles, and use of decision models customized for a domain. That is, the coordination mechanism should only help assess the implications for other domains of decisions made in one domain and should not explicitly make decisions for the other domains. This would require propagation of decision-related information from one domain to another through some form of inter-domain mapping, and transparency (to all domains) of decisions made in any domain.

Using Linkages and Domain Transparency

The linkages L_I, L_F, and L_K can clearly be adapted to satisfy the requirements of distributed coordination. Independence of the three linkages from one another is ideally suited for the requirement that control over decision making be retained in individual domains. The intra-domain linkages L_F and L_K are independent of each other (and of L_I). Decisions made using L_F, for example, permit the creation and use of all relevant innovations in material, processes, and scheduling. Shop floor personnel can be involved in decision making if they want to, and they can try out various decision models such as simulation, optimization, and expert system, customized for manufacturing domain decisions in areas such as scheduling and capacity management.

As mentioned earlier, the linkage L_I must be adapted to address the requirement of decision transparency between domain pairs. To understand what is implied by transparency, consider an example. Assume that the marketing domain is evaluating the policy of quicker delivery of products to customers as a way of increasing its market share. A part of the time reduction in the order-to-delivery cycle must also come from a reduction in manufacturing cycle time. There are several ways to cut the interval between the arrival of an order and departure of a physical product. First, inventory could be held at a higher level in the BOM diagram, enabling products to be assembled more quickly using inventoried subassemblies. If subassemblies are not inventoried, time to assemble the components needed to make the subassemblies would have to be added to the lead time. Because of the value-added effect, holding inventories at higher levels in a BOM diagram would certainly increase cost. To build an inventory, a larger-than-required number of subassemblies would have to be manufactured until the inventory buildup is complete. This would cause a disturbance in the manufacturing system (Chakravarty and Balakrishnan

1998), as manufacture of some other components would have to be held back (or the components would have to be procured from subcontractors). Second, the product in question and its associated components could be assigned a high priority in the manufacturing system. This will create a disturbance similar to the one mentioned above. Third, manufacturing equipment can be upgraded to make the processes faster, or flexible manufacturing equipment can be acquired to better utilize capacity—all requiring additional investment. Finally, the product itself could be redesigned to make it more manufacturable by reducing manufacturing steps and/or decreasing time at each step by doing away with special setups requiring jigs and fixtures. Redesign will, however, cause a one-time delay and may incur additional cost.

Decision aids (quantitative models, expert system, etc.) are available in the manufacturing domain to assess costs and other impacts of each of the four options above. For the system to be transparent, these decision aids must be made accessible and user-friendly so that marketing personnel can run them themselves. All they would need to do is to identify the product and the desired percentage reduction in its lead time. The output from the decision models would be in the form of cost and other impacts such as excess inventory and time needed for disturbances (if any) to settle.

Thus augmentation of L_I would be in the form of its capability to access and run decision models that are not in the same domain that needs the results. Data must also be retrieved from other domains where the models are actually run and results obtained. Such capabilities can be created in the system using an intranet. Some software vendors such as SAP, 12 Technologies, and Ariba, also claim to have built such features into their currently available software. The augmented (i.e., adapted) linkage L_I, together with linkages L_F and L_K, will thus be endowed with all the properties required for a successful distributed coordination.

Architecture of Coordination

The coordination architecture, presented below, helps in the implementation of the three linkages L_I, L_F, and L_K. The flow of control has been clearly identified in the context of domain and interface entities. It also helps to clarify the issue of independence of the linkages. Let

MF = an index set of entities in the manufacturing domain

MK = an index set of entities in the marketing domain

I = an index set of entities at the marketing-manufacturing interface

IF = an index set of entities that is a subset of both I and MF

IK = an index set of entities that is a subset of both I and MK

That is,

$$IF \; = \; I \wedge MF$$

$$IK \; = \; I \wedge MK$$

The linkages can now be shown, as in Figure 2.6.

Note the following:

$$j \in IF \Rightarrow j \in I \text{ and } j \in MF$$

$$j \in IK \Rightarrow j \in I \text{ and } j \in MK$$

$$j \in IF \Rightarrow j \notin MF - IF, \text{ and vice versa}$$

$$j \in IK \Rightarrow j \notin MK - IK, \text{ and vice versa}$$

The coordination architecture will ensure the following:

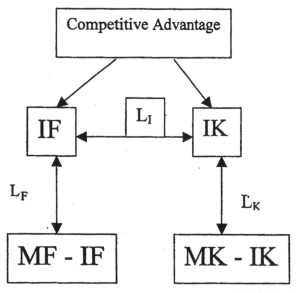

FIGURE 2.6: DOMAINS, LINKAGES, AND INTERFACE

- If competitive advantage (innovation, variety, response time, quality, and cost) needs to be strengthened, it will be communicated to the relevant domain entities that belong to the interface set I.
- On receiving this information, each domain carries out two sets of parallel activities. With entity values in the set IF as inputs, manufacturing domain fires the two linkages L_I and L_F sequentially.
- Linkage L_F helps to assess the intra-domain impact of decisions made in manufacturing, using decision models customized for domain MF (manufacturing).
- Given the inter-domain transparency of decisions, linkage L_I then accesses the decision models customized for domain MK (marketing). It evaluates the impact on domain MK of decisions made in domain MF.
- MF then aggregates the two sets of impacts obtained by L_I and L_F into a value called Π_F.
- MK, starting from information in set IF, carries out two sets of impact analysis, using linkages L_K and L_I. These activities are symmetrical to those described in the context of domain MF.
- MK aggregates the impacts into a value called Π_K.
- The values of Π_F and Π_K are communicated to the corporate planner. If $\Pi_K > \Pi_F$, the manufacturing domain is asked to try again to come up with a Π_F value exceeding Π_K. It follows that if $\Pi_K < \Pi_F$, marketing needs to redo its plan.
- The process of attempting to do better than the other domain continues until the values of Π_F and Π_K converge.

Implementation

We next consider how improvement in competitive advantage will be realized in two different scenarios: reduction in product delivery time to customer, and increase in product variety.

Example 1: Reduction in Delivery Time

Assume that the delivery time is to be reduced (as a competitive advantage) by T time periods, and assume that T_F and T_K are the corresponding reductions in manufacturing cycle time and sales cycle time.

Step 1: Value of T is communicated to the interface set I

Step 2: The inter-domain linkage L_I is stated as $T = T_F + T_K$

Step 3: The intra-domain linkage L_F, in manufacturing, is stated as

$T_F = R_1$ (inventory holding policy)

$T_F = R_2$ (scheduling priority)

$T_F = R_3$ (advanced manufacturing equipment)

$T_F = R_4$ (design for manufacturability)

Observe that cost in the manufacturing domain will increase with T_F. Correspondingly, we may write a cost expression as

$$C(T_F) = \text{Minimum } (C_1(T_F), C_2(T_F), C_3(T_F), C_4(T_F))$$

where

$C_1(T_F) = R_5$ (inventory holding policy)

$C_2(T_F) = R_6$ (scheduling priority)

$C_3(T_F) = R_7$ (advanced manufacturing equipment)

$C_4(T_F) = R_8$ (design for manufacturability)

As an illustration, we show a set of possible cost curves in Figure 2.7.

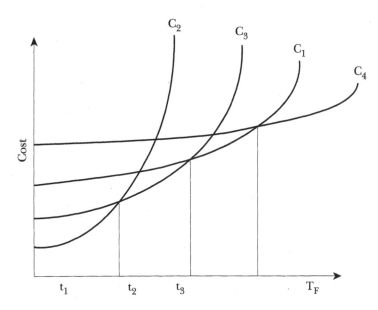

FIGURE 2.7: VARIATION OF COST WITH MANUFACTURING CYCLE TIME T_F

Observe in Figure 2.7 that if the reduction in lead time is to increase from T_F to $T_F + \delta$, the manufacturing domain will choose an action (from the four possible actions listed) that increases its cost by a minimum amount. It therefore follows that manufacturing cycle time reduction options will be chosen as follows:

$T_F \leq t_1 \Rightarrow$ assign a higher priority to the product in scheduling

$t_1 < T_F \leq t_2 \Rightarrow$ invest in manufacturing equipment

$t_2 < T_F \leq t_3 \Rightarrow$ minimize in-process waiting time

$t_3 < T_F \Rightarrow$ improve design for manufacturability

Similar cost curves can be assessed for the sales cycle time T_K, and the two cost curves can be shown, as in Figure 2.8.

Observe that the best values of $C_F(T_F)$ and $C_K(T_K)$ are being determined in their respective domains independently of each other. Hence both domains have the same opportunity to be proactive. The point at

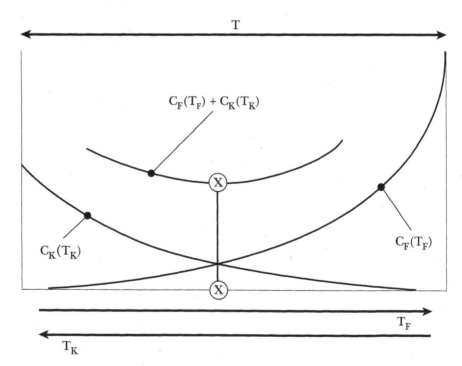

FIGURE 2.8: TOTAL COST AND BENEFIT

which the total cost $C_F(T_F) + C_K(T_K)$ is minimum would occur at a point such as $T_F = T_F^*$ (Figure 2.8). Hence

$$C(T) = C_F(T_F^*) + C_K(T - T_F^*)$$

If maximization of $C(T)$ with respect to T is desired, we may continue our analysis as described below.

Assume that an increase in market revenue for any T is given by $Q(T)$. We may express the net benefit Π of delivery time reduction as

$$\Pi(T) = Q(T) - C(T)$$

Now T is varied, and $\Pi(T)$ is determined by repeating the sequence described above (i.e., finding T_F^* for each value of T). $\Pi(T)$ can be graphed against T, and the value of T that maximizes $\Pi(T)$ can be ascertained.

Example 2: Increasing Product Variety

Variety can be increased by enlarging the set of admissible customer-desired product attributes. The inter-domain linkage L_I cannot be stated as a simple equation, as in the previous example, because the mapping of customer-desired product attributes to engineering features of the product is multidimensional and interactive.

Remember that in example 1, each domain was permitted to determine the best values of $C_F(T_F)$ and $C_K(T_K)$ to preserve proactivity. However, the transparency capability of L_I was not used, as the relationship was mathematically transparent in the linkage equation $T = T_F + T_K$. Since, in the case of product variety, the mapping function between I_F and I_K cannot be expressed as a simple equation, an inter-domain mapping (discussed in Chapter 6) must be found to transform variety V_F, used in manufacturing, to variety V_K for marketing, and vice versa. For the time being, we assume that such a mapping exists. In Chapter 6 we will see how such an inter-domain mapping can be conceptualized. The system profits can now be assessed corresponding to a product variety decision, V_F or V_K. Since profit Π is the difference between revenue and cost, we have

$$\Pi_F = Q_F(V_F) - C_F(V_F) \tag{2.1}$$

$$\Pi_K = Q_K(V_K) - C_K(V_K) \tag{2.2}$$

where Π_F (Π_K) is the profit based on the product decision in the manufacturing domain (or marketing domain), $Q_F(x)$ ($Q_K(x)$) is the market revenue corresponding to the product decision in the manufacturing domain (marketing domain), corresponding to a product variety level of x, and $C_F(x)$ ($C_K(x)$) is the manufacturing cost of the product decision made in the manufacturing domain (marketing domain), corresponding to a product variety level of x.

Both manufacturing and marketing domains would decide on the respective values (V_F and V_K) of product variety as domain decisions. To use linkage L_I, manufacturing will map V_F to V_K (using inter-domain mapping) before running decision models in the marketing domain. The marketing domain, likewise, will map V_K to V_F before using manufacturing domain models. The domains are required to determine the values of Π_F or Π_K (as appropriate), as in equations 2.1 and 2.2. Note that the manufacturing domain cannot determine $Q_F(V_F)$, as in equation 2.1, unless the linkage L_I is augmented with transparency capability, as discussed earlier. Thus in terms of mapped values we have $Q_F(V_F) = Q_K(V_K')$. Similarly, the marketing domain will require transparency in L_I to determine the value of C_K (V_K). That is, $C_K(V_K) = C_F(V_F')$.

It is obvious that Π_F and Π_K are not likely to be equal. Hence we would need several rounds of iteration, as explained in the coordination architecture. For this purpose we use a superscript r to denote that the values relate to the rth round of iteration. If in round 1 $\Pi_F^{(1)} < \Pi_K^{(1)}$, the manufacturing domain will be asked to go back to the drawing board and come up with a solution such that $\Pi_F^{(2)} \geq \Pi_K^{(1)}$. If the manufacturing domain does not redesign its system plan and just settles for $\Pi_F^{(2)} = \Pi_K^{(1)}$, it will be not in the domain interest, as in determining $\Pi_K^{(1)}$ the manufacturing domain is not being allowed to be proactive.

Domain Bids and Incentives

Before we proceed further with the iterations, let us try to understand the theory behind these iterations. It is clear that as variety (V_F or V_K) increases, both manufacturing cost and market revenue will go up. Revenue will rise quickly at first and then taper off as the market becomes saturated with variety. Cost will rise slowly at first but will rise rapidly as it becomes harder to increase variety, as shown in Figure 2.9. Since revenue is a con-

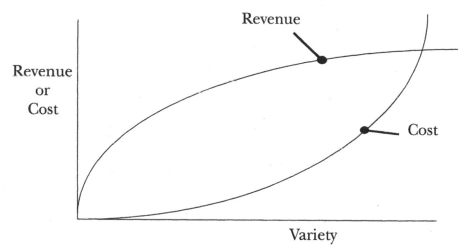

FIGURE 2.9: COST AND REVENUE WITH VARIETY

cave function of variety and cost is a convex function of variety, it can be easily shown that revenue will be a concave function of cost, as shown in Figure 2.10.

Since manufacturing aims at minimizing cost, it will locate itself at a point such as A_1, in Figure 2.10, in round 1. Marketing, on the other hand, aims at maximizing revenue, and would therefore locate at point B_1 in round 1. The values of costs (the horizontal axis in Figure 2.10) at A_1 and B_1 can be used in the learning process in respective domains, as discussed below. Although the equation of the profit curve is not known, it is clear that it would have a shape as shown in Figure 2.10. It is clear from Figure 2.10 that in round 2, the cost in the manufacturing solution would increase and the manufacturing cost in the marketing solution would decrease. If $\Pi_F^{(1)} < \Pi_K^{(1)}$, in round 2 the manufacturing domain will be asked to increase market revenue $Q_F(V_F)$ by permitting the manufacturing cost, $C_F(V_F)$, to increase. The manufacturing domain will thus be forced to choose a different value of V_F in round 2. If $\Pi_K^{(1)} < \Pi_F^{(1)}$, the manufacturing domain, similarly, will be asked to reduce the manufacturing cost $C_K(V_K)$ in round 2 so as to increase $\Pi_K^{(2)}$, even though the revenue $Q_K(V_K)$ may drop. The marketing domain will have to decide on a different value for V_K to move to round 2. Observe, in general, that if $\Pi_F^{(n)} > \Pi_K^{(m)}$, the manufacturing domain may not be able to produce a solution in round $r+1$, where $r = $ maximum (n,m). Therefore, the domain that is strongly proac-

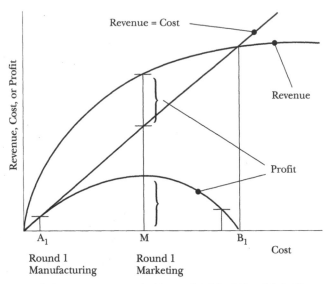

FIGURE 2.10: REVENUE, COST, AND PROFIT

tive and comes up with a good solution (high value of Π) does not have to keep altering its plans in the process of finding system synergy. Learning occurs from the transparency of cost and revenue values (from one domain to another), which tells the low-performing domain the direction (higher or lower) in which its manufacturing cost and market revenue need to move.

Observe in Figure 2.10 that the shape of the profit curve is such that a move by the domains toward each other in terms of manufacturing costs will also ensure an increase in system profit. The only ways this framework can fail to deliver results are (1) if one of the domains fails to be proactive and adopts the solution generated by the other domain (free ride) and (2) if both domains fail to be proactive by colluding to operate at a low profit level. Scenario (1) can occur when $\Pi_F^{(1)} < \Pi_K^{(1)}$, if the manufacturing domain in round 2 settles for a solution identical to that of the marketing domain without investigating other solutions. Notice that the solution corresponding to $\Pi_K^{(1)}$ is generated by the marketing domain, assuming the manufacturing domain to be reactive. It is obvious that the transparency property per se would not make manufacturing proactive; it would only facilitate the use of decision models to evaluate the solution proposed by marketing. Scenario (2) is similar to the first except that both domains may collude in not being proactive and settle on a solution that would be

below the profit curve in Figure 2.10. We shall discuss how incentive programs can be designed for domain decision makers to guard against such
possibilities.

Mathematical Analysis

We can perform simple mathematical analysis if the revenue curve as
shown in Figure 2.10 is strictly concave. We can estimate a concave curve
by mathematical functions of the type

$$z = a\left(1 - e^{-bx}\right)$$

where, in this case, z is the revenue and x is the cost. We can thus express
profit as

$$y = a\left(1 - e^{-bx}\right) - x \tag{2.1}$$

$$\frac{dy}{dx} = a\,e^{-bx} - 1$$

$$\frac{d^2y}{dx^2} = -a\,b\,e^{-bx} < 0$$

Hence the profit curve in Figure 2.10 is concave, and it will have a minimum at

$$x = \frac{1}{b}\ln\left(ab\right)$$

To see how the values of the constants a and b can be estimated, let us assume that the manufacturing and marketing domains obtain values of x
and y in round 1 as shown in Table 2.2.

Manufacturing	Marketing
$x_F^{(1)} = 200$	$x_K^{(1)} = 1{,}000$
$y_F^{(1)} = 1{,}300$	$y_K^{(1)} = 1{,}600$

Table 2.2: Domain Bids

Equation 2.1, after substitution of appropriate values, will generate two equations with two unknowns as shown below:

$$1,500 = a\left(1 - e^{-200b}\right)$$

$$2,600 = a\left(1 - e^{-1,000b}\right)$$

Solving iteratively for a and b, $a = 2,653$, $b = 1/240$, so the profit equation would be

$$y = 2,653\left(1 - e^{-x/240}\right) - x$$

and the maximum value of profit will be $y^* = 1,836$, $x^* = 576.67$. Since the mathematical equation of the profit curve is only an estimation of the reality, the value of $y^* = 1,836$ at $x^* = 576.67$ may not be realizable. But the value of x^* can be used to guide the domains in the general direction in which they should look to find better values. With respect to the values in Table 2.2, since $x_F^{(1)} < x^* < x_K^{(1)}$, it follows that in round 2, the cost values should have the relationship $x_F^{(2)} > x_F^{(1)}$ and $x_K^{(2)} < x_K^{(1)}$.

Note that after every round, the number of available data points will increase, so the estimate of x^* can be improved. For example, for two sets of data, we may rewrite equation 2.1 as

$$y = \frac{a_1}{2}\left(1 - e^{-b_1 x}\right) + \frac{a_2}{2}\left(1 - e^{-b_2 x}\right) - x$$

Now there would be four equations and four unknowns to be solved. It is clear that since the equations are not linear, there will be a limit to how many unknowns we can solve for. In such cases, other methods of curve fitting can be used. One such method is linear programming, which would minimize the maximum deviation between y (actual) and y (from the curve), that is, $y_F^{(n)} - a(1 - e^{-bx_F^{(n)}}) + x_F^{(n)}$, for all n. The decision variables would be a and b.

Incentive Schemes

Porteus and Whang (1991) have suggested an incentive scheme, based on the principal-agent paradigm, to maximize coordination between manufacturing and marketing. Although there is no provision in their model for distributed coordination, their ideas are interesting. They treat the manufacturing and marketing domains as independent agents of the company who pay a fixed amount to the company for the privilege to be a part of the system (franchise). In turn, manufacturing is paid an amount proportional to their production, and marketing is allowed to retain all of their revenue from the market. We can adapt this incentive scheme to our scenario, but since there is more information available in distributed coordination, implementation would be simpler in our case. As in Porteus and Whang, we require each domain to pay a fixed amount to the corporation, but unlike payments based on production or sales, manufacturing and marketing would get paid according to the profits Π_F and Π_K, respectively. This will solve the problem of collusion, but the manufacturing domain can still "abdicate" and settle for the free-ride solution Π_K. If this happens, or if the manufacturing domain cannot surpass Π_K several rounds after Π_K is established, it would have the unusual implication that marketing personnel can run the manufacturing domain better than manufacturing can itself. This would suggest serious problems in manufacturing.

2.5 INFORMATION FLOWS FOR COORDINATION

We have seen the roles of the linkages L_I, L_F, and L_K in interfacing manufacturing and marketing domains. Two types of information flows were mentioned. These were (1) inter-domain—within the interface set stipulated by linkage L_I between set IF (manufacturing) and set IK (marketing), and between domains for providing domain transparency, and (2) intra-domain—for making decisions within a domain, controlled by the linkages L_F and L_K in their respective domains. Information flows controlled by L_I relate to what we have defined in Chapter 1 as social knowledge, whereas the information flows in L_F and L_K involve both social and tacit knowledge. Thus, adopting Anand and Mendelson's (1997) terminology for domain interface, we may define

$$\tilde{x}_F, \tilde{x}_K = \text{vectors of social knowledge in manufacturing}$$
$$\text{and marketing domains, respectively}$$

$$\tilde{\tau}_F, \tilde{\tau}_K = \text{vectors of tacit knowledge in the two domains, respectively}$$

We further stratify the social knowledge (in a domain) into two groups—a group denoted by I for interfacing purpose in linkage L_I, and a group denoted by F or K for decision making in respective manufacturing and marketing domain. We write

$$\tilde{x}_F = (\tilde{x}_{IF}, \tilde{x}_{FF})$$

$$\tilde{x}_K = (\tilde{x}_{IK}, \tilde{x}_{KK})$$

Note that $\tilde{x}_F$ and $\tilde{x}_K$ will be determined for a specific view corresponding to the competitive advantage variable of interest. For example, for reducing delivery lead time, as we have seen in section 2.4,

$$\tilde{x}_{FF} = \text{(inventory holding policy, scheduling priority,}$$
$$\text{advanced manufacturing equipment, design for manufacturability)}$$

The flow of information between the domains will now be

$$Interdomain\ flow = (\tilde{x}_{IF}, \tilde{x}_{IK})$$

That is, it is not necessary to transfer all social knowledge between domains for distributed coordination. This is so because $\tilde{x}_{FF}$ has no use in the marketing domain and $\tilde{x}_{KK}$ has no use in the manufacturing domain. This reduction in information flow simplifies the implementation of distributed coordination, which would not have been possible otherwise.

The information used for decision making in a domain will be

$$\left.\begin{array}{c} Intradomain\ flows\ in \\ manufacturing\ domain \end{array}\right\} = (\tilde{x}_{FF}, \tilde{\tau}_F)$$

$$\left.\begin{array}{c} Intradomain\ flows\ in \\ marketing\ domain \end{array}\right\} = (\tilde{x}_{KK}, \tilde{\tau}_K)$$

In the manufacturing domain the information $\tilde{x}_{FF}$ and $\tilde{\tau}_F$ would be used to create decision models, procedures, and rules. Let $\tilde{y}_F$ denote a vector of the above decision models, procedures, and rules in the manufacturing domain. So

$$\widetilde{y}_F = relation_F\left(\widetilde{x}_{FF}, \widetilde{t}_F\right)$$

$$\widetilde{y}_K = relation_K\left(\widetilde{x}_{KK}, \widetilde{t}_K\right)$$

Therefore, information flows for domain transparency can be stated as

$$\left.\begin{array}{c} Information\ flow\ for \\ domain\ transparency \end{array}\right\} = \left(\widetilde{y}_F, \widetilde{y}_K\right)$$

REFERENCES

Anand, K., and H. Mendelson (1997), "Information and Organization for Horizontal Multimarket Coordination," *Management Science*, vol. 43, no. 12, pp. 1609–1627.

Chakravarty, A. (1997), "A Model For Switching Dispatching Rules in Real Time in a Flexible Manufacturing Cell," *Production and Operations Management*, vol. 6, no. 4, pp. 398–418.

Chakravarty, A., and N. Balakrishnan (1998), "Reacting in Real Time to Production Contingencies in a Capacitated Flexible Cell," *European Journal of Operational Research*, vol. 110, pp. 1–19.

Chakravarty, A., and S. Ghose (1992), "Supporting Manufacturing Marketing Interface Management with a Partially Automated Knowledge Base: A Precursor to Computer Integrated Business Enterprise," *Journal of Intelligent Manufacturing*, vol. 3., pp. 347–362.

Chang, Y., T. Sueyoshi, and R. Sullivan (1996), "Ranking Dispatching Rules by Data Evelopment Analysis in a Job Shop Environment," *IIE Transactions*, vol. 28, pp. 631–642.

Chang, T., and R. Wysk (1985), *An Introduction to Automated Process Planning Systems*, Prentice Hall, Englewood Cliffs, N.J.

Christensen, C. (1992), "Quantum Corporation: Business and Product Teams," Harvard Business School Case 9-692-023.

Garvin, D. (1984), "What Does Product Quality Really Mean?" *Sloan Management Review*, vol. 26, no. 1, pp. 25–43.

Kotler, P. (1988), *Marketing Management: Analysis, Planning, Implementation, and Control*, Prentice Hall, Englewood Cliffs, N.J.

Krishnan, V., S. Eppinger, and D. Whitney (1997), "A Model-Based Framework to Overlap Product Development Activities," *Management Science*, vol. 43, no. 4, pp. 437–451.

Lilien G., and P. Kotler (1983), *Marketing Decision Making: A Model Building Approach*, Harper and Row, New York.

Loch, C., and C. Terwiesch (1998), "Communication and Uncertainty in Concurrent Engineering," *Management Science*, vol. 44, no. 8, pp. 1032–1048.

Murakoshi, T. (1994), "Customer-Driven Manufacturing in Japan," *International Journal of Production Economics,* vol. 37, pp. 63–72.

Pisano, G. (1994), "ITT Automotive: Global Manufacturing Strategies," Harvard Business School Case 9-695-002.

Porter, M. (1985), *Competitive Advantage: Creating and Sustaining Superior Performance,* The Free Press, New York.

Porteus, E., and S. Whang (1991), "On Manufacturing/Marketing Incentives," *Management Science,* vol. 37, no. 9, pp. 1166–1181.

Stalk, G., and T. Hout (1990), *Competing Against Time: How Time Based Competition Is Reshaping Global Markets,* The Free Press, New York.

Wheelwright, S., and K. Clark (1992), *Revolutionizing Product Development,* The Free Press, New York.

3

Knowledge Organization
for Domain Decisions

3.1 INFORMATION FOR DECISIONS

Decision making is choosing one of many feasible alternatives. An alternative is usually a course of action that may comprise a single activity or a set of interrelated activities. An objective choice assumes that a value of benefits (payoff) can be associated with each alternative considered. For example, deciding how many units of which products should be produced and sold in a planning horizon can be based on evaluating each feasible production mix (itself based on production capacity and product demand) for its contribution to overall profit. The two elements of profit-cost and revenue, must be known or easily determinable. To assess the total cost, the cost estimates for each subactivity of a production activity are aggregated. Similarly, the revenues associated with different product mixes are estimated based on customer preferences, market segments, substitute products, price sensitivity, and competition.

To estimate the cost of an activity, one needs to first determine the constituents of the enabling process: resources, skills, timeliness, schedules, raw material, inventories, redundancies, layout, and information processing. The revenue-generating activities, on the other hand, include order processing, pricing, advertising, product warranty servicing, and quality control. The decision maker will need information, directly or indirectly, from all related activities that help shape the product, deliver it to the customer, and provide after-sale service. Direct information will come from activity-related data stored in the database. Indirect information will be obtained from data analysis using spreadsheet, statistics, and optimization models. The decision maker would use this information to aid a mental model, built up from his/her empirical observations, called experience (Lillien, Kotler, and Murthy 1990). The decision-making sequence can thus be represented as in Figure 3.1.

The two direct constituents of decision making, therefore, are a mental model and insights (or knowledge) related to interactions of variables. Insights could be garnered by running some of the analytical models and analyzing results. To appreciate the difference between a mental model and insights, consider an insight that says, "There is a 90% chance of improving customer service to an acceptable level by increasing manufacturing flexibility by 15%." The mental model, on the other hand, might say that the 10% risk of not satisfying customers, even after additional investment to increase flexibility by as much as 15%, is unacceptable. Insights in the marketing domain may offer explanations if sales vary excessively between territories, and if the unavailability of certain products increases a company's overall profits. Finally, insights such as "As the market for a product becomes more competitive, the manufacturer will be willing to pay more to the suppliers for components of the product" help link marketing and manufacturing domains.

3.2 DOMAIN KNOWLEDGE

Insights in a domain can accrue from running decision models, whether prescriptive or descriptive. Insights can also emerge from repeated observations of past behavior of a system. A few examples of such insights from manufacturing and marketing domains are listed (Chakravarty and Ghose

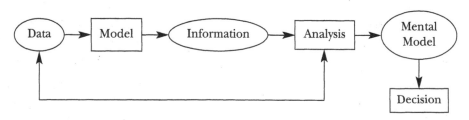

FIGURE 3.1: DECISION-MAKING SEQUENCE

Postulates	Description
F_1	Actual quality engineered in a product increases with technology sophistication and workers' skill, and decreases with process complexity.
F_2	Process complexity increases with the number and level (strength) of product features; the increase is more than proportional and hence nonlinear with features.
F_3	The number of products per standard component generally exceeds the corresponding number for unique components.
F_4	The total number of unique processes increases faster with unique components than with standard components.
F_5	The higher the intensity of competition, the shorter the lead times and the higher the variations in lead times.
F_6	The higher the intensity of competition, the higher the number of new processes and products.
F_7	The number of schedule changes increases with variations in lead times.
F_8	The number of schedule changes increases with the number of new products and processes.
F_9	The greater the flexibility in manufacturing, the greater the ability to absorb schedule-change "shocks."
F_{10}	The lower the manufacturing flexibility and the greater the schedule variability, the larger the in-process inventories.
F_{11}	The nature and sequence of manufacturing operations (i.e., process plan) for a product can vary with the technology of the manufacturing equipment.
F_{12}	A change in process plan will cause a change in schedule.

Table 3.1: Postulates in Manufacturing Domain

1993), as postulates F_1 to F_{12} and K_1 to K_{11} (F for manufacturing domain and K for marketing domain), as in Tables 3.1 and 3.2.

Observe that the postulates listed in Table 3.1 do not directly help in a specific decision such as planning production quantities or specifying which product a machine should be processing at any give time. As we shall see shortly, these postulates can be used in a strategic way in organizing domain knowledge so that decision makers can quickly access the relevant entities and postulates to come to a quick decision.

For a decision maker to arrive at a specific decision a different kind of domain knowledge is required. Consider, for example, machine scheduling, which may require information from entities such as product, process, process plan, and machine flexibility. Since processing capacity of a machine is limited, a procedure for sequencing operations on machines will

Postulates	Description
K_1	The closer the advertised product attributes are to the customer-desired attributes, the more effective the advertising campaign will be in raising the product's quality perception.
K_2	The greater the extent of after-sale service, the higher the product's quality perception.
K_3	Product demand will increase with market size, sales promotions, and quality perception, and will decrease with price.
K_4	The greater the product unavailability, the lower the customers' preference for it in subsequent periods.
K_5	The higher the importance of attributes in a competitor's products, the greater the likelihood that customers will switch to that competitor's products.
K_6	The higher the product variety, the higher the sales.
K_7	The higher the complexity of a product, the longer the time to market.
K_8	Product promotions increase market share but not necessarily revenue.
K_9	Product promotions can increase manufacturing cost more than proportionally to the increase in sales.
K_{10}	The higher the product quality, the lower the cost of warranty.
K_{11}	It is less expensive to increase the attractiveness of a warranty package as product quality increases.

Table 3.2: Postulates in Marketing Domains

be needed that would ensure that product due dates are satisfied and machine utilization is maximized. Many sequencing rules using simulation, optimization, and commonsense heuristics have been developed for different problem scenarios. Also note that domain knowledge used for different problems may differ substantially. Models commonly known by three-letter acronyms such as MPS, for master production scheduling; MRP, for materials requirement planning; ERP, for enterprise resource planning (Davenport, 1998); and JIT, for just-in-time scheduling are used in different problem scenarios. Most of them (except perhaps JIT) are available as commercial software that can be linked to a company's transaction-processing system so as to work directly with the most current data.

Similar domain knowledge and appropriate models have been established in marketing for well-structured problems. These include consumer brand choice (Gensch, Arersa, and Moore 1990); product pricing, markup, and discounting (Dobson and Kalish 1988); product attribute bundling, packaging, and after-sale service (Green and Srinivasan 1990); advertising and sales promotion (Blattberg and Nelson 1990); and channels, sales territories, and distribution (McGuire and Staelin 1986). Like manufacturing domain knowledge, marketing domain knowledge addresses specific marketing questions. For example, domain knowledge in advertising can be used to determine how much should be spent on an advertising program, what the content of an advertising message should be, and what media should be used.

Linking Knowledge in Context

Observe that each of the postulates listed interlinks several entities, such as quality, technology, product, process, and production control, to describe a relationship. Using such postulates, one can create a linked graph of important entities. Consider, for example, postulates F_5 to F_{12}. In Figure 3.2, the nine entities linked by these eight postulates are shown as nodes. The arcs represent postulates or relationships, and the arrows on the arcs indicate the direction of influence between a pair of nodes.

Observe in Figure 3.2 that to make a decision on a schedule change, information will be required on planned manufacturing lead time, throughput of various processes, task times of manufacturing operations, process plan, and the flexibility of the manufacturing system. It is clear that to trigger a scheduling change, it is not necessary for *all* six incident

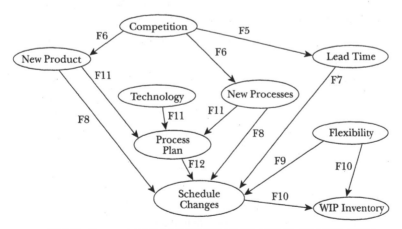

FIGURE 3.2: LINKAGE VARIABLES WITH POSTULATES

(Source: Chakravarty and Ghose 1993. Reprinted with permission from the Production and Operations Management Society.)

arcs at the schedule change node to be active (i.e., carry change-related information). For example, a change in manufacturing lead times may require a schedule change, even though there may not be any changes in products, processes, process plans, or manufacturing flexibility. Thus, the incident arcs at a node are clearly not inclusive/AND type. Neither are they exclusive/OR. They can best be described as AND type for the arcs belonging to individual subsets of the incident arcs. Thus for the five incident arcs at the schedule change node, there would be $^5C_1 + {}^5C_2 + {}^5C_3 + {}^5C_4 + {}^5C_5 = 31$ possible subsets (of arcs) not all of which would be meaningful for schedule changes. For example, Lead Time and New Products would be meaningful in the same subset if the interaction between these two entities causes a schedule change that cannot be captured by these entities independently. Note that each such subset will correspond to a relationship for which a postulate must exist. As shown in Figure 3.2, there are only four meaningful subsets corresponding to the postulates F_7, F_8, F_9, and F_{12}, as shown in Table 3.3.

It is clear that a diagram of linkages as shown in Figure 3.2 is an effective way of organizing data and information required by a decision maker for a particular decision. However, identifying a subset of nodes (variables), as in Table 3.1, does not by itself complete a decision-making process. If information on predecessor nodes is readily available in the database, the schedule change can be determined (resolved) unambiguously. Certain nodes, such as Process-Plan, may not have such information in the database. The required information must therefore be extracted from the next generation of predecessor nodes—New Products, New Processes, and Technology. The search must be continued from one generation of nodes to the next until all needed information is identified in the database. Resolution of a node may become enormously complicated because (1) it is not known a priori which relationship (out of a possible 31 at the node Schedule Change) corresponding to a subset of predeces-

Postulate	Subset
F_7	{Lead Time, Schedule Changes}
F_8	{New Products, New Processes, Schedule Changes}
F_9	{Manufacturing Flexibility, Schedule Changes}
F_{12}	{Process Plan, Schedule Changes}

Table 3.3 Subsets of Variables Defined by Postulates

sor nodes should be "fired," and (2) it is not known how many generations of predecessor nodes would need to be processed to obtain a resolution. We discuss this search problem next to see how knowledge and information can be organized to simplify acquisition of information related to a decision. We first look at a scenario where only a single relationship exists between a node and other nodes linked to it (that is, there is no flexibility in how a particular node is processed).

3.3 LINKING NODES IN A CHAIN

Consider an inventory control scenario with constant demand. Assume that inventory is replenished by ordering a fixed reorder quantity (ROQ) whenever the inventory level drops to a specified reorder point (ROP). Using basic economic concepts, mathematical formulas have been developed for expressing ROQ and ROP in terms of system parameters, ordering cost (OCS), holding cost (HCS), mean demand (MDM), standard deviation of demand (SDM), lead time (LT), and customer service level (SVL). The values of these six system parameters are therefore assumed to be available in the database. We can express the inventory concepts in terms of postulates that describe the relationship of the system parameters with ROQ, ROP, and three other variables of interest: average inventory level (AIL), average inventory cost (AIC), and average number of orders (ANO) (all averaging is done over several periods of time). Five postulates are listed in functional form below:

$$ROQ = f_1\,(HCS,\ OCS,\ MDM)$$

$$ANO = f_2\,(MDM,\ ROQ)$$

$$ROP = f_3\,(MDM,\ LT,\ ROQ,\ SDM,\ SVL)$$

$$AIL = f_4\,(LT,\ MDM,\ ROQ,\ ROP)$$

$$AIC = f_5\,(HCS,\ OCS,\ MDM,\ LT,\ ROQ,\ ROP)$$

The fact that a single relationship suffices for determining each of the five dependent variables tells us that decision making in a constant-demand scenario is very structured. A diagram linking the above variables is shown in Figure 3.3.

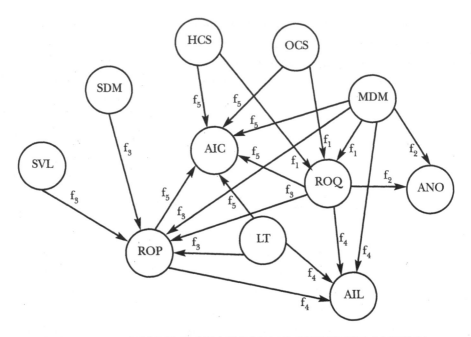

FIGURE 3.3: INFORMATION LINKAGE IN INVENTORY CONTROL

Given the above structure, it is now easy to identify the information chain necessary to determine the values of any of the dependent variables. Suppose we want to find the value of the average inventory level (AIL). Since relationship f_4 is required to resolve AIL, the variables in the argument of f_4, LT, MDM, ROQ, and ROP, need to be resolved. The values of LT and MDM are available in the database. To resolve ROQ, function f_1 needs to be evaluated, which requires values of HCS, OCS, and MDM, all of which are available in the database. Next, to resolve ROP, function f_3 is required, which in turn requires the values of MDM, LT, ROQ, SDM, and SVC; all but ROQ are available in the database. Since ROQ is already resolved, the chain is completed. Formation of this resolution chain is shown as a tree diagram in Figure 3.4.

The sequence of information processing can be described in three steps.

> Step 1: ROQ is determined using the database values of HCS, OCS, and MDM.
> Step 2: ROP is resolved using the database values of MDM, LT, SDM, SVL, and the value of ROQ from step 1.

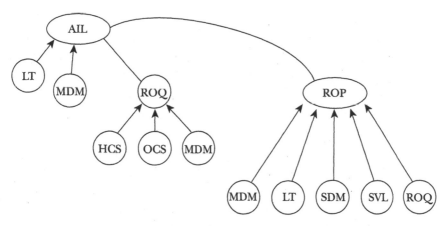

FIGURE 3.4: DECISION RESOLUTION TREE

Step 3: AIL is resolved using the database values of LT and MDM, the value of ROQ from step 1, and the value of ROP from step 2.

All data, models, and relationships required can thus be made available to the decision maker in order to determine average inventory level, AIL. In this particular example, since mathematical formulas are available to describe each relationship, a computer program can easily be written that would use the relevant database information to compute AIL. In a more complex situation, when such mathematical formulas are not available, the domain decision maker can extrapolate from mental models after studying the data and/or information categories made available to him/her for each node. Presenting information to the decision maker in this format is like giving him/her clues in solving a puzzle. For example, by associating LT, MDM, ROQ, and ROP with AIL, we are making sure that the decision maker is not starting from a blank slate. He/she can be confident that it is not necessary to look beyond the four variables mentioned above to come up with a mental model for AIL.

Consider now a more complex situation in which more than one mathematical (or logical) expression may be required to compute ROQ and ROP, depending on market and/or business conditions. Those conditions may be stipulated as (1) whether demand is constant, is random, or varies with time, (2) whether the product is perishable, (3) whether the product is purchased in lots or produced in-house at a given rate, (4) whether price discounts based on purchase quantities are available, (5) whether inventory position is monitored continuously, and (6) whether replenishment orders are placed (a) at fixed time intervals, (b) for fixed quantities, and (c) when triggered by inventory position (Chikan 1990). Ghiaseddin, Matta, and Sinha (1994) have created 16 scenarios based on the above conditions, as shown in Figure 3.5. In the context of our relationship diagram in Figure 3.3, there would now be several functions linking a node to other nodes, but only one is required for processing the node. It is not certain which of the multiple functions should be chosen at a node to construct a resolution tree, as in Figure 3.4. The choice of a function will depend on database values, on conditions (1) to (6), above, that identify a scenario. Ghiaseddin and colleagues, instead of starting from a decision to be resolved (a goal), start from database values to test condi-

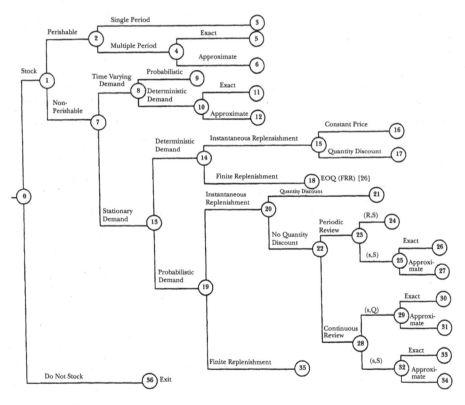

FIGURE 3.5: INVENTORY REPLENISHMENT SCENARIOS

(Source: Ghiaseddin, Matta, and Sinha 1994.) Adapted and reprinted with permission of INFORMS, Linthicum, Maryland.

tions (1) to (6) in order to identify a realized scenario. Once the scenario is identified, the relationship diagram will have exactly one function defining each node, and so the resolution tree is easily constructed. Ghiaseddin and colleagues were able to use domain knowledge to limit their analysis to 16 scenarios. In general, with two values to a condition and eight conditions (as in their analysis), there can be as many as 256 scenarios.

Note that the inventory replenishment problem discussed above belongs to a special class of problems where there exists a set of conditions (1) to (6) that generate scenarios for all decision variables such as ROQ and ROP. In general, decision scenarios for any two decision variables that are related to each another (such as ROQ and ROP) may not be identical. In such cases, the scenarios for each variable will have to be instantiated individually, and since the variables may be related, as in Figure 3.3, the number of scenarios can multiply to an astronomical figure quickly. We first look at some such semi-structured situations, and then discuss some solution approaches.

3.4 SEMI-STRUCTURED DECISIONS

Consider the problem of setting a new sales quota (NQ) of books for salespersons (Holsapple and Whinston 1987) as shown in Figure 3.6. This quota depends on a base amount (base), an economic factor (EF), local factor (LF), a product factor (PF), and an adjustment factor (AF). The base amount is determined on the basis of the previous period's quota and actual sales. The economic factor is determined by growth rate and employment. The relationships governing the economic factor, however, depend on the state of the economy: good, fair, or poor. Thus one function for each state of the economy is required for computing EF. Characterization of the state of the economy (EC) is based on the growth rate and unemployment level. Four possible combinations of growth rate and unemployment, requiring four functions, generate these three states of economy. Local factor (LF) depends on the state of the economy and local advertising, and three possible combinations of these variables produce three values of LF. Product factor (PF) depends on the strength of the product line and the number of old and new book titles sold, and can have three possible values, generated by three different functions. The adjustment factor (AF) can have three possible values depending on whether

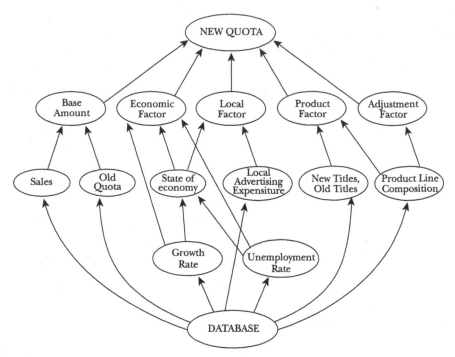

FIGURE 3.6: RELATIONSHIP DIAGRAM FOR NEW SALES QUOTA

the product line is strong, weak, or neither. Computation of the new quota (NQ) requires three different functions, as the adjustment factor (AF) is discrete. The relationship diagram is shown in Figure 3.6. The values of sales, old quota, growth rate, unemployment rate, local advertising expenditure, number of new and old book titles, and product line composition are available in the database but would, obviously, differ for different salespersons.

Eighteen relationships—three for new quota, three for product factor, three for local factor, four for state of the economy, three for economic factor, and two for base amount—are shown in Figure 3.7.

As there is more than one relationship linking a variable to other variables, and as the market and business conditions that create different scenarios are not identical for all variables, it is clear that the procedures used in conjunction with Figures 3.3 and 3.5 are inadequate to calculate the new sales quota. As alluded to earlier, complexity of analysis increases as the relationship for resolving a variable node must be determined in real time. If the correct set of relationships is not chosen and if database values turn out to be incompatible with the relationships chosen backtracking will be required. A number of approximations (heuristics) have been suggested to overcome this problem (Holsapple and Whinston 1987). We will discuss a different approach for this problem, a variant of which was proposed by Chakravarty and Sinha (1990) and used by Chakravarty and Ghose (1993).

Examples from Domains

To appreciate how interlinked intra-domain decisions are, two other decision scenarios from two domains are described. In the first scenario we explore how the quality of a product delivered to customers is affected by decisions made in the manufacturing domain and how such decisions are themselves interlinked.

As shown in the relationship diagram in Figure 3.8, at the topmost level of relationship it is postulated that the quality level in delivered products depends upon manufactured quality (attained in manufacturing) and SQC (which screens out substandard products before they are delivered to the customer). The manufactured quality of products increases with better manufacturing technology (equipment such as flexible manufacturing and CAD/CAM computer aided manufacturing) and better-skilled workers but decreases with process complexity. Process complexity, in turn, increases with the complexity of the process plan (number of steps, sequence of operations, specification, etc.), and whether new processes such as laser cutting, electronic foundry, electromechanical etching, and the like are put in place. Process plans and process technology from research and development (R&D) or from vendors determine the extent of new processes or upgrading of existing processes required. The cost of new processes or process upgrading, together with the customer-desired product attributes, determines the level of actual product attributes. Finally, the development of manufacturing and process technology is determined by the extent of in-house innovations.

The relationship diagram in Figure 3.8 combines the input of "customer-desired attributes" from marketing with manufacturing domain knowledge (inclusive of engineering design) such as process planning, characteristics of manufacturing processes, worker skill, and manufacturing equipment. This knowledge needs to be expressed as sets of postulates

R1: BASE
IF: SALES > 1.15 * QUOTA
THEN: BASE = QUOTA + (SALES – 1.15 * QUOTE)

R2: BASE
IF: SALES < = 1.15 * QUOTA
THEN: BASE = QUOTA

R3: EFACTOR
IF: ECONOMY = "good" AND KNOWN ("GROWTH")
THEN: EFACTOR = GROWTH

R4: EFACTOR
IF: ECONOMY = "fair" AND KNOWN ("LOCALADS")
AND KNOWN ("GROWTH")
THEN: EFACTOR = GROWTH/3; LAFACTOR = LOCALADS/120,000

R5: EFACTOR
IF: ECONOMY = "poor" AND KNOWN ("GROWTH")
AND KNOWN FACTOR ("UNEMPLOYMENT")
THEN: EFACTOR = MIN (GROWTH, .085 – UNEMPLOYMENT)

R6: ECONOMY
IF: GROWTH > = .04 AND UNEMPLOYMENT < .076
THEN: ECONOMY = "good"

R7: ECONOMY
IF: GROWTH > = .02 AND GROWTH < .04
AND UNEMPLOYMENT < .055
THEN: ECONOMY = "good"

R8: ECONOMY
IF: GROWTH > = .02 AND GROWTH < .04
AND UNEMPLOYMENT > = .055 AND UNEMPLOYMENT < .082
THEN: ECONOMY = "fair"

R9: ECONOMY
IF: GROWTH < .02 OR UNEMPLOYMENT > = .082
THEN: ECONOMY = "poor"

R10: LFACTOR
IF: ECONOMY = "good" AND LOCALADS > 2,000
THEN: LAFACTOR = LOCALADS/100,000

R11: LFACTOR
IF: ECONOMY = "poor" AND LOCALADS < 1,500
THEN: LAFACTOR = –.015

R12: LFACTOR
IF: (ECONOMY = "poor" AND LOCALADS > = 1,500)
OR (ECONOMY = "good" AND LOCALADS < = 2,000)
THEN: LAFACTOR = 0

R13: PFACTOR, STRONG, WEAK
IF: PROD IN ["computer," "romance," "scifi"]
THEN: PFACTOR = (NEWTITLES + OLDTITLES)/OLDTITLES – 1
STRONG = FALSE
WEAK = FALSE

R14: PFACTOR, STRONG, WEAK
IF: PROD IN ["reference," biography," "psychology," "sports"]
THEN: PFACTOR = .75* (NEWSTITLES + OLDTITLES)/OLDTITLES – 1)
STRONG = FALSE
WEAK = FALSE

R15: PFACTOR, STRONG, WEAK
IF: NOT (PROD IN ["computer," "romance," scifi," reference," biography,"
"psychology," sports"])
THEN: PFACTOR = .45* (NEWSTITLES + OLDTITLES)/OLDTITLES – 1)
WEAK = TRUE; STRONG = FALSE

R16: NEWQUOTA
IF: STRONG AND KNOWN ("BASE") AND KNOWN ("EFACTOR")
AND KNOWN ("PFACTOR") AND KNOWN ("LAFACTOR")
THEN: INPUT RISE NUM \
WITH "Enter estimate of percentage sales increase" \
+ "due to rising interest in" + PROD
NEWQUOTA = BASE* (1 + EFACTOR + LAFACTOR + \
PFACTOR + RISE/100)

R17: NEWQUOTA
IF: WEAK AND KNOWN ("BASE") AND KNOWN ("EFACTOR") AND
KNOWN ("PFACTOR") AND KNOWN ("LAFACTOR")
THEN: INPUT FALL NUM \
WITH "Enter estimate of percentage sales decrease" \
+ "due to falling interest in" + PROD
NEW QUOTA = BASE* (1 + EFACTOR + LAFACTOR + 1 \
PFACTOR – FALL/100)

R18: NEWQUOTA
IF: NOT (WEAK OR STRONG) AND KNOWN ("BASE")
AND KNOWN ("EFACTOR") AND KNOWN ("PFACTOR")
AND KNOWN ("LFACTOR")
THEN: NEWQUOTA = BASE*(1 + EFACTOR + LAFACTOR + PFACTOR)

FIGURE 3.7: RELATIONSHIP FOR NEW SALES QUOTA

(Source: Holsapple and Whinston 1987)
Reproduced with the permission of McGraw Hill.

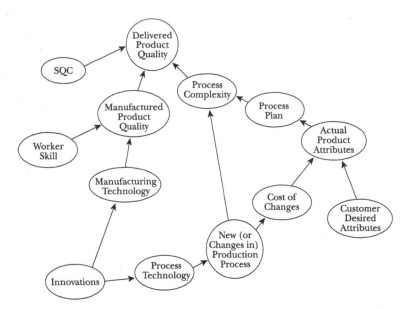

FIGURE 3.8: RELATIONSHIP DIAGRAM FOR PRODUCT QUALITY

related to the nodes in the diagram. It is not easy to find functional forms of relationships that would be valid for all ranges of values of input and output variables. For example, the relationship between product attributes and process plans cannot be expressed as mathematical equations that would be valid for all combinations of attributes and their levels (strengths). More likely they will be presented as logical statements, each of which may be valid for a specific range of values. These would be similar to some of the relationships used in Figure 3.7, where the range of validity was controlled by IF conditions associated with relationships.

In Figure 3.9 the relationship diagram establishes how the customer's perception of product quality is shaped by his/her perception of the existence of specific attributes of the product and the importance of these attributes along with performance of product (delivery quality), price, warranty package, and after-sale service. The advertising campaign, together with experience in using the product (delivered quality), shapes perception of attributes in the product. Ad campaigns can influence individual customers' perception of the importance of particular product attributes. The advertising campaign itself is shaped by the need to narrow the gap between actual and customer-desired attributes of products and by the need to improve the company's competitive position.

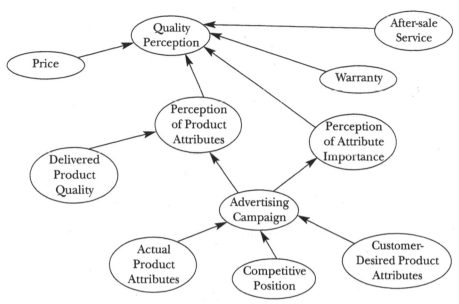

FIGURE 3.9: PERCEPTION OF PRODUCT QUALITY

As we described in previous examples, postulates that define relationships among the nodes in Figure 3.9 will be derived from marketing knowledge built up over many years in individual organizations or from common knowledge available in the literature. For example, customer expectations from the product (customer-desired attributes) and customer satisfaction with competing products will have a strong impact on the contents of ad messages. Similarly, the actual attributes of the product will determine what claims an ad can make about features present in the product.

3.5 RELATIONSHIP CHAIN CONSTRUCTION

If a chain of relations linking entities (to be evaluated) with the database can be found, the decision maker can concentrate on this subset of relations to come to a final decision. In Figure 3.6 such a chain linking the new quota to the database must link several other variables in its path. To provide the advantage of focus to the decision maker and to minimize the cost of data processing, the number of relations used in such a chain must be the minimum possible. In the context of Figures 3.6 and 3.7, a mathematical programming formulation that minimizes the number of relations used is discussed next.

Let the relations as in Figure 3.7, including another seven relations for database retrieval, be numbered 1 to 25. Let the nodes (called variables) also be numbered consecutively, starting from the new quota node and ending in the database node.

Assume that node j is to be resolved, with R $(r \in R)$ being a set of relations and I $(i \in I)$ being a set of nodes (variables). Let

$$x_r = 1, \text{ if relation r is used in a chain (0 otherwise)}$$

$$a_{ri} = 1, \text{ if node i is an input to relation r (0 otherwise)}$$

$$b_{ri} = 1, \text{ if node i is an output of relation r (0 otherwise)}$$

Objective Function

Since the total number of relations to be used is to be minimized, the objective function would be

$$\text{Minimize} \sum_{r \in R} x_r \tag{3.1}$$

Constraints

Since node j is to be resolved, the system must make sure that at least one of the relations used $(x_r \neq 0)$ has node j as its output node. That is,

$$\sum_{r \in R} b_{rj} x_r \geq 1; \ \forall_j, j \in I \tag{3.2}$$

Similarly, to make sure that no more than one relation is used to resolve a node, we would have

$$\sum_{r \in R} b_{ri} x_r \leq 1; \ \forall_i, i \in I \tag{3.3}$$

Next, note that if a relation uses node k, node k must be an output node from at least one used relation. That is,

$$\sum_{r \in R} a_{rk} x_r \leq M \sum_{r \in R} b_{rk} x_r; \ \forall_k, k \in I \qquad (3.4)$$

Observe that the objective function 3.1 with constraints 3.2, 3.3, and 3.4 would complete the formulation if there were no conditions attached to the firing of the relationships. Observe in Figure 3.7 that all relations that do not use database values have IF conditions attached to them. Nine of the IF conditions, corresponding to relations 1, 2, 6, 7, 8, 9, 13, 14, and 15, are expressed in terms of database values. The other nine are expressed in terms of values at intermediate nodes. These two types of IF conditions need to be treated separately, as discussed below.

IF–Database

Consider relations R_1 and R_2. The associated IF conditions can be evaluated using the values of sales and old quota, both of which are obtained from the database. These IF conditions can be modeled by placing constraints on x_1 and x_2 that control triggering of R_1 and R_2, respectively, as shown below.

$$x_1 + x_2 = 1$$

Letting node i = sales and node k = old quota,

$$x_1 \leq \frac{1}{M}(V_i - 1.15\ V_k) \qquad (3.5)$$

$$x_2 \leq \frac{1}{M}(1.15\ V_k - V_i) \qquad (3.6)$$

where M is a large positive number, and V_i is the database value of the variable at node i.

IF–Intermediate Node

Consider the relations R_3, R_4, and R_5, which use the node called "state of the economy" as a condition for triggering one of the three rules. Three values of the state of the economy—good, fair, and poor—are used for stating conditions. We first identify the relations that generate a specific state of economy as an output node. These are

State of the Economy	Relation
Good	R_6, R_7
Fair	R_8
Poor	R_9

Now consider R_3. Since R_3 is triggered when the state of the economy is good, we may write

$$x_3 + x_4 + x_5 = 1$$

$$x_3 \leq x_6 + x_7$$

For R_4 and R_5, we may similarly write

$$x_4 \leq x_8$$

$$x_5 \leq x_9$$

Finally, consider a compound IF condition such as the one associated with relation R_{12}. It is clear that the IF condition can be modeled by the following constraint set:

$$x_{10} + x_{11} + x_{12} = 1$$

$$x_{12} \leq y_1 + y_2$$

$$y_1 + y_2 = 1$$

$$y_1 \leq x_9$$

$$y_1 \leq \frac{1}{M}(V_i - 1,500) + 1$$

$$y_2 \leq x_6 + x_7$$

$$y_2 \leq \frac{1}{M}(2,000 - V_i) + 1$$

where node i = local advertising and

y_1 = 1, if the condition "economy is poor and local ads $\leq$ 1,500" is true; y_1 = 0 otherwise

y_2 = 1, if the condition "economy is good and local ads $\leq$ 2,000" is true; y_2 = 0 otherwise

3.6 KNOWLEDGE MODULARIZATION

A decision maker should be able to easily retrieve information he/she needs during the process of decision making. That is, in the context of Figures 3.3 and 3.4, he/she may want to know the value of AIL, or in the context of Figure 3.5, he/she may want to know the value of new quota. In each instance the decision maker will send a query to the knowledge base asking for AIL or new quota. Observe that to resolve a query about AIL (Figure 3.4) only three of five possible relations were used. Similarly, to resolve the query about new quota, only a subset of 25 possible relations is used. The relations required to resolve queries on AIL, AIC, and ANO from Figure 3.4 are shown in Table 3.4.

In the context of organizing the knowledge base, one of the issues is whether all five relations should be stored in one location or whether they should be split and stored by query type. In the first case, a cost will be incurred in searching for the relevant relations every time a query is made. In the second case, there will be no search cost, but now there will be duplication in storage, as five storage locations will be needed instead of just one. Planning for the differentials in query frequency is the second issue. That is, if AIL needs to be resolved 100 times for every time AIC or AON is resolved, it may be preferable not to group AIL with AIC or AON. As a query may share some of the relations with other queries, *flexible knowledge modules,* which can be used by more than one query, become attractive. This does not require a query to be totally resolved in a single module, but

Query	Relevant Relations
AIL	R_1, R_3, R_4
AIC	R_1, R_3, R_5
ANO	R_1, R_2

Table 3.4: Required Relations for Query Resolution

a good knowledge organization will minimize the number of modules of relations retrieved for query resolution in a given period. If the modules are too large (as would be the case in having all five relations in a single module), the cost of an intra-module search for relevant relations will be high. If, on the other hand, the modules are too small (such as having one relation per module), the inter-module retrieval cost will be high, as each query may require several modules. The organization of relations into modules for given query probabilities is therefore a nontrivial problem.

Chakrvarty and Sinha (1990) have used a mathematical programming formulation of the type shown below to create an optimal number of modules of relations.

Model for Relation Modules

Let N ($n \in N$) be a set of modules created. Let

$$z_{kn} = 1 \text{ if query } k \text{ uses module } n \text{ (0 otherwise)}$$

$$y_{rn} = 1 \text{ if relation } r \text{ is assigned to module } n \text{ (0 otherwise)}$$

$$u_{kn} = \text{number of relations not used for query } k \text{ in module } n$$

$$x_{rk} = 1 \text{ if relation } r \text{ is required in resolving query } k$$

The value of x_{rk} is known, and u_{kn} is a continuous variable.

$$A = \text{relative cost of inter-module search to intra-module search}$$

$$p_k = \text{probability of query k occurring}$$

The model can be stated as:

$$\text{Minimize} \sum_{k \in K} \sum_{n \in N} A \, p_k \, z_{kn} + \sum_{k \in K} \sum_{n \in N} p_k \, u_{kn} \tag{3.7}$$

subject to

$$\sum_{n \in N} y_{rn} = 1, \ \forall_r, r \in R \tag{3.8}$$

$$\sum_{r \in R} x_{rk} \, y_{rn} - M z_{kn} \leq 0, \ \forall_k, \ \forall_n, k \in K, n \in N \tag{3.9}$$

$$\sum_{r \in R} y_{rn}(1 - x_{rk}) + M z_{kn} - u_{kn} \leq M, \ \forall_k, \ \forall_n \qquad (3.10)$$

$u_{kn} \geq 0; \ y_{rn} = 0, 1; \ x_{rk} = 0, 1 \text{ (known)}; \ z_{kn} = 0, 1,$ and

M is a large positive number

Observe that objective function 3.7 is the expected inter- and intra-module search cost. Constraint 3.8 ensures that no relation is left unassigned to a module. Constraint 3.9 ensures that if query k does not use module n, then none of the relations relevant to resolving query k will be assigned to module n. Constraint 3.10 ensures that the number of relations unused in module n is $y_{rn}(1-x_{rk})$ if query k requires module n, and is zero otherwise.

Using a value of $A = 2$, and using query probabilities of 0.1, 0.3, and 0.6 for queries AIC, AIL, and AON, respectively, Chakravarty and Sinha solve the above model; the optimal modules and queries are shown in Figure 3.10. One relation is unutilized for each of queries 1 and 2 of module 1, and 1 relation is unutilized for query 1 in module 3.

Note that if query probabilities change, the module composition may change. In fact, if the query probabilities are changed to 0.6, 0.3, and 0.1

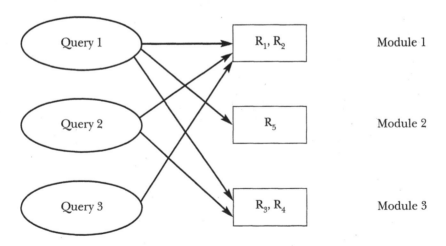

FIGURE 3.10: OPTIMAL RELATION–MODULES AND QUERIES

(Source: Chakravarty and Sinha 1990. Adapted and printed with permission of INFORMS, Linthicum, Maryland.)

for queries 1, 2, and 3, respectively, relations R_3 and R_5 migrate from modules 3 and 2, respectively, to module 1, and R_2 migrates from module 1 to module 2.

3.7 ENTERPRISE RESOURCE PLANNING

There are many routine decisions that must be made in any organization. These decisions can be very structured, such as the decision to pay a worker based on number of hours worked or the decision to place an inventory replenishment order based on current inventory status. Such decisions usually relate to processing of a mass of data on entities such as employee time accounting, asset accounting, vendor performance tracking, and monitoring product quality. The objective is to categorize the processed data (information) in such a way that any one of the categories may be used to arrive at a decision. Enterprise resource planning (ERP), a term used to describe software from companies software such as SAP, Manugistic, Baan, and PeopleSoft, attempts to rationalize the capture and processing on a mass scale of data that cut across functional boundaries.

Functions supported by typical ERP systems include finance, operations and logistics, human resources, and sales and marketing. Davenport (1998) lists the business processes supported by SAP's R/3 package as

Financials
Accounts receivable and payable
Asset accounting
Cash management and forecasting
Cost-element and cost-center accounting
Executive information system
Financial consolidation
General ledger
Product-cost accounting
Profitability analysis
Profit-center accounting
Standard and period-related costing

Human Resources
Human resources time accouting
Payroll

Personnel planning
Travel expenses

Operations and Logistics
Inventory management
Material requirements planning
Materials management
Plant maintenance
Production planning
Project management
Purchasing
Quality management
Routing management
Shipping
Vendor evaluation

Sales and Marketing
Order management
Pricing
Sales management
Sales planning

Scheer (1994) provides a different view of entities that need to be coordinated in a manufacturing plant. Figure 3.11 lists the entities in five broad groups, called order release, detailed scheduling, engineering, factory data entry, and data analysis. In designing an ERP, identification of system entities, as in Figure 3.4, would be the starting point. The entities are then linked by information flows in a way similar to that in entity relationship diagrams (Chen and Associates 1988).

At the core of an ERP lies a comprehensive database, a part of which may be similar to what is shown in Figure 3.11. The strength of any ERP lies in its ability to collect data from different points in the organization and feed them to different application modules, commonly known as business processes. A typical system can be configured to produce sales contracts for customers based on salesperson's inputs, production schedules, bills of material, routing of components in a facility, and shipping schedules. Thus, starting from a sales order, it can track and update all value chain functions, ending in updates of the corporate balance sheet, accounts payable, accounts receivable, cost center accounts, and corporate

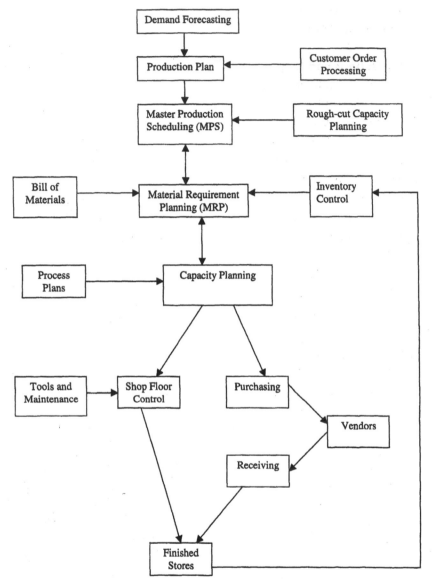

FIGURE 3.11: BUILDING BLOCKS IN AN ERP

cash balance. It can therefore provide real-time information to decision makers rapidly.

As Davenport points out, ERP systems may suffer from a lack of flexibility. Although such systems may provide in excess of 3,000 configuration tables so that organizations can customize the software to their business practices, the options are still limited. As a result, a company may be forced to abandon a competitive niche it might have created from its own unique business practices. The ideal solution, therefore, would involve selectivity in the choice of which business processes are configured on the ERP and which processes are retained independent of ERP for competitive advantage. Compaq Computer, for example, decided to develop its own proprietary order management system when it adopted ERP for the switch to a build-to-order system. Using multiple systems (ERP and a proprietary system), however, raises other problems of integration. Companies must now invest in interface software to create a linkage among multiple systems. They must also decide which processes should be made common throughout the company and be put on the ERP, and which processes (in which units of the company) must be retained untouched for their competitive advantage. This is the so-called federalist model. Davenport reports that companies such as Monsanto, Union Carbide, Dow Chemical, and Owens-Corning have adopted ERP as a centralized system, whereas Hewlett-Packard has implemented it as a federalist system.

Nokia Telecommunications, on the other hand, decided to develop its own ERP, as they saw a tremendous pressure from a demand-driven system that led to an increase in customization at local (country) levels. They created an entity called accounts management as an interface among customer demand, area management, product divisions, and customer service.

Nokia Telecommunications had a severe problem in coordinating product line managers with global customers (Jarvenppa and Tuomi 1995). Production of product lines was based on demand forecasts. With finished products in hand, salespeople frantically looked for customers, who often needed products different from what was being supplied. In 1995 Nokia designed and implemented a new system based on the notion of customer orders triggering component production in different product lines.

The redesigned system was based on three core business processes: product process, customer commitment process, and management and support process. The product process extends from the determination of business needs, at one end, to the development and launching of new systems, products, and features, at the other. The customer commitment process extends from tendering the order to order delivery, implementation, and after-sale service. The support process includes financials, human resource management, and direction setting.

It is clear that Nokia has created its own federalist system. While an ERP system provides a menu of configuration tables to choose from, Nokia's system permits development of new modules at the local levels and their integration with the whole system. It follows that a good ERP system must provide for optimization of the relationship chain, as we discussed in section 3.5. But today's ERPs come with predetermined knowledge modules (configuration tables). Nokia's system, on the other hand, allows optimal modularization of knowledge at local levels, before their integration.

3.8 IMPLEMENTATION

Database and Data Warehouse

The entity called "database" in the relationship diagram (Figure 3.6) is a traditional database used to support transaction processing. It is clear that because of the emphasis on relations (or postulates), a relational database (McFadden and Hoffer 1991) would be most appropriate. As discussed in section 3.7, the ERP databases are of this type. There are many similarities between entity relationship diagrams (Chen and Associates 1988) used in a relational database and the relationship diagram we have used to support decision making. Most relational databases permit the setting up of query tables. Such queries, however, relate to the evaluation of status variables using simple transactions or individual logical rules. Queries in a semi-structured decision-making environment, as we have seen, are not aimed at assessing the status values (truth values or numerical values) of specific variables (entities). The objective is to determine the set of knowledge entities that would be required by the decision maker to resolve specific queries. Thus the current relational database would need augmentation to provide this capability. Other approaches, such as knowledge frames (Bu-Huliga and Chakravarty 1988) and object-oriented knowledge representation (Chakravarty et al. 1997), may be more appropriate, as they have certain specific characteristics that may be uniquely suited to our current needs. For example, in a knowledge frame, entities can be entered as an "is a" or "if needed" type. The is-a characterization permits inheritance and can be resolved directly from database values. The if-needed characterization, on the other hand, permits running a model or evaluation of a *set* of logical statements. In our context, entities such as new quota or ROP will have an if-needed characterization and will be evaluated by running the mathematical programming model for relationship chain construction, described earlier. Bu-Huliga and Chakravarty (1988) show how a model for retooling of a flexible machine can be triggered using the if-needed characterization. In object-oriented representation, both entities and relations can be declared as objects. Chakravarty and colleagues (1997) show how mathematical heuristics and/or mathematical models can be manipulated and used by an object-oriented system.

Tacit Knowledge and Organization

Organization of domain knowledge must, in addition to addressing the issues of domain decisions, respond to system needs for transparency in the context of distributed coordination. While the modular organization of knowledge and chains of relations corresponding to a decision strengthen the L_F and L_k linkages (discussed in Chapter 2), the augmented L_I linkage should enable the decision aids in one domain to be run by another domain, for transparency. Well-structured decision aids help transparency, but the existence of tacit knowledge in a domain hampers it. To see this, let us revisit the sales quota problem in Figure 3.7. The IF condition in relation R_1 requires sales to exceed $1.15 \times$ quota. The constant 1.15 is a manifestation of tacit (or local) knowledge. Similarly, relation R_4 stipulates that efactor = growth/3, the constant ⅓ being derived from tacit knowledge. Note that these values of constants, once stipulated in a rule, can be easily retrieved by linkage L_I. However, as the domains are required to be proactive, the values of the constants, or the entire postulate corresponding to a relation, can change. Thus there may be a time lag between the relationship used in a domain (L_F or L_K) and that retrieved by another domain through linkage L_I. Thus, if the manufacturing domain is consistently more proactive than marketing, marketing will suffer from this lag effect. But this is how it is supposed to be; the domain that is more proactive will be rewarded. The second issue concerns structured and semi-structured postulates. It should be appreciated that the decision scenario in Figure 3.7 is semi-structured in that the best choice of the chain of relations is not apparent. But the postulates (corresponding to a relation) are well structured, as they are expressible mathematically. It is obvious that well-structured postulates provide better transparency. Most domain decision makers would claim that it is not always possible to come up with mathematical functions to describe a relationship. But the point being lost in this argument is that if a decision maker can arrive at a decision, however arduously, he/she should be able to express the mental model as a set of logical statements, using as many constants for tacit knowledge (as in the relations in Figure 3.7) as he/she sees fit. Remember, as emphasized earlier, that the constants and even entire postulates can be updated as more learning accrues in the system.

The other form of tacit knowledge relates to the assessment of query probabilities in a domain and the restructuring of relation modules as these probabilities change. Observe that, other than the lag effect discussed earlier, it does not affect distributed coordination, since other domains do not need to know how a particular domain organizes its knowledge modules. Finally, the issues of accessibility, protocols, and database integrity need to be addressed in linking an augmented L_I with L_F or L_K. Many such hurdles may be overcome by using the hyperlink format available in any intranet designed for an organization.

REFERENCES

Blattberg, R., and S. Nelson (1990), *Sales Promotion: Concepts, Methods, and Strategies,* Prentice Hall, Englewood Cliffs, N.J.

Bu-Huliga, M., and A. Chakravarty (1988), "An Object Oriented Knowledge Representation for Hierarchical Real Time Control of Flexible Manufacturing," *International Journal of Production Research,* vol. 36, pp. 777–793.

Chakravarty, A., and S. Ghose (1993), "Tracking Product-Process Interactions: A Research Paradigm," *Production and Operations Management,* vol. 2, no. 2, pp. 72–93.

Chakravarty, A., H. Jain, J. Liu, and D. Nazareth (1997), "Object Oriented Domain Analysis for Flexible Manufacturing Systems," *Integrated Computer Aided Engineering,* winter, pp. 290–309.

Chakravarty, A., and D. Sinha (1990), "Knowledge Modularization for Adaptive Decision Modeling," *ORSA Journal on Computing,* vol. 2, no. 4, pp. 312–324.

Chen and Associates (1988), *E-R Designer Reference Manual,* Chen and Associates Baton Rouge, La.

Chikan, A. (1990), *Inventory Models,* Kluwer Academic Publishers, Dordrecht, The Netherlands.

Davenport, T. (1998), "Putting the Enterprise into the Enterprise System," *Harvard Business Review,* July–August, pp. 121–131.

Dobson, G., and S. Kalish (1988), "Positioning and Pricing a Product Line," *Marketing Science,* vol. 7., no. 2, pp. 107–125.

Gensch, D., N. Arersa, and S. Moore (1990), "A Choice Modeling Market Information System that Enabled ABB Electric to Expand Its Market Share," *Interface*, vol. 20, no. 1, pp. 6–25.

Ghiaseddin, N., K. Matta, and D. Sinha (1994), "A Structured Expert System for Model Management in Inventory Control," *ORSA Journal on Computing*, vol. 6, no. 4, pp. 409–422.

Green, P., and V. Srinivasan (1990), "Conjoint Analysis in Marketing Research: New Developments and Directions," *Journal of Marketing*, vol. 54, no. 4, pp. 3–19.

Holsapple, C., and A. Whinston (1987), *Business Expert Systems*, Irwin, New York.

Jarvenpaa, S., and I. Tuomi (1995), "Nokia Telecommunications: Redesign of International Logistics," Harvard Business School Case 9-996-006.

Lillien, G., P. Kotler, and K. Murthy (1992), *Marketing Models*, Prentice Hall, Englewood Cliffs, N.J.

McFadden, F., and J. Hoffer (1991), *Database Management*, Benjamin/Cummings, Redwood City, Calif.

McGuire, T., and R. Staelin (1986), "Channel Efficiency Incentive Compatibility, Transfer Pricing, and Market Structure: An Equilibrium Analysis of Channel Relationships," in *Research in Marketing*, vol. 8, Bucklin (ed.), JAI Press, New York.

Scheer, A. (1994), *Business Process Reengineering*, Springer-Verlag, New York.

Product Design and Time-to-market

Assessing the relative values of product features that enhance market share but also increase manufacturing cost. Understanding how product introduction to markets may be hastened and how it may impact competition and profits.

4

Marketing Approaches to Product Design

In Chapter 2 we discussed how the interface between marketing and manufacturing could be operationalized by a "bidding" scheme. Both manufacturing and marketing domains are asked (separately) to maximize company-wide profit accruing from manufacture and sale of innovative products. As shown in Figure 2.10, the unique aspect of this framework is that bidding does not terminate with a set of one-time bids from domains; the parties concerned are allowed to improve their bids over several rounds by exploiting the transparency of decision making in both domains.

It is likely that in round 1 the marketing domain will maximize a company's revenues without regard to manufacturing costs, and the manufacturing domain will minimize manufacturing costs while ignoring market revenues. In subsequent rounds marketing will try to hold manufacturing cost below a threshold, equal to manufacturing cost in the previous round, while maximizing revenues. Thus emphasis in the marketing domain is to maximize revenue. In what follows in this chapter, we explore the revenue maximization objective, ignoring manufacturing costs.

Market revenue, a figure arrived at by multiplying unit price and sales, may be increased by increasing either unit price or sales. Sales, in turn, can be increased by bundling desirable attributes in products, by advertising and promotion, and by timely distribution. These, together with unit price, form the five elements of marketing mix that determine a company's marketing strategy. These five elements have direct and indirect implications for manufacturing cost and practices and will be explored further in later chapters.

In this chapter we discuss how a product design, in terms of bundling attributes, affects market share and market revenue. We divide this discussion into five sections: product positioning in attribute space, discrete choice model, part worth and conjoint model, deterministic model, and mass customization.

4.1 PRODUCT POSITIONING

Customers evaluate products on several dimensions. In Figure 3.1 the major entities that impact quality perception of a product are shown. These include the weights assigned to the importance of attributes, warranty, after-sale service, and price. Although entities such as warranty and after-sale service may be considered as attributes of a product, we differentiate such "soft" attributes, which are not engineered into the product, from "hard" attributes. For example, attributes such as speed and acceleration (car), cooling rate (refrigerator), wash cycles (washers), and memory (computer) are hard attributes, as they are specifically designed into the product by design engineers and determine the degree of difficulty in product manufacturing.

Consider two common attributes of a car: gas mileage (in mpg, miles per gallon), and time to accelerate (in seconds). Assume that car A is designed for 45 mpg and takes 18 seconds to accelerate from 0 to 60 mph, and car B is designed for 30 mpg and takes 12 seconds to accelerate from 0 to 60 mph. These two cars can be positioned in a two-dimensional attribute space, as shown in Figure 4.1, along with two other cars (C and D). Observe that cars such as A, B, and C lie on an efficient frontier. Car D, on the other hand, is not on the efficient frontier, as it is outperformed by car A on gas mileage and by car B on acceleration time. Car D does not outperform A or B on any attribute.

Notice that there is only a partial dominance between a car such as B and any other car on the efficient frontier. That is, one car may dominate another in some attributes but not in all. If gas mileage is very important to a customer, he/she will choose A, and if acceleration is very important, C will be chosen. If, on the other hand, both attributes are important, the

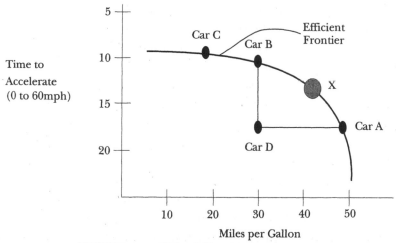

FIGURE 4.1: PRODUCT ATTRIBUTE SPACE

customer will choose *B*. Thus a customer, in general, will choose a car on the efficient frontier depending on the importance he/she assigns to the two attributes. This, however, does not imply that car *A* is ideal for the customer who chooses *A;* it implies only that this customer prefers *A* to *B* or *C*. If other choices were available, he/she probably would choose a different car. Assume now that a customer's ideal car corresponds to point *x* on the efficient frontier. Obviously, this customer will assess cars *A* and *B* in terms of the distance of point *x* (in the attribute space) from points *A* and *B*, and the customer's assigned importance weights for attributes.

If point *x* is close to *A*, the customer would buy car *A*, and if it is close to *B*, he/she would buy car *B*. It follows that there would exist a point on the efficient frontier between points *A* and *B* where a customer will be indifferent between cars *A* and *B*. Let this point be $\hat{x}$. Thus customers whose ideal points fall between *A* and their individual $\hat{x}$ would buy car A, and those with ideal points between their $\hat{x}$ and *B* would buy car *B*. To estimate how many customers buy car *A* (or car *B*), we would need to know the statistical distribution of customers' ideal points on the efficient frontier. Sales revenues can then be estimated for given product prices, and the best position of car *A* would be the one that maximizes its estimated profit. Given the distribution of customer ideal points, the decision variables are the attribute strengths (which determine the product's position in the attribute space) and its unit price.

The attribute strengths of a product, relative to a customer's ideal product, determine the utility of the product to the customer. Utility can therefore be measured as the distance between the product's positioning and the customer's ideal point. It is well known that if attribute dimensions are orthogonal to one another (as in Figure 4.1), distance between any two points in a multi-attribute space can be measured in a Euclidean sense. That is, the squared values of the difference between attribute strengths can be added.

Consider a single product in a market (i.e., no competition). Since customers have no other recourse, the product may be positioned anywhere. If cost is not a factor, the seller may want to position the product so as to maximize the sum of utilities of all customers, an abstract measure of customer satisfaction. However, if there is a competitive product, customer satisfaction would not be just an abstraction; it would be directly translated into market share. To ensure that the customer buys the product, the company would position the product so that the customer's ideal point lies between the product's positioning and his/her point of indifference. When there are several customers in the market with their ideal points distributed widely, it is not possible to ensure that every customer's ideal point will lie between their individual points of indifference and the product's positioning. In such a case product positioning must ensure that a maximum number of customers have their ideal points between the above bounds. This would be tantamount to maximizing market share.

Several procedure have been suggested for establishing a product's positioning as some "central tendency" of customer ideal points, in a profit-maximization context. These procedures tend to be simple or complex based on how many attribute dimensions are included in the analysis, and how the relative importance of attribute dimensions are established. The important assumption in all such positioning procedures is that a new product can be positioned (technologically) anywhere in an n-dimensional attribute space. That is, the number of potential locations is assumed to approach infinity. This assumption permits the product's utility to be measured as a continuous function of the product's positioning (measured by its coordinates). Hotelling's positioning model (Hotelling 1929) was perhaps the first of such procedures. It is not of much practical value, as only a single attribute is considered. It is, however, very helpful in bringing about conceptual clarity.

In what follows, we first discuss Hotelling's model to see how sales revenues are impacted by product positioning, and how the revenues can be maximized. We follow this with a discussion of the LINMAP model, where multiple attributes are permitted and where the importance weights of attributes are assessed based on choices customers make. Finally, we discuss the DEFENDER model, which translates distance measure to angles (at the origin) and points of indifference to preferred angles. This enables a graphical analysis and permits a quick interpretation and assessment of a new product as to whether it is an "attack" product.

Hotelling Positioning Model

We discuss a simple example where only a specific attribute is considered for repositioning a product. The importance weight of the attribute is not included in this analysis, as it would be irrelevant in a single-attribute scenario. We consider a competitive scenario where the company is interested in the optimal attribute-strength values of its product, as in Lilien, Kotler, and Murthy (1992). Let the attribute strength of the company's product be denoted by x_1 and that of the competing product by x_2. Let the corresponding unit prices be p_1 and p_2, respectively. Let the ideal points (in attribute strength) of customers be denoted by x, a random variable with probability density function $f(x)$. To simplify our exposition, we assume that $f(x)$ is uniformly distributed, $a \leq x \leq b$. Assume that the maximum utility of either product is A and is realized by those customers whose ideal points coincide with the positioning of either product. Since utility decreases with the distance from the ideal point, we can define the utility u_i of product i to a customer whose ideal point is at x as

$$u_i(x) = A - (x - x_i)^2$$

and the customer's surplus defined as utility – unit price, as

$$s_i(x) = A - (x - x_i)^2 - p_i$$

As discussed earlier, product i will be positioned to maximize its sales revenue (R_i) where

$$R_i = m_i p_i S \tag{4.1}$$

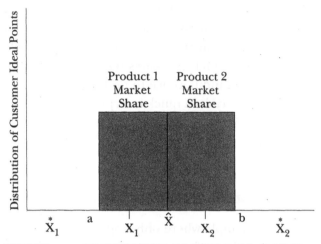

FIGURE 4.2: COMPETITIVE POSITIONING MODEL

(Source: Lilien et al. 1992. Adapted and printed with permission
from Prentice Hall, Inc.)

S is the market size and m_i is the market share of product i. The market
share of product 1 (assuming $x_1 < x_2$) may be expressed as (see Figure 4.2)

$$m_1 = (\hat{x} - a)/(b - a) \tag{4.2}$$

where $\hat{x}$, the point of indifference between products 1 and 2, is defined by
$s_1(\hat{x}) = s_2(\hat{x})$. That is,

$$A - (\hat{x} - x_1)^2 - p_1 = A - (\hat{x} - x_2)^2 - p_2$$

which simplifies to

$$\hat{x} = \frac{x_1 + x_2}{2} + \frac{p_2 - p_1}{2(x_2 - x_1)} \tag{4.3}$$

Substituting the value of $\hat{x}$ from equation 4.3, and m_1 from equation 4.2
into equation 4.1, we have

$$R_1 = \frac{p_1}{2} \left\{ \frac{(x_1 + x_2 - 2a) + \left(\dfrac{p_2 - p_1}{x_2 - x_1}\right)}{(b - a)} \right\} \cdot S \tag{4.4}$$

Similarly, we may write for $i = 2$,

$$R_2 = \frac{p_2}{2} \left\{ \frac{(2b - x_1 - x_2) - \left(\dfrac{p_2 - p_1}{x_2 - x_1}\right)}{(b - a)} \right\} \cdot S \qquad (4.5)$$

Observe that R_1 is concave in p_1 and convex in x_1. Similarly, R_2 is concave in p_2 and convex in x_2. Since R_1 and R_2 are being maximized, we may only set $\dfrac{\partial R_1}{\partial p_1} = 0$ and $\dfrac{\partial R_2}{\partial p_2} = 0$. That is,

$$\frac{\partial R_1}{\partial p_1} = \frac{S}{2(b - a)} \left\{ x_1 + x_2 - 2a + \frac{p_2 - 2p_1}{x_2 - x_1} \right\} = 0 \qquad (4.6)$$

$$\frac{\partial R_2}{\partial p_2} = \frac{S}{2(b - a)} \left\{ 2b - x_1 - x_2 - \frac{2p_2 - p_1}{x_2 - x_1} \right\} = 0 \qquad (4.7)$$

Solving equations 4.6 and 4.7 for p_1 and p_2, we have

$$p_1^* = \left(\frac{x_2 - x_1}{3} \right) (x_1 + x_2 + 2b - 4a) \qquad (4.8)$$

$$p_2^* = \left(\frac{x_2 - x_1}{3} \right) \{ -(x_1 + x_2) + 4b - 2a\} \qquad (4.9)$$

Substituting p_1^* and p_2^* in R_1 and R_2, we have

$$R_1 = \frac{x_2 - x_1}{18(b - a)} (x_1 + x_2 + 2b - 4a)^2 \cdot S \qquad (4.10)$$

$$R_2 = \frac{x_2 - x_1}{18(b - a)} \{ -(x_1 + x_2) + 4b - 2a\}^2 \cdot S \qquad (4.11)$$

R_1 can now be shown to be concave in x_1 (and R_2 concave in x_2).
  For maximization, $\frac{\partial R_1}{2x_1} = 0$ and $\frac{\partial R_2}{\partial x_2} = 0$. After simplification, we would have

$$x_1^* = \frac{x_2^* - 2(b - 2a)}{3}$$

$$x_2^* = \frac{x_1^* - 2(2b - a)}{3}$$

That is,

$$x_1^* = \frac{5a - b}{4}$$

$$x_2^* = \frac{5b - a}{4}$$

The corresponding prices would be

$$p_1^* = 3(b - a)^2/2 = p_2^*$$

The point of indifference, $\hat{x} = \dfrac{p_2 - p_1}{2(x_2 - x_1)} + \dfrac{x_1 + x_2}{2}$, can be simplified to

$$\hat{x} = (a + b)/2$$

The optimal revenues would be

$$R_1^* = R_2^* = 3S(b - a)^2/4$$

Notice that $x_1^* < a$ and $x_2^* > b$, since a $<$ b. As customer ideal points lie in the interval from a to b, it implies that neither of the two products exactly corresponds to the ideal points of any customer, as shown in Figure 4.2.

It can be shown that

$$a - x_1^* = x_2^* - b = \delta$$

Hence

$$x_1^* + x_2^* = a - \delta + b + \delta = a + b$$

$$x_1^* - x_2^* = a - b - 2\delta$$

Therefore, it follows from equations 4.8 and 4.9 that both p_1^* and p_2^* increase with δ. Since market share of the products shown in Figure 4.2 is not affected when $\delta > 0$, it follows that in the optimal solution (for $\delta > 0$), the prices p_1 and p_2 can be increased with δ, increasing both companies, revenues. Prices, of course, cannot be increased indefinitely, as customer surplus will become negative eventually.

It is interesting to compare the above results with what would happen in a monopoly. It is clear that the monopolist will not gain by a price differential between products. It can be shown that the monopolist will position the products at $x_1^* = (3a + b)/4$ and $x_2^* = (3b + a)/4$ and price both products at $p^* = A - \frac{(b-a)^2}{4}$ so that $R^* = p^* = A - \frac{(b-a)^2}{4}$. Comparing the monopoly with duopoly, we see that

$$x_2^* - x_1^* = (b - a) / 2 \text{ in a monopoly}$$

$$x_2^* - x_1^* = 3(b - a) / 2 \text{ in a duopoly}$$

That is, price competition induces the two companies to move their products farther apart than what a monopolist would.

Eliashberg and Manrai (1992) consider a scenario with two attributes (denoted as x and y), two customer segments, and a utility function of the type $\{(x_a - x)^2 + (y_a - y)^2\}^{-1}$, where x_a, y_a is the location of the ideal product. They show that (1) if the customer segments are not very heterogeneous or if the utility of existing product(s) is low, the new product would be positioned to serve both segments equally; otherwise it would primarily serve a single segment, and (2) in the early stages of a product's life cycle, when technology evolution is in its early stages, a single-segment product focus would be optimal.

Example (Hotelling Model)

Let $b = 80$, $a = 30$, $S = 100$.

Optimal product positioning

Product 1	$x_1^* = \dfrac{5a - b}{4} = \dfrac{150 - 80}{4}$	$= 17.5$
Product 2	$x_2^* = \dfrac{5b - a}{4} = \dfrac{400 - 30}{4}$	$= 92.5$

Optimal price

Product 1 $\qquad p_1^* = \dfrac{3(b-a)^2}{2} = \dfrac{3(80-30)^2}{2} = 3{,}750$

Product 2 $\qquad\qquad p_2^* = p_1^* = 3{,}750$

Optimal revenue

Product 1 $\qquad\qquad R_1^* = \dfrac{3S(b-a)^2}{4}$

$$= \dfrac{3 \times 100 \times (80-30)^2}{4} = 187{,}500$$

Product 2 $\qquad\qquad R_2^* = R_1^* = 187{,}500$

Arbitrary positioning

Consider now variation of price and revenue with x_1 and x_2.
From equation 4.8 we have

$$p_1 = \left(\frac{x_2 - x_1}{3}\right)(x_1 + x_2 + 40) = \{(x_2^2 + 40x_2) - (x_1^2 + 40x_1)\} / 3$$

Similarly, from equation 4.9 we have

$$p_2 = \{(x_1^2 - 260x_1) - (x_2^2 - 260x_2)\} / 3$$

Note the following:

$$\frac{\partial p_1}{\partial x_1} = \frac{-(2x_1 + 40)}{3}, \frac{\partial p_1}{\partial x_2} = \frac{2x_2 + 40}{3}$$

$$\frac{\partial p_2}{\partial x_1} = \frac{2x_1 - 260}{3}, \frac{\partial p_2}{\partial x_2} = \frac{-(2x_2 - 260)}{3}$$

It is clear that p_1 is a decreasing concave function of x_1 and an increasing convex function of x_2. Similarly, p_2 is a convex function of x_1 (stationary point: $x_1 = 130$) and a concave function of x_2 (stationary point: $x_2 = 130$). That is, if the two products get closer in their positioning, p_1 will decrease, but p_2 may increase or decrease.

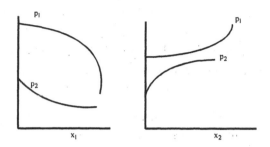

Observe that if x_1 is increased from its optimal value of 17.5, both p_1 and p_2 will drop, but p_1 will drop more slowly than p_2 initially. For example if x_1 is increased to 32.5 while holding x_2 at 92.5, p_1 will drop to 3,300 (from 3,750), and p_2 will drop to 2,700 (from 3,750). Revenue of the first product, R_1, will decrease slightly to 181,500 (from 187,500), and R_2 will decrease drastically to 67,500 (from 187,500). Note that there will be a net drop of 126,000 in total revenue, $R_1 + R_2$.

LINMAP Model

Positioning a product in a multi-attribute space becomes much more complex, as the number of dimensions increases and the importance weights of attributes are incorporated. Shocker and Srinivasan (1974) propose a linear programming model called LINMAP that jointly establishes individual ideal points and attribute-importance weights for customer responses.

The model has two distinct stages. In stage 1 they determine ideal products and attribute-importance weights for all customers. In stage 2 they use this information to locate the position (X_n) of a new product that maximizes the company's revenue. The following notation is used in stage 1.

Stage 1 Formulation

$I\,(i \in I)$ is a set of all individual customers

$J\,(j \in J)$ is a set of products

$N\,(n \in N)$ is a set of product-attributes

Note that attribute n is the same as dimension n in the attribute space.

x_{jn} = position of product j in dimension n in the attribute space; x_{jn} is measured as the strength of the nth attribute in product j (known)

w_{in} = importance weight of attribute n (dimension n), as imputed from customer i (unknown)

X_{in} = position of customer's ideal product in dimension n (unknown)

x_{kn} = position of the new product k in dimension n (unknown)

y_{ij} = distance of product j from customer i's ideal product (unknown)

$P\,(p \in P)$ is a set of all product pairs.

Customer i is asked to rank two products in pair p as preferred and not preferred. Using this information, the authors define

$u_{i,p1}$ = distance of customer i's preferred product in pair p from his/her ideal product

$u_{i,p2}$ = distance of customer i's non-preferred product from his/her ideal point

Thus

$$u_{i,p1} < u_{i,p2}, \quad \forall p \in P \tag{4.12}$$

Using appropriate values of j (consistent with $p1$ and $p2$), the authors define a distance measure

$$u_{ip1} \text{ (or } u_{ip2}) = y_{ij} = \sum_{n \in N} (x_{jn} - X_{in})^2 \, w_{in} \tag{4.13}$$

where $j = p1$ or $p2$, that is, the first or second product in pair p.

As an example, assume that the sixth product pair comprises products 2 and 5, where customer i's preference is product 5. Then the value of j in equation 4.13 corresponding to $p1$ would be 5, and for $p2$ it would be 2. Observe that for any given values of X_{in} and w_{in} ($n \in N$), values of $u_{i,p1}$

and $u_{i,p2}$ can be determined from equation 4.13, and the violation of condition 4.12, measured as $u_{i,p1} - u_{i,p2}$, is determined. Obviously, if condition 4.12 is not violated, the violation measure will have a negative value. Clearly, for optimal values of X_{in} and w_{in}, the sum of the violation measures over all product pairs would be minimized. Since a positive value of violation in a pair cannot be traded off with a negative value, all negative values of violation in the above summation are set equal to zero. Mathematically we express this as determine X_{in} and w_{in} to

$$\text{Minimize} \sum_{p \in P} z_{ip} \qquad (4.14)$$

such that

$$z_{ip} \geq u_{i,p1} - u_{i,p2}, \forall\, p \in P \qquad (4.15)$$

The authors use a unique scaling for normalization, so that

$$\sum_{p \in P} (u_{i,p2} - u_{i,p1}) = 1 \qquad (4.16)$$

The formulation of expressions 4.14 to 4.16 is solved individually for each customer i. Note that the model is not linear because of the nonlinearity in equation 4.13. The authors suggest several variable transformations to transform the entire model to a linear form. Solution of the model provides values of X_{in}, w_{in} (for all n), and y_{ij} from equation 4.13.

The probability of customer i purchasing product j is estimated from

$$\Pi_{ij} = \frac{a_i}{y_{ij}^b}, \sum_j \Pi_{ij} = 1$$

where a_i is a constant and the parameter b is estimated separately from the historical data of multiple customers.

Stage 2: Positioning of a New Product

Let $q_i = Prob$ (customer i buys a product from the company after the new product k is introduced).

Therefore,

$$q_i = \frac{\Pi_{ik} + \sum\limits_{j \in J} \Pi_{ij}}{\Pi_{ik} + \sum\limits_{j \in J} \Pi_{ij} + \sum\limits_{r \in J} \Pi_{ir}}$$

where $\bar{J}$ is the set of competing products.

Letting $\sum\limits_{j \in J} \Pi_{ij} = h_i$ and Q_i = annual purchase of customer i, we can express incremental sales to customer i as $(q_i - h_i) Q_i$. Further, if we let $C(x_{k1}, x_{k2} \dots x_{km})$ as the unit cost of producing the new product, and p its unit price, we can write incremental profit as

$$\psi = \sum_{i \in I} (q_i - h_i) \{p - C(x_{k1}, x_{k2} \dots x_{km})\} Q_i \qquad (4.17)$$

Maximization of ψ determines x_{kn} ($n = 1$ to m), for the new product k. Inclusion of the cost term in equation 4.17, which is a multidimensional nonlinear function, makes solution of the above formulation very hard (Shocker and Srinivasan do not solve the model).

DEFENDER Model

The DEFENDER model was proposed by Hauser and Shugan (1983). It is a directional model in that it establishes the direction in which a product must be repositioned in the attribute space after a competitor introduces a new product to attack the company's product. In Figure 4.3, positioning of four products in attribute space is shown. All attribute values are rescaled to attribute-value per dollar.

Let

$$p_j = \text{unit price of product } j$$

$$x_{jn} = \frac{y_{jn}}{p_j} = \text{position of product } j \text{ on attribute } n, \text{ where } y_{jn} = \text{strength}$$

$$\text{of attribute } n \text{ in product } j$$

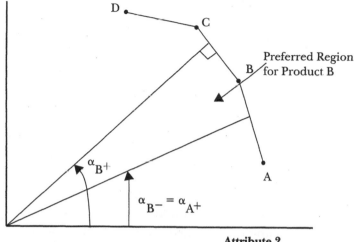

FIGURE 4.3: PREFERRED ANGLES IN ATTRIBUTE / $ SPACE

w_{in} = importance weight of attribute n to customer i.

u_{ij} = utility of product j to customer i

α_i = preferred angle for customer i

Thus

$$u_{ij} = \sum_n w_{in} x_{jn} = \sum_n w_{in} y_{jn} / p_j \qquad (4.18)$$

To understand the notion of preferred angle, consider a customer i who is indifferent between products A and B. Therefore, $u_{iA} = u_{iB}$. That is, with two attributes per product,

$$\sum_{n=1}^{2} w_{in} y_{An} / p_A = \sum_{n=1}^{2} w_{in} y_{Bn} / p_B$$

That is,

$$w_{i1}\left(\frac{y_{A1}}{p_A} - \frac{y_{B1}}{p_B}\right) = w_{i2}\left(\frac{y_{B2}}{p_B} - \frac{y_{A2}}{p_A}\right)$$

Which simplifies to

$$\frac{\left(\dfrac{y_{A1}}{p_A} - \dfrac{y_{B1}}{p_B}\right)}{\left(\dfrac{y_{B2}}{p_B} - \dfrac{y_{A2}}{p_A}\right)} - \frac{w_{i2}}{w_{i1}} = tan\,(\alpha_i) \qquad (4.19)$$

Consider a line drawn from the origin, orthogonal to the line AB, as shown in Figure 4.3. It can be shown that if customer i's ideal product is located on the orthogonal line from the origin, customer i will be indifferent between products A and B, and the orthogonal line will be at an angle α_i with respect to attribute 1. This would happen only if the ratio w_{i2}/w_{i1} equals tan (α_i). This angle α_i is called the preferred angle for customer i in the context of products A and B and is denoted as α_{A+} or α_{B-} (Figure 4.3). Assuming w_{i1} and w_{i2} vary widely among customers, the preferred angle will be distributed between 90° and 0° in the attribute space. We can compute the expected number of customers whose preferred angles fall between α_{B+} and α_{B-}, between products A and C, as shown in Figure 4.3. The region bounded by α_{B+} and α_{B-} is known as the preferred region for product B over all customers as a whole. Thus the concept of preferred angle is somewhat similar to that of an ideal product, but it is much more powerful, as it refers to an unlimited number of ideal points lying on the orthogonal line.

Hauser and Shugan define a competitor's product E as an "attack" product in regard to B if E is positioned adjacent to B in the attribute space. Adjacency requires that E is positioned between B and C or B and A, as shown in Figure 4.4. If it is positioned beyond C or below A, it will have no effect on B's sales. Let α_{B+} be denoted as $\alpha_{B+,E}$ after E is introduced.

It is clear from Figure 4.4 that $\alpha_{B+,E} < \alpha_{B+}$, and hence some of B's market share will be lost to E, as shown in the figure. The company can respond by repositioning B to location B' so as to increase the value of the preferred angle $\alpha_{B+,E}$.

Hauser and Shugan show that the optimal response to attack from E would be to reposition B to B' and cut its unit price simultaneously.

4.2 DISCRETE CHOICE MODELS

Observe that in determining product positioning, customers are involved only in terms of identifying their individual ideal products. Utility (or dis-

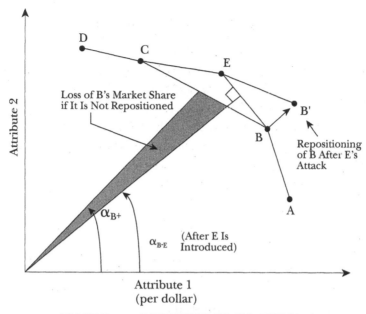

FIGURE 4.4: RESPONSE TO E'S ATTACH

utility) of a product is measured in terms of the distance between the position of the product and that of the customer's ideal product. Attribute strength is treated as a continuous variable, but the mathematical expression for market share of a new product with an arbitrary attribute bundling becomes too cumbersome (mathematically) to use in decision making. Although the graphic capabilities of positioning models are valuable for generating insights, their usefulness is limited to a scenario with only a few products, a few product attributes, and few customers (aggregated into segments for tractability).

In choice-based market analysis, on the other hand, attribute strength is treated as a discrete variable. Customers are asked to provide information on their attribute-importance weights and on their ranking of new products that have specified attribute levels. Utility models have been constructed that exploit this rich source of data from customers. Since customer response data may suffer from low reliability, utility models allow for random errors in observations. The choice mechanism (decision to buy), however, is deterministic. The resulting market share model (such as the logit model), unlike the positioning model, is extremely amenable to mathematical manipulations and is easier (compared to positioning models) to use for parameter estimation. Discrete choice models, therefore, have found increasing use, especially in designing a product line.

Logit and Probit Models

Consider the problem of making a choice of one of two products, j and k. The process of making a choice is discrete (either j or k) and is called the binary choice problem (Ben-Akiva and Lerman 1985). Assume that utility of product j to customer i is u_{ij}. Since the utility value is obtained by getting the customer to express how much he/she values the product, there is always an element of uncertainty about the solicited value of utility. The true value of utility u_{ij} is therefore expressed as the sum of its assessed value v_{ij} and an assessment error $\in_{ij}$. That is,

$$u_{ij} = v_{ij} + \in_{ij}$$

According to the theory of need gratification, customer i will choose product j over product i if $u_{ij} \geq u_{ik}$. Therefore, the probability of customer i purchasing product j can be expressed as

$$P_{ij} = Prob\left(u_{ij} \geq u_{ik}\right)$$

which can be rewritten as

$$P_{ij} = Prob\left(\in_{ik} - \in_{ij} \leq v_{ij} - v_{ik}\right) \qquad (4.20)$$

where $\in_{ij}$ and $\in_{ik}$ are independent random variables.

Let $\in_i = \in_{ik} - \in_{ij}$. Obviously, the form of the distribution of $\in_i$ must be known for P_{ij} to be determined. Different forms of $\in_i$'s distribution will lead to different choice models. A logistic distribution will lead to the logit model and a normal distribution will lead to the probit model. Consider the logistics distribution first. The cumulative probability, $Prob\left(\in_i \leq x\right)$ can be expressed as

$$F(x) = \{1 + exp\left(-\beta x\right)\}^{-1}$$

Therefore,

$$Prob\left(\in_i < v_{ij} - v_{ik}\right) = \left[1 + exp\left\{-\beta(v_{ij} - v_{ik})\right\}\right]^{-1}$$

Thus

$$P_{ij} = Prob\,(\in_i < v_{ij} - v_{ik}) = \frac{exp\,(\beta\,v_{ij})}{exp\,(\beta\,v_{ij}) + exp\,(\beta\,v_{ik})} \qquad (4.21)$$

which is the binary logit model.

We obtain the same result as in equation 4.21 if the distribution of $\in_i$ is double-exponential. That is,

$$Prob\,(\in_i \leq x) = F(x) = exp\,\{-exp\,(-\beta x)\}$$

If, on the other hand, $\in_i$ is distributed as $N\,(0, \sigma^2)$, it would lead to the probit model.

Thus,

$$Prob\,(\in_i \leq x) = F(x) = \int_{-\infty}^{x} \frac{1}{(\sigma\sqrt{2\Pi})}\, exp\left\{\frac{-(\in_i/\sigma)^2}{2}\right\} d\in_i$$

Therefore

$$P_{ij} = Prob\,(\in_i < v_{ij} - v_{ik}) = \int_{-\infty}^{v_{ij}-v_{ik}} \frac{1}{(\sigma\sqrt{2\Pi})}\, exp\left\{\frac{-(x/\sigma)^2}{2}\right\} dx \qquad (4.22)$$

where $x = v_{ij} - v_{ik}$.

The probit model, as in equation 4.22, lacks a closed-form solution, and hence it is computationally very inefficient.

The binary logit model can be easily extended to a choice problem involving more than two products. Now the probability of choosing product j over product i will be expressed as

$$P_{ij} = Prob\,(u_{ij} \geq u_{ik}), \text{ for all } k \in S_i$$

where S_i is the set of products that customer i is considering for purchase.

In terms of earlier notations we have

$$P_{ij} = \prod_{k \neq j} \{Prob\,(\in_{ik} \leq v_{ij} - v_{ik} + \in_{ij})\}, \text{ for all } k \in S_i$$

where

$$S_i = S \text{ for all } i$$

Using a double-exponential distribution, we can write

$$Prob\left(\in_{i1} \le x_1 \text{ and } \in_{i2} \le x_2 \text{ and } \cdots \in_{im} \le x_m\right) = \prod_{k \ne j} exp\left\{e^{-\beta(v_{ij} - v_{ik} + \in_{ij})}\right\}$$

Integrating $\in_{ij}$ out of the above equation, we get

$$P_{ij} = \frac{exp\left(\beta\, v_{ij}\right)}{\sum_{k \in S} exp\left(\beta\, v_{ik}\right)} \tag{4.23}$$

The utility v_{ij} is, of course, influenced by the attribute strengths of the product. For different application scenarios, different expressions for v_{ij} have been used. The most common application has been that of the linear compensatory value model

$$v_{ij} = \sum_n w_{in}\, x_{ijn} \tag{4.24}$$

where

x_{ijn} is the respondent i's evaluation of product j with respect to attribute n

w_{in} is the importance weight assigned to attribute n, by customer i

Substituting equation 4.24 into 4.23 and letting $\beta = 1$, we get

$$P_{ij} = \frac{exp\left(\sum_n w_{in}\, x_{ijn}\right)}{\sum_{k \in S} exp\left(\sum_n w_{in}\, x_{ikn}\right)} \tag{4.25}$$

Clearly equation 4.25 is a special form of Luce's axiom. The importance weight w_{in} can be estimated using procedures such as conjoint analysis, and will be discussed later.

Consider the example in Chakravarty and Baum (1992), which uses the multinomial logit model to compute market share of new products. They consider two attributes and two possible levels (strengths) of each attribute. The various combinations of attributes and their strengths lead to eight possible products, shown in Table 4.1.

The importance weights are representative values (expected values) of the entire set of customers (over all i). Consider a product line consisting of the first three products. Using equation 4.24, the utility values of products would be

$$v_1 = 0.4, \qquad v_2 = 0.6, \qquad v_3 = 1.0$$

Assume that the utility of the status quo product(s) is 2.0. Since the market share of product j equals the probability that it would be bought, we can write the market shares m_j of product j (using equation 4.25) as

$$m_1 = \frac{e^{0.4}}{e^{0.4} + e^{0.6} + e^{1.0} + e^{2.0}} = 0.1113 = 11.13\%$$

$$m_2 = \frac{e^{0.6}}{e^{0.4} + e^{0.6} + e^{1.0} + e^{2.0}} = 0.1357 = 13.57\%$$

$$m_3 = \frac{e^{1.0}}{e^{0.4} + e^{0.6} + e^{1.0} + e^{2.0}} = 0.2022 = 20.22\%$$

m_s (market share of status-quo product) = 55.08%

Assume that a fourth product that has only attribute 1 but at level 2—that is $v_4 = 0.8$—is to be introduced. The denominator would equal $e^{0.4} + e^{0.6} + $

| Attribute | Importance | Attribute Levels | | | | | | | |
(n)	Weight w_{in}	Product 1	Product 2	Product 3	Product 4	Product 5	Product 6	Product 7	Product 8
1	0.4	1	0	1	2	2	2	0	1
2	0.6	0	1	1	0	1	2	2	2

Table 4.1 Attribute Bundling in Product Design

(Source: Chakravarty and Baum 1992.)

$e^{1.0} + e^{0.8} + e^{2.0} = 15.647$, and the revised market share would equal $m'_1 = 9.55\%$, $m'_2 = 11.64\%$, $m'_3 = 17.34\%$, $m'_4 = 14.22\%$, $m'_s = 47.25\%$ Observe that all three old products and the status quo product lose market share to the new product. It is implicitly assumed that the low values of one attribute can be compensated for by the high values of another. This may not hold in all situations, as explained below.

Noncompensatory utility models have been proposed (Beltman 1979) for situations when an extreme value of one attribute may be used to decide for the entire product. There are two types of noncompensatory models, conjunctive and disjunctive. Conjunctive models are applicable in a scenario where an unsatisfactory value for any single attribute can disqualify the entire product. For example, a safety-conscious customer may not compromise on this attribute no matter how superior other attributes of a car happen to be. A threshold value T_{ijn} for attribute n and product j is determined for each customer i, so that customer i would not buy the product if the strength of attribute n is below this threshold. Thus, we can define a binary variable δ_{ijn} as

$$\delta_{ijn} = 1, \text{ if } x_{ijn} \geq T_{ijn}, \text{ and}$$
$$\delta_{ijn} = 0, \text{ if } x_{ijn} < T_{ijn}$$

The probability P_{ij} of customer i purchasing product j is therefore expressed as

$$P_{ij} = \prod_{n} (\delta_{ijn}) \tag{4.26}$$

The flip side of this scenario, where no single attribute can disqualify a product, is modeled as a disjunctive noncompensatory model. This implies that all attributes of the product have to be below their respective threshold values for the product to be disqualified. We can model this simply by modifying the definition of p_{ij} as in equation 4.26, to

$$P_{ij} = \text{Minimum}\left(\sum_{n} \delta_{ijn}, 1\right) \tag{4.27}$$

Gensch, Arersa, and Moore (1990) have used a combination of compensatory and noncompensatory logit models to design attribute bundles of products for ABB Electric. Their model consists of two stages. In stage 1, they rescale the attribute values to zero or the difference from a threshold value, as

$$\delta_{ijn} = \text{Maximum } (0, x_{ijn} - T_{ijn}), \text{ and}$$

$$y_{ij} = \prod_{n} (\delta_{ijn}) \tag{4.28}$$

The above noncompensatory model eliminates any product with at least one attribute below the threshold value. In stage 2 of the model, they use a compensatory utility model as in equations 4.24 and 4.25 for products remaining in the choice set. They report improvement in prediction accuracy by a factor of three, in comparison to using a single-stage (compensatory or noncompensatory) model.

4.3 CONJOINT ANALYSIS

Conjoint analysis is an experimental procedure for assessing values of w_{in}, which are also called part-worths of attributes. Respondents are asked to rate each product (bundle of attributes, with attributes at different levels) on a 0 to 100 scale. The group of products evaluated by customers is chosen carefully, so that none of the products dominates in terms of attribute levels (partial dominance is okay). Thus, for the example in Table 4.1, products 1, 2, and 3 are the only non-dominated products. A regression equation relating the part-worths to product rating v_{ij} is set up as

$$v_{ij} = \sum_{n=1}^{N} \sum_{m=1}^{Mn} a_{inm} x_{jnm} + \epsilon_{ij} \tag{4.29}$$

where

v_{ij} = respondent i's rating of product concept j (known)

a_{inm} = part-worth of attribute n at level m to respondent i (unknown)

x_{jnm}= 1, if the mth level of the nth attribute is present
in product concept j, and 0 otherwise (known)

M_n = number of levels of attribute n (known)

N = number of attributes (possible) of the product (known)

$\in_{ij}$ = random error with $N(0, \sigma^2)$ distribution

The values of a_{inm} can be estimated using a regression-type program, and the utilities with the estimated part-worths $\hat{a}_{inm}$ can be calculated from

$$v_{ij} = \sum_{n=1}^{N} \sum_{m=1}^{Mn} \hat{a}_{inm} \, x_{jnm}$$

Pizza Design

Lilien and Rangaswamy (1998) describe the use of conjoint analysis in designing pizza with five attributes: crust, topping, cheese type, cheese amount, and price. The levels of each attribute and the part-worths (from conjoint analysis) are shown in Table 4.2.

Using the above part-worth information, new product concepts can be easily evaluated. Table 4.3 shows three possible new products, called Aloha Special, Meat-Lover's Treat, and Veggie Delite.

To estimate the 16 values of a_{inm} in regression (equation 4.29), 16

Attribute Type	Level 1 Description	Level 1 Part-Worth	Level 2 Description	Level 2 Part-Worth	Level 3 Description	Level 3 Part-Worth	Level 4 Description	Level 4 Part-Worth
Crust	Pan	0	Thin	10	Thick	15		
Topping	Pine-apple	0	Veggie	10	Sausage	25	Pepperoni	30
Cheese Type	Romano	0	Mixed Cheese	3	Mozzarella	10		
Cheese Amount	2 oz.	0	4 oz.	8	6 oz.	10		
Price	$9.99	0	$8.99	20	$7.99	35		

Table 4.2: Conjoint Analysis for Pizza Design

(Source: Lilien and Rangaswamy 1998. Adapted and printed with permission from Addison Wesley.)

Products	Attributes										Utility
	Crust		Topping		Cheese Type		Cheese Amount		Price		
	Level	Part-Worth	Level	Part-Worth	Level	Part-Worth	Level	Part-Worth	Level	Part-Worth	
Aloha	Pan	0	Pine-apple	0	Moz-zarella	10	4 oz.	8	$8.99	20	38
Special Meat-Lover's Treat	Thick	15	Pep-peroni	30	Mixed Cheeses	3	6 oz.	10	$9.99	0	58
Veggie Delite	Thin	10	Veggie	10	Romano	0	2 oz.	0	$7.99	35	55

Table 4.3: Evaluation of New Product Concepts

(Source: Lilien and Rangaswamy 1998. Adapted and printed
with permission from Addison Wesley.)

product bundles (one for each product concept) were presented to the customers for their preference rating. Note that with five attributes—four at three levels each and one at four levels—the number of possible product bundles are $3^4 \times 4 \times 3 \times 3 \times 3 = 324$. However, it is not necessary to rate all 324 product bundles in the experiment, since there are only 16 unknowns (a_{inm}) in the regression equation. These 16 orthogonal product concepts, along with customers' preference scores, are shown in Table 4.4.

A regression analysis software can be used to estimate the 16 part-worth values by solving simultaneous equations, one for each row in Table 4.4.

Food Processor Design

Page and Rosenbaum (1987) have applied conjoint analysis to redesign an existing food-processor product line at Sunbeam Corporation. They identified 12 attributes of the product, as shown in Table 4.5. See Page and Rosenbaum (1987) for details in tables 4.5 to 4.7.

Based on part-worths of attribute levels from a conjoint analysis (one for each attribute), the most desirable attribute level for each attribute type is shown with asterisks in Table 4.5. However, since the relative importances of the attributes to the customers are not equal, the desirable attribute levels may not constitute the optimal choice for the food processor

Product Bundle	Crust	Topping	Type of Cheese	Amount of Cheese	Price	Performance Score
1	Pan	Pineapple	Romano	2 oz.	$9.99	0
2	Thin	Pineapple	Mixed	6 oz.	$8.99	43
3	Thick	Pineapple	Mozzarella	4 oz	$8.99	53
4	Thin	Pineapple	Mixed	4 oz.	$7.99	56
5	Pan	Veggie	Mixed	4 oz.	$8.99	41
6	Thin	Veggie	Romano	4 oz.	$7.99	63
7	Thick	Veggie	Mixed	6 oz.	$9.99	38
8	Thin	Veggie	Mozzarella	2 oz.	$8.99	53
9	Thick	Pepperoni	Mozzarella	6 oz.	$7.99	68
10	Thin	Pepperoni	Mixed	2 oz.	$8.99	46
11	Pan	Pepperoni	Romano	4 oz.	$8.99	80
12	Thin	Pepperoni	Mixed	4 oz.	$9.99	58
13	Pan	Sausage	Mixed	4 oz.	$8.99	61
14	Thin	Sausage	Mozzarella	4 oz.	$9.99	57
15	Thick	Sausage	Mixed	2 oz.	$7.99	83
16	Thin	Sausage	Romano	6 oz.	$8.99	70

Table 4.4: Orthogonal Products for Conjoint Analysis

(Source: Lilien and Rangaswamy 1998. Adapted and printed with permission from Addison Wesley.)

Attributes	Level 1	Level 2	Level 3
Price	$49.99*	$99.99	$199.99
Motor	Regular	Heavy-duty*	Professional
Number of blades	3	5	7*
Bowl size	1.5 quarts	2.5 quarts*	4 quarts
Number of speeds	1	2*	7
Other use	None*	Blender	Mixer
Configuration (motor and bowl)	Side by side*	Bowl on top	In a cabinet
Bowl type	Regular*	Side discharge	—
Feed tube pusher	Regular—solid*	3 interchangeable components	—
Feed tube size	Regular	Large*	—
Bowl shape	Cylindrical*	Spherical	—
Pouring spout	Present*	Absent	—

Table 4.5: Attributes of a Food Processor

(Source: Page and Rosenbaum 1987. Reprinted with permission from Elsevier Science.)

as a whole. The importance weights were determined as price (13.8%), number of speeds (4.2%), bowl shape (7.6%), bowl size (11.7%), motor power (17.7%), other uses (10.6%), configuration (5.4%), bowl type (4.4%), feed tube pusher (2.5%), size of tube (1.4%), number of blades (18.4%), and pouring spout (2.3%). Thus an average buyer would trade a large feed tube for a low price or a larger number of blades.

To evaluate new product lines, the managers created a select few attribute profiles from 69,984 ($3^7 \times 2^5$) possibilities and used simulation to ascertain the utility of each profile. Twelve utility functions, one for each attribute type, were used in this simulation. With these utility values, they then used an appropriate procedure, which they called MANANOVA, to compute the market share of each profile. The product profiles chosen by Sunbeam are shown in Table 4.6.

The market share (computed values) with five new models under two different market scenarios are shown in Table 4.7. The combined market share of new models, at about 25%, exceeds the market share of Sunbeam's old line by 18%. After additional trials of new product concepts, Sunbeam chose a product line with three models, and increased its market share by 10% within a year.

	Model 1	Model 2	Model 3	Model 4	Model 5
I. Price	$199.99	$45.99	$99.99	$99.99	$49.99
II. Motor power (1, 2, or 3)	3	1	2	2	2
III. Number of blades	5	3	5	5	5
IV. Bowl size	4	2.5	1.5	1.5	1.5
V. Number of speeds	1	1	7	7	2
VI. Other uses (1, 2, or 3)	1	1	1	1	1
VII. Configuration (1, 2, or 3)	2	1	2	2	2
VIII. Bowl type (1 or 2)	1	1	1	2	1
IX. Type of feed tube pusher (1 or 2)	2	1	2	2	1
X. Size of feed tube (1 or 2)	2	1	2	2	2
XI. Bowl shape (1 or 2)	1	1	1	1	1
XII. Pouring spout (1 or 2)	2	1	2	2	2

Table 4.6: Input for Market Share Simulations

(Source: Page and Rosenbaum 1987. Reprinted with permission from Elsevier Science.)

Brands and Models	Predicted Market Shares (%)	
	Scenario 1	Scenario 2
Sunbeam		
Model 1—Professional ($179)	7.1	7.0
Model 2—14056 ($30)	3.3	3.2
Model 3—Five speeds ($59)	5.8	4.0
Model 4—Seven speeds ($79)	8.7	8.7
Model 5—Two speeds ($49)	—	3.9
Brand Total	24.9	26.8
Cuisinart		
DLC-7 Pro	1.1	1.1
DLC-8F	5.0	5.0
DLC-10E	0.0	0.0
DLCX	19.7	18.9
Brand Total	25.8	25.0
General Electric		
FP 1	5.1	4.7
FP 3	13.0	13.0
FP 6	3.4	3.4
Brand Total	21.5	21.1
Hamilton Beach		
702	1.8	1.6
736	3.4	3.0
738	3.8	3.6
2002	0.7	0.7
Brand Total	9.5	8.9
KitchenAid		
KFP 700	0.2	0.2
Moulinex		
LM 2	1.0	0.9
LM 5	15.0	15.0
Brand Total	16.0	15.9
Robot Coupe		
RC 2000	0.1	0.0
RC 2100W	0.1	0.1
RC 3500	0.1	0.1
RC 3600	0.2	0.2
RC 2800	0.2	0.2
Brand Total	0.7	0.6

Table 4.7: Market Shares

(Source: Page and Rosenbaum 1987. Reprinted with permission from Elsevier Science.)

4.4 DETERMINISTIC MODELS

Choice models, discussed in section 4.2, are very useful in a scenario where the number of customers is large. Assumptions of normality and independence of error in observing customer utilities can be easily justified. Notice that choice models can be used only if attribute part-worths are known. In addition, as in the Sunbeam example, an exhaustive simulation is required for picking the attribute combination with maximum profit. It is clear that the optimal product design may be missed if the search process in simulation is not carried out on a fine grid of multidimensional attribute strengths. Because of the simulation's run time and cost, there would be a limit to how fine the search grid can be.

In situations where the number of customers is not large or where they can be divided into a small number of homogeneous segments, and where assessment of attribute part-worths is error-free, more direct optimization approaches can be utilized. We discuss two such approaches, both involving mathematical programming. The first approach is versatile in that it can optimize a variety of objective functions, but the attribute part-worths must be established using experiments such as conjoint analysis. The second approach directly uses the part-worth of attributes at various levels (strengths) to rank products.

Profit or Welfare Maximization

Green and Krieger (1985) model profit/welfare maximization by assuming that (1) a set of products of interest with different attribute-level combinations can be identified, and (2) these products can be presented to all customers for their evaluation. In the model,

u_{ij} = utility of product j to customer i

x_{ij} = 1, if customer i chooses product j; 0 otherwise

y_j = 1, if product j is offered to customers; 0 otherwise

w_{ij} = seller's utility (profit contribution) from selling product j to customer i

The welfare maximization objective can be written as

$$\text{Maximize} \qquad W = \sum_i \sum_j u_{ij} x_{ij} \qquad (4.30)$$

Similarly, the seller's objective can be written as

$$\text{Maximize} \qquad \Pi = \sum_i \sum_j w_{ij} x_{ij} \qquad (4.31)$$

The choice variable, to be feasible, must ensure (1) customer i chooses no more than one product, (2) he/she cannot choose a product if it is not offered, and (3) he/she would choose the product, from those offered, to derive the highest utility from it. These can be expressed by the following set of mathematical constraints:

$$\sum_i x_{ij} \leq 1, \ \forall_i \qquad (4.32)$$

$$x_{ij} \leq y_j, \ \forall_{i \text{ and } j} \qquad (4.33)$$

$$u_{ij} y_j \leq \sum_k u_{ik} x_{ik}, \ \forall_{i \text{ and } j} \qquad (4.34)$$

Since no more than k products are to be chosen, we have the final constraint,

$$\sum_j y_j \leq k \qquad (4.35)$$

Green and Krieger develop an efficient heuristic, called the greedy interchange heuristic, to solve for the welfare maximization objective equation 4.30. Solving for the seller's objective (equation 4.31) is much more difficult, however,

McBride and Zufryden (1988) make an additional assumption to solve the harder problem (equation 4.31). They assume $w_{ij} = w_i$. That is, the seller's utility from selling the products is independent of product type. Now, since a customer's choice does not affect the weights in the objective function, constraint 4.34 becomes redundant, and the problem can be solved by an efficient set-covering procedure.

Dobson and Kalish (1988) add other dimensions to this problem by including unit price as a decision variable and accounting for the fixed cost of offering a product (product development cost). At this time we consider the Dobson/Kalish model without fixed costs, and discuss the full model in Chapter 7. Assuming p_j as the unit price for product j, the profit maximization model is stated as

$$\sum_j y_j \leq k \qquad\qquad (4.36)$$

Constraints 4.32 and 4.33 will be unchanged, and 4.34 will be replaced by,

$$\sum_k (u_{ik} - p_k)\, x_{ik} \geq (u_{ij} - p_k) y_j, \ \forall_{i,j} \qquad\qquad (4.37)$$

For the special case where products offered (y_j) and customer choices (x_{ij}) are known, optimal prices can be determined by solving the dual formulation of the linear program as a set of shortest-path problems. To see the interactions between price and utility, Dobson and Kalish provide the following two-product (A and B), two-customer (1, 2) example:

Let the utilities be expressed as

$$u_{1A} = \$10 \qquad u_{1B} = \$8$$

$$u_{2A} = \$12 \qquad u_{2B} = \$15$$

$$c_A = c_B = 1$$

$$x_{1A} = 1 \qquad \text{and} \qquad x_{2B} = 1$$

To obtain maximum profit, the seller would like to charge $10 for A and $15 for B. However, in that case customer 2 will switch to A, as he/she will have a utility surplus of $2 (12 − 10). The solution will not therefore be feasible. For feasibility, the seller will have to make sure that customer 2 obtains a minimum of $2 surplus from purchase of B. This can be done by charging $13 ($15 − $2) for B.

4.5 PRODUCT RANKING

In conjoint analysis part-worths of attributes are established by treating them as dependent variables (unknown), customer rating of products as independent variables (known), and attribute levels as known constants. For any attribute, the customer ratings form "observations," one for each customer. Note that part-worths are kind of average values over the range of customers. Thus important information on individual customers is not used in the next phase, where simulation is used to evaluate new (proposed) products.

Kohli and Krishnamurti (1987) suggest a procedure that overcomes this difficulty in that (1) individual customer information is used, up to the point of final decision making, and (2) an optimization procedure is used instead of simulation to rank products. Consider N attributes and L levels for each attribute. There would be L^N possible product profiles. To determine optimal product profiles, the simulation procedure would need to evaluate all NL profiles. Kohli and Krishnamuri suggest a dynamic heuristic that needs to consider only $(N-1)\,L^2$ of the possible L^N profiles.

As an example of this procedure, let w_{ijn} = part-worth of attribute j, level n (relative to status quo product) to customer i. Consider the part-worths for three customers, three attributes with two levels per attribute, as in Table 4.8. The first chart in Table 4.9 is constructed by adding the part-worths of attribute 2/level 1 (A_2L_1) to the corresponding part-worths of attribute 1/level 1 (A_1L_1) and attribute 1/level 2 (A_1L_2), individually for each customer. There are three non-negative elements in column 1 and two non-negative elements in column 2 (chart 1 in Table 4.9).

Hence the attribute combination $A_1L_1 + A_2L_1$ will be preferred to $A_1L_2 + A_2L_1$. In chart 2, we see that there is a tie in the number of non-negative elements. Adding the non-negative elements, the tie is broken, as the sum in column 2 exceeds that in column 1. Hence $A_1L_2 + A_2L_2$ will be preferred to $A_1L_1 + A_2L_2$. Thus, the two choices $A_1L_1 + A_2L_1$ and $A_1L_2 + A_2L_2$ will be carried to the next stage, where attribute 3's part-worths are added. The effect of adding A_3L_1 to the two chosen products is shown in chart 1, Table 4.10, and that of adding A_3L_2 in chart 2, Table 4.10. Counting the non-

Customer	Attribute 1		Attribute 2		Attribute 3	
	Level 1 (1,1)	Level 2 (1,2)	Level 1 (2,1)	Level 2 (2,2)	Level 1 (3,1)	Level 2 (3,2)
1	100	110	80	85	−200	−190
2	80	60	−70	−100	130	100
3	70	90	90	110	140	140

Table 4.8: Part-Worths of Attributes

Customer	$A_1 L_1 + A_2 L_1$	$A_1 L_2 + A_2 L_1$	Customer	$A_1 L_1 + A_2 L_2$	$A_1 L_2 + A_2 L_2$
1	180	190	1	185	195
2	10	–10	2	–20	–40
3	160	180	3	180	200
Number of Non-negative Elements	3*	2	Number of Non-negative Elements	2	2
			Sum of Non-negative Elements	365	395*

Table 4.9: Adding Attribute 2 to Attribute 1

negative elements and adding the nonnegative elements in Table 4.10, as before, we see that the product $A_1 L_2 + A_2 L_2 + A_3 L_3$ will be the top-ranked product, followed by $A_1 L_1 + A_2 L_1 + A_3 L_1$. Note that all three customers will purchase the top-ranked product, whereas only two of three customers will buy the second-ranked product.

Customer	$A_1 L_1 + A_2 L_1 + A_3 L_1$	$A_1 L_2 + A_2 L_2 + A_3 L_1$
1	–20	–5
2	140	90
3	300	340
	2	2
	440*	430

Chart 1

Customer	$A_1 L_1 + A_2 L_1 + A_3 L_2$	$A_1 L_2 + A_2 L_2 + A_3 L_2$
1	–10	5
2	110	60
3	300	340
	2	3*

Chart 2

Table 4.10: Adding Attribute 3

4.6 MASS CUSTOMIZATION

As we have seen, product line design is usually a choice of k products from n possibilities ($k << n$). This implies that not every customer will get to have his/her top-ranked product. Mass customization is the response to the customer's desire to buy his/her top-ranked product. It is aimed at providing a large number of customers their top choices, creating a competitive advantage for the company. While simple cosmetic variations in products may provide a sense of mass customization, new technology and/or novel uses of old technology can create sustainable dimensions of mass customization for competitive advantage.

Gilmore and Pine (1997) describe four types of mass customization, which they call adaptive, cosmetic, collaborative, and transparent. As mentioned earlier, except for cosmetic customization, all such customization depends heavily on technology for realization. Adaptive customization depend on products being made flexible so that they can be used for multiple applications. Collaborative customization is made possible by advanced technology using digital data processing that conveys customer-desired product attributes to designers in real time. Transparent customization requires a system learning capability that can analyze the patterns of customers' preference as expressed in actual purchases and match the attributes of individualized products to those patterns.

The four approaches to customization can be positioned in a two-dimensional space, as shown in Figure 4.5. Product customization can be obtained by tailoring them to customer specifications (as ordered) or by creating a large number of product platforms, as shown in Figure 4.5. It is clear that for a made-to-order policy, the manufacturing domain would be

		Made to Order	
		Few Products	Many Products
Product Platforms	Many	Transparent (Revealed Preference)	Collaborative (Made-to-Order)
	Few	Adaptive (Flexible Product)	Cosmetic (Peripheral)

FIGURE 4.5: MASS CUSTOMIZATION

expected to be in a reactive mode, while in a platform-based system, manufacturing can be proactive and can produce to a schedule.

A totally made-to-order system is perhaps the form of customization that is best understood. If in such a system the number of product platforms is kept low, customization can only be of a peripheral nature, where only minor changes can be made to the engineering features of the product. Companies attempt to compensate for this by creating product differentiation through marketing means. These may include (1) packaging, labeling, dispenser features, and storage, (2) sales brochures, flyers, video/audio tapes, and testimonials, (3) delivery frequency, home delivery, delivery lead time, and 24-hour delivery, (4) price, promotion, payment terms, warranty, after-sale service, and ordering policy, and (5) brand name, club membership, and brand loyalty privileges.

With a large number of platforms, made-to-order systems can be taken to a new dimension, that of real-time customization. That is, the product specifications are determined jointly by the manufacturer and the customer. Such a system has been put in practice by Matsushita Electric for bicycle manufacturing (Murakoshi 1994), described in Chapter 2. Bicycle specifications are determined when the customer visits a showroom and tries out a flexible frame. The customized attribute values of the bicycle are captured directly by an information system. Another novel application is reported in Japan from the home-building industry. Sophisticated virtual reality is used where customers in a showroom put on headsets and gloves to "walk through" a kitchen and try out different layouts in real time until they find a satisfactory one, which is recorded and used for order placement. In both examples customers design their own products, but they do not start from scratch. In the bicycle example, attributes of the product are predetermined by the manufacturer, and customers choose the attribute levels they prefer. Similarly, in the virtual reality example, attributes of the room being remodeled are determined by the service provider. In both examples, highly sophisticated technology, including CAD software and artificial intelligence, is necessary to enable customization in real time. Observe that since product attributes are predefined, a large number of attributes must be provided for, and since each combination of attribute levels may lead to a new product, the manufacturing system must allow for a large number of product platforms.

Consider now the case of products not made to order. That is, production may take place in batches, or products may be configured based on a customer's revealed preferences, although there may not be an explicit order from a customer.

The question that arises naturally is how products can be customized if they are produced in batches for unspecified customers. The answer, of course, lies in technology that enables customers to reconfigure products to a different shape or form after purchase. This is becoming a common practice in most industries, where products are made with certain degrees of flexibility built in them. Prime examples are food processors that can be quickly adapted to be used as blenders and mixers (using a common motor for drive), sport-utility vehicles that can be transformed from two-wheel drive to four-wheel drive, home air-conditioning/heating systems that can be programmed for different temperature settings at different times of the day according to the customer's preferences, and toys that can be transformed to different forms and shapes (e.g., from a monster to a race car). To endow products with such flexibility, programmable technology such as microprocessors or complex mechanical movements that can be adapted by the customer must be built into the product. It is clear that since customers do the adaptations on the products themselves, only a limited number of platforms would suffice. Each product or platform will, however, have to be more complex, almost intelligent, to transform simple adaptations by the customer into a sequence of complex maneuvers to alter product characteristics.

Physically realizing a product that the customer did not know he/she wanted and never placed an order for is perhaps the ultimate form of customization. Companies that can do it effectively will, of course, gain a tremendous competitive advantage. This would require an advanced technology that would track a customer's purchases and use a learning process such as artificial intelligence (AI) to reveal his/her preferences. The advent of the Internet and electronic commerce is beginning to make such customization commercially viable. Electronic commerce makes it possible to track customer purchases online. Companies such as Amazon.com have taken it a step further, using AI to discern patterns in customers' reading habits. As new books emerge, the company matches the attributes of individual books with customers' revealed reading preferences and alerts customers to the new book if a match exists.

It is obvious that in the context of a manufacturing company, a large number of product platforms must be maintained so that they may be delivered without delay as matches with customers are established.

REFERENCES

Beltman, J. (1979), *An Information Processing Theory of Consumer Choice*, Addison-Wesley, Reading, Mass.

Ben-Akiva, M., and R. Lerman (1985), *Discrete Choice Analysis: Theory and Application to Travel Demand*, MIT Press, Cambridge, Mass.

Chakravarty, A., and J. Baum (1992), "Coordinated Planning for Competitive Products and Their Manufacturing Operations," *International Journal of Production Research*, vol. 30, no. 10, pp. 2293–2311.

Dobson, G., and S. Kalish (1988), "Positioning and Pricing a Product Line," *Marketing Science*, vol. 7., no. 2, pp. 107–125.

Dobson, G., and S. Kalish (1993), "Heuristics for Pricing and Positioning a Product-Line Conjoint and Cost Data," *Management Science*, vol. 39, no. 2, pp. 160–175.

Eliashberg, J., and A. Manrai (1992), "Optimal Positioning of New Product Concepts: Some Analytical Implications and Empirical Results," *European Journal of Operational Research*, vol. 63, pp. 376–397.

Gensch, D., N. Arersa, and S. Moore (1990), "A Choice Modeling Market Information System that Enabled ABB Electric to Expand Its Market Share," *Interfaces*, vol. 20, no. 1, pp. 6–25.

Gilmore, J., and J. Pine (1997), "The Four Faces of Mass Customization," *Harvard Business Review*, Jan./Feb., pp. 91–101.

Green, P., and A. Krieger (1985), "Models and Heuristics for Product Line Selection," *Marketing Science*, vol. 4, no. 1, pp. 1–19.

Hauser, J., and S. Shugan (1983), "Defensive Marketing Strategies," *Marketing Science*, vol. 3, fall, pp. 327–351.

Hotelling, H. (1929), "Stability in Competition," *Economic Journal*, vol. 39, pp. 41–57.

Kohli, R., and R. Krishnamurti (1987), "A Heuristic Approach to Product Design," *Management Science*, vol. 33, no. 12, pp. 1523–1533.

Lilien, G., P. Kotler, and S. Murthy (1992), *Marketing Models*, Prentice Hall, Englewood Cliffs, N.J.

Lilien, G., and A. Rangaswamy (1998), *Marketing Engineering*, Addison-Wesley, Reading, Mass.

McBride, R., and F. Zufryden (1988), "An Integer Programming Approach to Optimal Product-Line Selection," *Marketing Science*, vol. 7, no. 2, pp. 126–140.

Murakoshi, T. (1994), "Customer Driven Manufacturing in Japan," *International Journal of Production Economics*, vol. 37, pp. 63–72.

Page, A., and H. Rosenbaum (1987), "Redesigning Product Lines with Conjoint Analysis: How Sunbeam Does It," *Journal of Product Innovation Management*, vol. 4, pp. 120–137.

Shocker, A., and V. Srinivasan (1974), "Multi-Attribute Approaches for Product Concept Evaluation and Generation: A Critical Review," *Journal of Marketing Research*, vol. 16, May, pp. 159–180.

CHAPTER

5

Design Engineering

5.1 THE DESIGN PROCESS

Product concepts in the marketing domain, as we have seen, emphasize market share and select attribute strengths of a product accordingly. The attributes in question are those that can be readily understood by customers, such as fuel consumption in a car. On the other hand, the configuration (components, materials and signals) that enables a product to attain a certain performance level is not a concern at this stage.

Determination of a product's physical features is the responsibility of design engineering. This involves decisions related to the number of components, the role of each component, the hierarchy among components, the layout of components and assembly, the choice of material, the choice of product processes, specifications and tolerances, signal processing protocols, detailed blueprints, and costs. Such decisions, obviously, depend upon the nature of relationship among customer-desired attributes of a product and its physical features. Such relationships can be complex, as they may not be explicit. That is, features such as components and material that affect product attributes may not be independent of one another. Consider, as an illustration, a design of four components, as shown in Figure 5.1.

The directed arc $A \rightarrow B$ tells us that the design of component B is dependent upon the design of component A. It is therefore clear from Figure 5.1 that while component D will be the last to be designed, the order of designing components A, B, and C is not sequential. To understand such complex relationships, design engineers attempt to divide the design

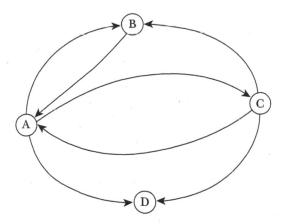

FIGURE 5.1: COMPONENT INTERDEPENDENCE

process into discrete design stages (Von Hippel 1990). Independent choices are then made for each design stage. This is called *conceptual design.*

The next phase of engineering design is called *form design* or embodiment design. In this phase the physical shapes of components and assemblies that perform specific functions are determined, their layout in relation to one another is established, and the dimensions (length, width, etc.) and tolerances are deduced. Detailed blueprints are then produced and production processes are finalized.

5.2 CONCEPTUAL DESIGN

A product concept can be formalized in terms of the functions it needs to perform to attain a certain level of output. Such functions may be interdependent, and so the method of coordinating them must also be determined. Thus the set of functions and their coordination mechanisms must be evaluated together in terms of performance and cost. Function design and engineering evaluation are therefore discussed next.

Function Design

It is not generally appreciated that even for a relatively simple product, the number of engineering or physics principles that need to be satisfied can be very large. Consider, for example, a dough-shaping machine. The dough must be fed in, then prepared by mixing in additives, and dispensed to appropriate shape forms. After the shapes are formed, they are separated by shape type. The activities of preparing, shaping, and separating generate waste, some of which can be recycled. Thus the main function, shaping dough, needs to be conceptualized in terms of seven different subfunctions: feed in, prepare, dispense, shape, separate, feed out, and recycle waste, as shown in Figure 5.2. Identification of all subsidiary functions that support the product's main function is the first step in product design. The functions can be sequential, as shown in Figure 5.2 They can also be in parallel or form a hierarchical structure, as shown in Figure 5.3.

In Figure 5.3, functions 2 and 3 are parallel to each other; functions 1 and 4 are sequential to functions 2 and 3, respectively. The main function is supported by three hierarchical levels of subfunctions. How far should a main function be exploded into subfunctions? The answer obviously depends on whether a physical entity can be constructed that would satisfy the engineering principle associated with a subfunction at the lowest level of the hierarchy. Consider, for example, the subfunction "dispense" in Figure 5.2 The objective of the function is to transfer a measured quantity of material to a shaping unit. Therefore, this function may have to be divided further into two subfunctions: "measure" and "transfer." If the dough is channeled through the system in a uniform tube (up to the shaper), slicing off a certain length of the dough will be analogous to measuring a quantity. Since a simple engineering device can be easily designed for the slicing subfunction, no further division of function is required.

All functions in a function structure must operate synergistically with one another to enable the main function to be efficient. Coordination of

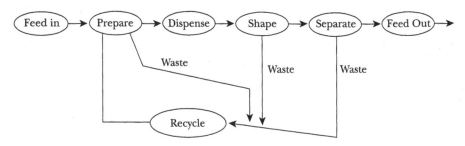

FIGURE 5.2: FUNCTION STRUCTURE FOR SHAPING DOUGH

(Source: Pahl and Beitz 1988.)

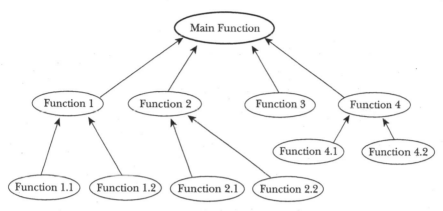

FIGURE 5.3: FUNCTION STRUCTURE

(Source: Pahl and Beitz 1988.)

function is therefore an important engineering design task. Consider, for example, transmission of torque in a car with a standard gearshift, as shown in Figure 5.4. A signal is sent by engaging the gear arm in different positions (reverse, first, second, third, etc.), and the clutch is engaged by releasing the clutch pedal. No signal will be sent if the gear arm is in the neutral position. As is clear from the decision table in Figure 5.5, torque transmission can take place only if a signal is sent and the clutch is engaged. That is, in any column, entries in the signal and engage clutch rows have to be YES for the torque transmission row to be a YES. A decision table is one way of expressing logical relationships among functions. Predicate calculus, petri nets, boolean algebra, and protocol diagrams are some other methods. Logical relations can usually be deduced from system requirements related to safety, reliability, fault prevention, and timeliness.

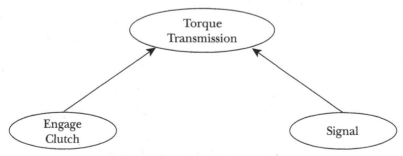

FIGURE 5.4: FUNCTION COORDINATION

Signal	YES	NO	NO	YES
Engage Clutch	NO	NO	YES	YES
Torque Transmission	NO	NO	NO	YES

FIGURE 5.5: DECISION TABLE FOR TORQUE TRANSMISSION

Limits in a relationship are determined by the constraints placed on the system.

Engineering Solutions

A function structure, as in Figure 5.3, provides a useful framework for identifying appropriate engineering solutions or devices. It is clear that engineering solutions must first be found for functions at the lowest level of the function structure hierarchy. These solutions must then be coordinated with one another as the designer moves up the hierarchy and designs for larger functions, culminating eventually at the main function.

Several engineering solutions may be permissible for a single function. Incompatibility among functions may, however, preclude some of these solutions. Consider a simple function structure comprising two subfunctions, with up to four possible solutions for each, as shown in Figure 5.6.

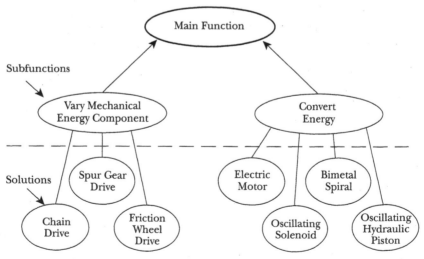

FIGURE 5.6: FUNCTION STRUCTURE

(Source: Pahl and Beitz 1988.)

Vary Mechanical Energy Component	Convert Energy			
	Electric Motor	Oscillating Solenoid	Bimetal Spiral	Oscillating Hydraulic Piston
Chain Drive	Yes, if chain drive can rotate	Yes, but slow motion	Yes	Yes, but requires additional linkage and low-speed piston
Spur Gear Drive	Yes	Yes, but additional elements required, and slow motion	Depends on angle of rotation	Yes but requires rack and pinion, low piston speed
Friction Wheel Drive	Yes	As in spur gear	Yes, but imprecise positioning	As in bimetal spiral

FIGURE 5.7: COMPATIBILITY OF ENGINEERING SOLUTIONS

(Source: Pahl and Beitz 1988.)

The compatibility matrix related to functions that vary the mechanical energy component and convert energy can be expressed as in Figure 5.7.

It is clear that the three compatible solution combinations are chain drive with bimetal spiral, spur gear drive with electric motor, and friction wheel drive with electric motor. Observe that other combinations, such as chain drive with oscillating solenoid, would be compatible but less desirable. In general, the compatible engineering solutions, from Figure 5.6 can be expressed in a tabular form as in Figure 5.8.

Functions	Engineering Solutions, j		
1	1	2	3
1	$E\,S_{1,1}$	$E\,S_{1,2}$	$E\,S_{1,3}$
2	$E\,S_{2,1}$	$E\,S_{2,2}$	$E\,S_{2,3}$

FIGURE 5.8: ENGINEERING SOLUTIONS

Thus $E S_{i,j}$ refers to the jth engineering solution for the ith function. In Figure 5.6, for the example, we have

$$E S_{1,1} \quad = \quad \text{Chain drive}$$

$$E S_{1,2} \quad = \quad \text{Spur gear drive}$$

$$E S_{1,3} \quad = \quad \text{Friction wheel drive}$$

$$E S_{2,1} \quad = \quad \text{Electrical motor}$$

$$E S_{2,2} \quad = \quad \text{Oscillatory solenoid}$$

$$E S_{2,3} \quad = \quad \text{Bimetal spiral}$$

The three compatible solutions are the pairs $(ES_{1,1}, ES_{2,3})$, $(ES_{1,2}, ES_{2,1})$, and $(ES_{1,3}, ES_{2,1})$. In general, if n_i engineering solutions are possible for the ith function, the number of possible combinations would be $\prod_{i=1}^{m} (n_i)$. A subset of these combinations will, of course, be incompatible. Determining the set of compatible combinations from the two-dimensional matrices, one of which is shown in Figure 5.7, will be an arduous task. The compatible combinations are called conceptual design variants, which are similar to the notion of design concepts used in conjoint analysis and discussed in Chapter 4. Singhal, Singhal, and Weeks (1988) have suggested a mathematical procedure for identifying compatible combinations for process design.

Evaluation of Design Variants

The set of compatible design variants may be very large. However, all such variants may not be equally attractive. We have seen how, in conjoint analysis, customers' ratings of product variants are solicited. The design variants likewise can be evaluated by designers in terms of product performance expectations. Unlike conjoint analysis, where for rating purposes the attributes of a product can be assumed to be independent of one another, the engineering solutions of a function can be highly interactive with those of another function, as shown in Figure 5.7. Hence the engineering solutions of a function cannot be rated independently of other engineering solutions. Instead, the entire set of design variants must be evaluated in terms of system objectives. Pahl and Beitz (1988) provide an example of a

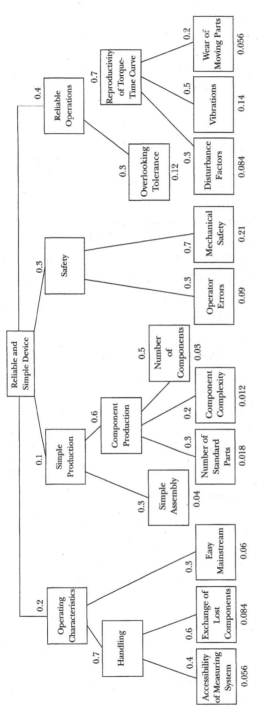

FIGURE 5.9: HIERARCHY OF SYSTEM ATTRIBUTES

hierarchy of system attributes of a product called an impulse-loading test rig, shown in Figure 5.9.

The overall system objective in Figure 5.9 is to design a reliable and simple test device. The four major system attributes are estimated to contribute to this objective in the following way:

Operating characteristics   20%
Production simplicity   10%
Safety   30%
Reliable operations   40%

At the next level, operating characteristics have contributions from ease of handling and ease of maintenance in the order of 70% and 30%. Contributions to production simplicity come from simple assembly (40%), component production (60%), and so on.

From Figure 5.9 we can easily deduce that the attribute component complexity will determine only 1.2% ($0.2 \times 0.6 \times 0.1$) of the overall system objective, as shown. The contributions of the 13 system attributes identified at the lowest level of the hierarchy are also shown.

In the next phase, the design variants are evaluated against each of the system attributes. However, since different system attributes are measured in different units, it will not be correct simply to use the magnitudes of these attributes as their attractiveness. One way to overcome this problem is to express the attribute magnitudes on a 0-to-10 scale of desirability. For example, the attributes, fuel consumption, component complexity, and life expectancy of a car may be scaled individually, as in Figure 5.10. Math-

	Attribute Magnitude		
Value Scale	Fuel Consumption	Component Complexity	Life Expectancy
10	60 miles/gal.	Elegantly simple	100,000 miles
9	55	Very simple	90,000
8	50	Simple (level 2)	80,000
7	45	Simple (level 1)	70,000
6	40	Somewhat simple	60,000
5	35	Average	50,000
4	30	Little complex	40,000
3	25	Complex (level 1)	30,000
2	20	Complex (level 2)	20,000
1	15	Very complex	10,000
0	10	Extremely complex	5,000

FIGURE 5.10: CONVERSION OF ATTRIBUTE MAGNITUDE TO VALUE

emetical functions, such as exponential and linear ones, have also been used for such scaling (Zangemeister 1970).

Pahl and Beitz (1988) use the above scheme to evaluate three design variants shown in Figure 5.11. It is clear that variant 2 is more desirable than either of other two variants, as it has the highest total score.

5.3 FORM DESIGN

Form design is the creation of a physical entity (or a set of physical features) that satisfies a function structure established in conceptual design with the least expenditure of resources in the form of energy, materials, and information. It is clear that for a product design to be realized, there must exist a mapping between the functions and the physical features.

Consider the functional structure in Figure 5.12 and the corresponding physical features in Figure 5.13. Notice the one-to-one correspondence between individual functions and physical features.

The rules that govern such mappings have come to be known as design rules. The design rules for efficient design can be stated as,

- Minimize total number of parts used
- Employ a modular design
- Maximize the number of standard components
- Maximize the number of parts that are multifunctional
- Maximize the number of parts that can be reused
- Maximize ease of fabrication
- Minimize use of fasteners
- Minimize assembly directions
- Maximize compliance
- Minimize handling

Design Matrix and Axioms

Mathematically, the mapping between functions x and features y can be represented as

$$x = A\,y$$

where x is a vector of functions, y is a vector of physical features, and A is a

Importance	Attributes	Variant 1			Variant 2			Variant 3		
		Magnitude	Value	Weighted value	Magnitude	Value	Weighted value	Magnitude	Value	Weighted value
0.056	Amount of wear	High	3	0.168	Low	6	0.336	Average	4	0.224
0.14	Vibration frequency	410/s	3	0.420	2,370/s	7	0.980	2,370/s	7	0.980
0.084	Factor disturbance	High	2	0.168	Low	7	0.588	Low	6	0.504
0.12	Overload reserve	5%	5	.600	10%	7	0.840	10%	7	0.840
0.21	Expected mechanical safety	Average	4	0.840	High	7	1.470	High	7	1.470
0.09	Possibilities of operator errors	High	3	0.270	Low	7	0.630	Low	6	0.540
0.03	No. of components	Average	5	0.150	Average	4	0.120	Average	4	0.120
0.012	Complexity of components	Low	6	0.072	Low	7	0.840	Average	5	0.060
0.018	Proportion of standard and bought-out components	Low	2	0.036	Average	6	0.108	Average	6	0.108
0.04	Simplicity of assembly	Low	3	0.120	Average	5	0.200	Average	5	0.200
0.06	Time and cost of maintenance	Average	4	0.240	Low	8	0.480	Low	7	0.420
0.084	Estimated time needed to exchange test connections	180 minutes	4	0.336	120 mins.	7	0.588	120 mins	7	0.588
0.056	Accessibility of measuring systems	Good	7	0.392	Good	7	0.392	Good	7	0.392
1.0	Total		51	3.812		85	6.816		78	6.446

FIGURE 5:11: EVALUATION OF DESIGN VARIANTS

(Source: Pahl and Beitz 1988.)

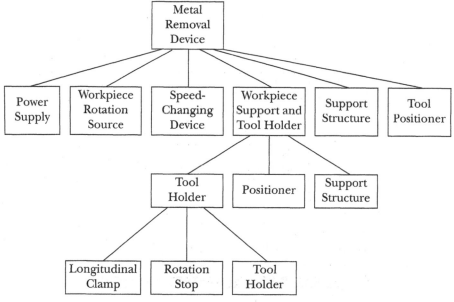

FIGURE 5.12: FUNCTION STRUCTURE

(Source: Suh 1990.)

design matrix defining the map. Therefore, for a linear system the ith function can be expressed as

$$x_i = \sum_j a_{ij} y_j$$

Thus, with three functions and three features we may write

$$\begin{bmatrix} x_1 \\ x_2 \\ x_3 \end{bmatrix} = \begin{bmatrix} a_{11} & a_{12} & a_{13} \\ a_{21} & a_{22} & a_{23} \\ a_{31} & a_{32} & a_{33} \end{bmatrix} \begin{bmatrix} y_1 \\ y_2 \\ y_3 \end{bmatrix} \qquad (5.1)$$

That is,

$$x_1 = a_{11}y_1 + a_{12}y_2 + a_{13}y_3 \qquad (5.2)$$

$$x_2 = a_{21}y_1 + a_{22}y_2 + a_{23}y_3 \qquad (5.3)$$

$$x_{31} = a_{31}y_1 + a_{32}y_2 + a_{33}y_3 \qquad (5.4)$$

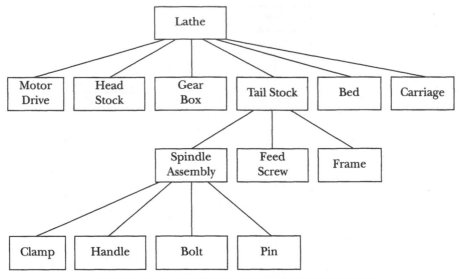

FIGURE 5.13: PHYSICAL FEATURE STRUCTURE

(Source: Suh 1990.)

Consider a triangular matrix of the form

$$A = \begin{bmatrix} a_{11} & 0 & 0 \\ a_{21} & a_{22} & 0 \\ a_{31} & a_{32} & a_{33} \end{bmatrix} \tag{5.5}$$

Since $y = A^{-1} x$, we have

$$y = \begin{bmatrix} 1/a_{11} & 0 & 0 \\ -a_{21}/a_{11} a_{22} & 1/a_{22} & 0 \\ -\left(a_{31} - \dfrac{a_{21}\, a_{32}}{a_{22}} \right) \Big/ a_{11} a_{33} & \dfrac{-a_{32}}{a_{22} a_{33}} & \dfrac{1}{a_{33}} \end{bmatrix} x \tag{5.6}$$

That is,

$$y_1 = x_1/a_{11} \tag{5.7}$$

$$y_2 = \left(x_2 - \frac{a_{21}}{a_{11}} x_1 \right) \Big/ a_{22} \tag{5.8}$$

$$y_3 = \left\{ x_3 - \frac{a_{32}x_2}{a_{22}} - \left(\frac{a_{31}}{a_{11}} - \frac{a_{21}a_{32}}{a_{11}a_{22}} \right) x_1 \right\} \Big/ a_{33} \qquad (5.9)$$

which can be rewritten as

$$y_1 = x_1/a_{11} \qquad (5.10)$$

$$y_2 = (x_2 - a_{21}y_1)/a_{22} \qquad (5.11)$$

$$y_3 = (x_3 - a_{32}y_2 - a_{31}y_1)/a_{33} \qquad (5.12)$$

It is clear that if $a_{21} = 0$, $a_{31} = 0$, variation in a specific y_j will affect only the value of the corresponding x_j. Such a system, where A is a diagonal matrix, is called an uncoupled system. That is, the functions x_1 and x_j are not interdependent. Next, consider the system with $a_{21} > 0$, $a_{31} > 0$, and $a_{32} > 0$. Assume y_2 and y_3 are fixed while y_1 is varied. It is clear from equations 5.10, 5.11, and 5.12 that not only x_1 but both x_2 and x_3 would have to be adjusted corresponding to any variation in y_1. It is, however, possible to vary y_2 instead of x_2 and y_3 instead of x_3, to maintain the validity of equations 5.11 and 5.12. Thus if the y_j are not required to be independent of one another, independence of x_j from one another can be ensured. Such a system is called a decoupled system.

Independence of x_j is a very desirable aspect of product design. Consider, for example, design of a refrigerator door. The consumer would like the refrigerator to function such that food is easily accessible when the door is open but there is no transfer of ambient heat into the refrigerator. An ideal door design would require that the functions, food accessibility, and heat transfer not affect each other.

Suh (1990) stated the above property of desirable design as the axiom of functional independence. A second axiom of Suh relates to the information content of the product, defined as the inverse of the probability of being able to manufacture the product as designed. Thus a product with high tolerance values will have a low information content. Obviously, it would be desirable to design products with low information content.

Observe that the axioms by themselves do not define a design process completely. That is, the axioms need to be interpreted in order to derive specific rules of design in different situations. For example, if the required product functions cannot be made independent, it may be possible to use the axioms in determining an altered product design. Suh (1990) has constructed seven corollaries, derived from the two axioms, that help the design process. Notice the similarities between the corollaries and design rules.

Axioms

1. In an optimal design, change in any feature would not affect more than one function.
2. The best design is a functionally uncoupled design that requires a minimum information content for each function (or feature).

Corollaries

1. Break up the function structure diagram into several groups of functions if the functions are interdependent.
2. Integrate physical features in a single physical part if functions are independently satisfied in the proposed solution.
3. Specify the largest allowable tolerance in stating functions.
4. Seek an uncoupled design that requires less information than coupled designs in satisfying a set of functions.
5. Minimize the number of functions and constraints.
6. Use standard parts if consistent with functions and constraints.
7. Use symmetrical shapes if they are consistent with functions and constraints.

The corollaries are related to the two axioms, as shown in Figure 5.14.

5.4 DESIGNING TOLERANCES

The designed specifications of physical features (or components) can be determined from matrices such as 5.1. Manufacturing a part to exact design specifications is, however, extremely difficult because of the uncertainties related to machines, material, and operator skill. Therefore, the manufactured components are allowed to vary a little from the desired values of specification. But, to be acceptable, the variations must be limited to a narrow band (akin to a statistical control chart). The maximum devia-

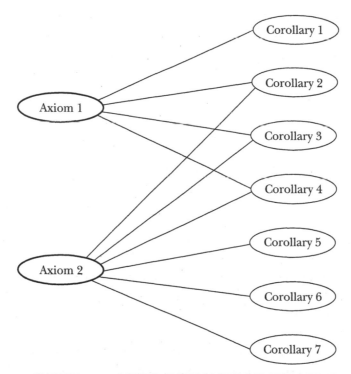

FIGURE 5.14: AXIOM-COROLLARY RELATIONSHIP

tions from the desired values within the band, are called tolerances of a specific part or component.

Consider an assembly comprising three components and their tolerances, as shown in Figure 5.15. To determine what the designed tolerances on the assembly should be, we need to understand how the part tolerances and assembly tolerances may interact.

Assume that deviations from the specified dimensions of components are random, and that the random variables are all normally distributed. Assume that component tolerances are designed based on a $\pm k \sigma$ limits

Component 1	Component 2	Component 3
15 ± 0.1	10 ± 0.08	20 ± 0.2
	$45 \pm t$	

FIGURE 5.15: ASSEMBLY TOLERANCE

(from the mean). The standard deviations of the component manufacturing processes would be $\frac{0.1}{k}$, $\frac{0.08}{k}$, and $\frac{0.2}{k}$ respectively. The standard deviation σ of the assembly dimension, assuming components are drawn at random, can be written as

$$\sigma^2 = \left(\frac{0.1}{k}\right)^2 + \left(\frac{0.8}{k}\right)^2 + \left(\frac{0.2}{k}\right)^2 \tag{5.13}$$

$$= 0.0564/k^2$$

Hence

$$\sigma = 0.238/k$$

Assuming a $k\,\sigma$ limit of tolerance for the assembly, it is clear that the tolerance on the assembly would be specified as $\pm$ 0.238. Observe that if the component tolerances were to be simply added, the range of assembly tolerance would be determined to be $\pm$ 0.38. Also note that the value of the assembly tolerance is independent of the value of k. High values of k imply smaller rejection rates and tighter manufacturing and assembly processes.

Next, consider the reverse problem of designing component tolerances given the tolerance on the final assembly. For a three-component assembly problem, shown in Figure 5.16, the tolerance on the final assembly is shown to be $\pm$ 0.5.

Again assuming a $\pm\,k\,\sigma$ limit for tolerance design, we have

$$\sigma^2 = \sigma_1^2 + \sigma_2^2 + \sigma_3^2$$

That is,

$$\left(\frac{0.5}{k}\right)^2 = \left(\frac{t_1}{k}\right)^2 + \left(\frac{t_2}{k}\right)^2 + \left(\frac{t_3}{k}\right)^2 \tag{5.14}$$

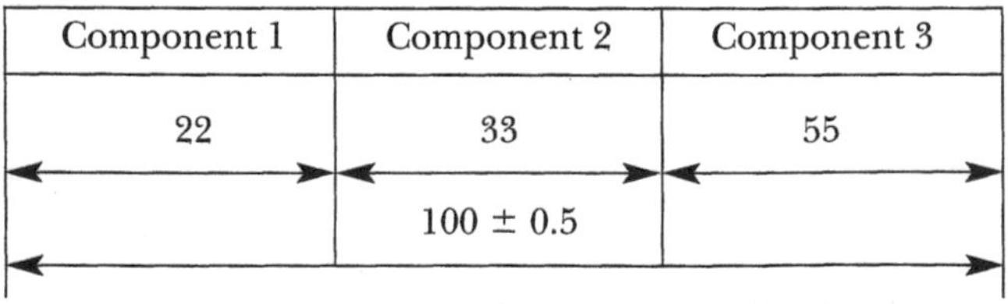

Component 1	Component 2	Component 3
22	33	55
	100 $\pm$ 0.5	

FIGURE 5.16: TOLERANCE DISTRIBUTION

where $\pm\, t_i$ is the tolerance on the ith component. It is clear that t_i cannot be determined uniquely, as the number of unknowns exceeds the number of equations. If we assume component tolerances to be equal to one another (t), we have from equation 5.14

$$3t^2 = 0.25$$

so that

$$t = \pm\, 0.29$$

There is, of course, no reason why the component tolerances should be equal. It is reasonable to assume that component tolerances would be assigned based on how hard it is to hold variation in individual component-manufacturing processes to the assigned tolerances. In general, however, we can say that the cost of manufacturing will decrease as tolerances for individual components increase. Consider a cost function of the type

$$Cost = a_i/tolerance$$

(Several other types of cost function have also been considered (Wu et al. 1988).) The decision problem can be restated as

$$\text{Minimize } \sum_i \frac{a_i}{t_i} \tag{5.15}$$

subject to

$$\left(\frac{t}{k}\right)^2 = \sum_i \left(\frac{t_i}{k}\right)^2$$

Using the Langrangian multiplier λ, and optimizing t_i, we have

$$t_i = \frac{t}{\sqrt{\sum_r (a_r)^{2/3}}} \cdot (a_i)^{1/3} \tag{5.16}$$

and

$$\lambda = \frac{1}{2} \cdot \left\{ \sum_r (a_r)^{2/3} \right\}^{3/2} \cdot k^3/t^3 \qquad (5.17)$$

Note that k has no effect on the value of t_i, but it does impact the value of λ. Since λ measures the marginal decrease in cost with t^2, the cost of component manufacturing will decrease at a faster rate with t if k increases. That is, the cost decreases faster as the tolerance band increases.

For the example in Figure 5.16, let $a_1 = 64$, $a_2 = 125$, and $a_3 = 216$. Substituting these values into equation 5.16, we have $t_1 = 0.228$, $t_2 = 0.285$, and $t_3 = 0.342$.

Robust Design

In situations where manufacturing process variability is a problem, it would be desirable to maximize the value of tolerances on components while holding product performance within certain bounds. This is called robust design (Hendrix, Mecking, and Hendriks 1996), as the product is not rejected even when the manufactured dimensions vary substantially from nominal dimensions (average values). Since both nominal dimension and tolerance need to be determined simultaneously, it must be ensured that the product's performance does not violate certain bounds.

Consider a product comprising two components (1 and 2), with two performance objectives of interest: cost and weight. Let the actual dimensions (as manufactured) of components be x_1 and x_2, respectively. Let the designed dimensions be μ_1 and μ_2, respectively, and let the designed tolerance be t. Let the performance values be determined as follows:

$$\text{Weight} = f(x_1, x_2) = 2x_1 + 3x_2$$

$$\text{Profit} = g(x_1, x_2) = 5x_1 + 2x_2$$

Let the upper limit on weight and the lower limit on profit be 5 and 8, respectively. Thus the performance bounds may be expressed as:

$$\left. \begin{array}{r} 2x_1 + 3x_2 \leq 5 \\ 5x_1 + 2x_2 \geq 8 \\ x_1 \geq 0 \\ x_2 \geq 0 \end{array} \right\} \qquad (5.17)$$

Letting $x_i = \mu_i + t$, we can rewrite the above set of inequalities as

$$\left.\begin{array}{r}2\mu_1 + 3\mu_2 + 5t \le 5 \\ 5\mu_1 + 2\mu_2 + 7t \ge 8 \\ \mu_1 \ \ge 0 \\ \mu_2 \ \ge 0\end{array}\right\} \qquad (5.18)$$

Note that t is to be maximized, but μ_1 and μ_2 are also decision variables. Since $\mu_i \ge 0$, the constraint $\mu_i + t \ge 0$ is equivalent to $\mu_i \ge 0$. Thus the set 5.18, where t is being maximized, is a linear program.

It can be easily verified that in the optimal solution $t = 9/11$, $\mu_1 = 5/11$, and $\mu_2 = 0$. Therefore, the optimal solution is not to use component 2 in the product. The designed dimension of the component would be $5/11 \pm 9/11$.

Observe that in formulation 5.18, it is implicitly assumed that the dimension of the manufactured component would not exceed $\mu_i + t$. However, as discussed earlier, manufacturing processes cannot be controlled with certainty. Therefore deviations from nominal dimensions must be considered random. Let x_1 and x_2 be normally distributed random variables with means μ_1 and μ_2 and standard deviations σ_1 and σ_2, respectively. Letting

$$y_1 = 2x_1 + 3x_2, y_2 = 5x_1 + 2x_2$$

we have

$$\left.\begin{array}{l}\mu(y_1) = 2\mu_1 + 3\mu_2 \, ; \mu(y_2) = 5\mu_1 + 2\mu_2 \\ \sigma(y_1) = (2\sigma_1^2 + 3\sigma_2^2)^{1/2} \, ; \sigma(y_2) = (5\sigma_1^2 + 2\sigma_2^2)^{1/2}\end{array}\right\} \qquad (5.20)$$

Assume that robustness of product (assembly of components 1 and 2) is defined such that

$$P(y_1 \le 5) = \alpha_1$$
$$P(y_2 \ge 8) = \alpha_2$$

Since

$$P(y_1 \leq 5) \Rightarrow P\left(\frac{y_1 - \mu(y_1)}{\sigma(y_1)} < \frac{5 - \mu(y_1)}{\sigma(y_1)}\right)$$

it follows that

$$\frac{5 - \mu(y_1)}{\sigma(y_1)} \geq z_{\alpha_1} = k \tag{5.21}$$

where z_α is the standard normal variate, so that

$$P(z \leq z_{\alpha_1}) = \alpha_1$$

Therefore, it follows from equation 5.21 that

$$5 - \mu(y_1) - k\sigma(y_1) \geq 0$$

Using equation 5.20, we rewrite the above as

$$2\mu_1 + 3\mu_2 + k(2\sigma_1^2 + 3\sigma_2^2)^{1/2} - 5 \leq 0 \tag{5.22}$$

For the variable y_2, we can similarly establish

$$5\mu_1 + 2\mu_2 + k(5\sigma_1^2 + 2\sigma_2^2)^{1/2} - 8 \geq 0 \tag{5.23}$$

In a similar way, the constraint

$$x_i \geq 0 \ (i = 1, 2)$$

can be shown to be equivalent to

$$\mu_i - k\sigma(y_i) \geq 0 \ (i = 1, 2)$$

Hence the two additional constraints are

$$\mu_1 - k(2\sigma_1^2 + 3\sigma_2^2)^{1/2} \geq 0 \tag{5.24}$$

$$\mu_2 - k(5\sigma_1^2 + 2\sigma_2^2)^{1/2} \geq 0 \tag{5.25}$$

Assuming equal tolerances for both components, we have $t = k\,\sigma_i$. The decision problem, from the expressions 5.22 to 5.25, can be stated as

$$\left.\begin{array}{l} \text{Maximize } t \\ \text{subject to} \\ \qquad t\sqrt{5} + 2\mu_1 + 3\mu_2 \leq 5 \\ \qquad t\sqrt{7} + 5\mu_1 + 2\mu_2 \geq 8 \end{array}\right\} \tag{5.26}$$

Note that equation 5.26 can be solved in a manner similar to 5.18. The solution is $t = 0.3396$, $\mu_1 = 1.165$, and $\mu_2 = 0.636$. The designed dimensions are

$$\text{Component 1} \longrightarrow 1.165 \pm 0.3396$$

$$\text{Component 2} \longrightarrow 0.636 \pm 0.3396$$

Comparing the solutions of equations 5.18 and 5.26, it is clear that inclusion of manufacturing uncertainty has reduced the tolerance (from 0.82 to 0.3396), has increased the nominal design dimensions of component 1, and has permitted use of component 2 in the product.

To keep our exposition simple, we have assumed that both components have equal tolerance ($t = t_i$). If this assumption is relaxed, the objective function will be a weighted sum of

$$t_i \left(i.e. \sum_i w_i t_i \right)$$

which remains linear. The constraints, however, become nonlinear. A procedure for solving nonlinear optimization (such as using Kuhn-Tucker conditions) will have to be explored for obtaining solutions.

The solution procedure for the linear case is very robust, since a standard linear programming package can be used. Observe that each additional performance objective adds a new constraint to the system, as in equation 5.19, and each additional component in the assembly adds a new variable.

5.5 DESIGN SEQUENCE

As we saw in equations 5.10 to 5.12, physical feature y_3 cannot be designed until y_2 and y_1 are known, and y_2 cannot be designed until y_1 is known. We

	y_1	y_2	y_3
y_1	×	1	1
y_2	0	×	1
$y_{.3}$	0	0	×

FIGURE 5.17: DESIGN PRECEDENCE MATRIX

can express this information in a design precedence matrix, shown in Figure 5.17. An entry of 1 in cell ij indicates that feature j cannot be designed until feature i is known. An entry of × indicates that the cell is redundant. The sequence of design, therefore, will be as in Figure 5.18.

In the above problem, the sequence is unambiguous because the matrix in Figure 5.15 is upper triangular, barring the diagonal elements. In general, however, the matrix may not be triangular. Therefore, to determine design sequence, the matrix must first be converted to a form that is as close as possible to being triangular. Steward (1981) suggested a partitioning procedure for this purpose and called the resulting matrix a design structure matrix (DSM). Black, Fine, and Sachs (1990) apply this procedure on a brake system design. The design precedence matrix and the design structure matrix for the brake system problem are shown in Figures 5.19 and 5.20.

Observe that Steward's partitioning procedure does not ensure conversion of a matrix to a triangular form, and as a result we have several entries of 1 on the cells below the diagonal of the matrix. This is not a weakness of Steward's procedure. Rather, it tells us that the matrix is not amenable to conversion to a triangular form implying that there does not exist a definite sequence for designing the physical features of the brake system. In other words, there must exist loops or circuits in the design process. Notice, for example, the loop in the design of the front piston and rear piston in Figure 5.20.

Observe, as in Figure 5.19, that the designs of the wheel and rear piston must precede the design of the front piston. The design of the rear piston, however, requires that the wheel and front piston be designed first. Clearly, there exists a loop between the front and rear pistons, meaning

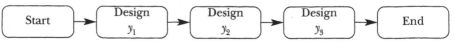

FIGURE 5.18: DESIGN SEQUENCE

		1	2	3	4	5	6	7	8
1	Wheel	0		1		1	1		1
2	Brake pedal		0	1	1			1	1
3	Rotor		1	0	1			1	1
4	Front lining			1	0				1
5	Rear piston		1	1	1	0	1	1	1
6	Front piston		1	1	1	1	0	1	1
7	Rear lining			1				0	1
8	Booster		1	1	1			1	0

FIGURE 5.19: DESIGN PRECEDENCE MATRIX

(Source: Black, Fine, and Sachs 1990.)

that they cannot be designed independently of each other. Several iterations may be required before completing the design of the front and rear pistons, as shown in Figure 5.21. The second group of features, comprising the brake pedal, rear lining, front lining, booster, and rotor, contains several such loops.

The presence of loops in a design sequence tells us that design of the group of physical features, shown as square blocks in Figure 5.20, would be highly interactive. Estimation of the time to complete a design becomes very complicated, as the point of time when the design process exits such a loop may be far from certain.

Consider the design process of three components (or physical features) *A*, *B*, and *C*. Let the intended design sequence be *A* followed by *B*

		1	6	5	2	7	4	8	3
1	Wheel	0	1	1				1	1
6	Front Piston		0	1	1	1	1	1	1
5	Rear Piston		1	0	1	1	1	1	1
2	Brake Pedal				0	1	1	1	1
7	Rear lining					0		1	1
4	Front lining						0	1	1
8	Booster				1	1	1	0	1
3	Rotor				1	1	1	1	0

FIGURE 5.20: STRUCTURED DESIGN MATRIX

(Source: Black, Fine, and Sachs 1990.)

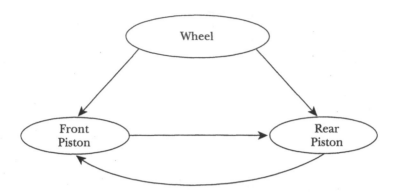

FIGURE 5.21: LOOP IN DESIGN SEQUENCE

followed by C. Design of B can obviously start after A is completed. However, it may not always be possible to obtain a design of B that would be compatible with A as already designed. In such a case we may have to redesign A. If A is now compatible with B (as designed), design of C can commence; if not we would have to redesign B. After designing C, B may require a redesign, which in turn may require A to be redesigned. If the redesign of A at this stage is compatible with both B and C, the design sequence would end. The sequence may also end after completing C or B if there is compatibility. Note that the uncertainty of whether or not a component would require redesign does not stay constant. For example, the frequency of redesigning A following design of B is not expected to be same as the frequency after a redesign of B (following design of C). Smith and Eppinger (1997) suggest the use of probability measures for redesign uncertainties. They define a stage r design as the work related to the design of component r, inclusive of all redesigns of components preceding component r that result from the design of component r. For the three-component example discussed above, we would thus have three design stages. Stage 1 is the design of A, stage 2 is the design of B and the possible redesign of A, and stage 3 is the design of C and the possible redesigns of B and A. An example of a three-stage design problem, adapted from Smith and Eppinger 1997, is shown in Figure 5.22.

The nodes in Figure 5.22 are numbered A_i, B_i, and C_i and indicate whether a design or redesign is performed on the respective components. In stage i component A is labeled as A_i, implying it would be the ith design of A (that is, the $i-1$th redesign), B is labeled B_{i-1}, and so on. The numbers on the arrows are probability measures. For example, the number 0.3

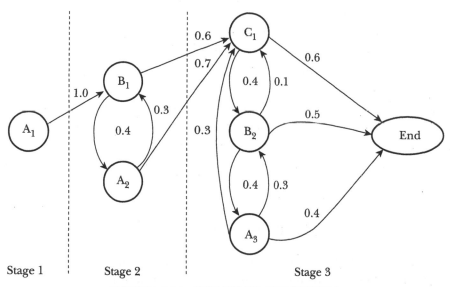

FIGURE 5.22: REDESIGN SEQUENCES

on the arrow joining A_3 to B_2 implies that there is a 30% chance of requiring a redesign of B after A is redesigned in stage 3. The time required to design or redesign components A, B, and C are assumed to be 8, 5, and 6, respectively.

Let T_{A_i}, T_{B_i} and T_{C_i} denote the expected time to completion of the design project (components A, B, and C), after reaching the nodes A_i, B_i, and C_i, respectively, in Figure 5.21. The following equations follow directly from Figure 5.22.

$$T_{A_1} = 8 + T_{B_1}$$

$$T_{B_1} = 5 + 0.4T_{A_2} + 0.6T_{C_1}$$

$$T_{A_2} = 8 + 0.3T_{B_1} + 0.7T_{C_1}$$

$$T_{C_1} = 6 + 0.4T_{B_2}$$

$$T_{B_2} = 5 + 0.1T_{C_1} + 0.4T_{A_3}$$

$$T_{A_3} = 8 + 0.3T_{B_2} + 0.3T_{C_1}$$

As there are six unknowns and six equations, the above set of equations can be solved simultaneously. The solution is $T_{C_1} = 10.8$, $T_{B_1} = 20.1$, and $T_{A_1} = 28.1$. Values of T_{B_2}, T_{A_2}, and T_{A_3} are not listed.

It is very likely that a subsequent redesign of a component would be of a shorter duration. For example, if the first and second redesigns of A require 6 and 3 units of time, and a redesign of B requires 4 units of time, the equations for T_{A_2}, T_{A_3}, and T_{B_2} can be adjusted accordingly, and the set of equations solved again.

Thus, for the intended sequence of ABC the expected completion time would be 28.1 units. To obtain the best sequence, the expected times of all other sequences, ACB, BAC, BCA, CAB, and CBA, would have to be computed.

5.6 COST OF DESIGN

According to Clark and Wheelwright (1993), as much as 70% of the cost (including manufacturing cost) of a product is determined before production commences. It is therefore essential to reduce the expected manufacturing cost before freezing the design. This requires that the costs of manufacturing processes and equipment be assessed for several design alternatives, and the one that results in the best compromise of cost and performance be chosen.

The first step in assessing the cost of a product as a function of its design would be to identify the cost drivers. The major cost drivers for a product are material, labor, production equipment, tooling, rework, and system overheads. Material and labor costs together define what is known as variable cost. While a change in product design may affect the quantity and type of material used, it may also add, delete, or modify manufacturing processes such as metal removal, assembly, and molding/casting. The choice of material and the complexity of the manufacturing processes also impact the choice of production equipment, tooling, and rework. A system cost, comprising assembly and fabrication of the production and its components, is also incurred. Ulrich, Sartorius, Pearson, and Jakiela (1993) include activities such as manufacturing engineering, industrial engineering, engineering changes, quality control, vendor support, master scheduling, MRP planning, purchasing, receiving, materials handling, production supervisors, and shipping in estimating system cost. They estimate system cost to be approximately 20% of the total cost of a product such as a camera.

Ulrich and Pearson (1998), in a study of coffee machines, identify 16 cost drivers:

Assembly content (hours)
Total purchase parts (U.S.)
Sheet metal use (kg)
Sheet metal processing time (hours)
Sheet metal press time (hours)
Tooling fabrication time (hours)
Plastic use (kg)
Molding processing time (hours)
Molding machine time (hours)
Total plastic mass (kg)
Mold fabrication time (hours)
Tooling lead time (weeks)
Number of parts
Number of molded parts
Number of unique parts
Number of fasteners

From their research the authors established numerical values for the above cost drivers corresponding to 18 different brands of coffee makers available in the market, as shown in Figure 5.23.

Using the above parameter values, the authors estimate the manufacturing cost of the products.

$$C = C_{assy} + C_{purchased\ parts} + C_{molded\ parts} + C_{sheet\ metal\ parts} + C_{tooling} + C_{supervision} + C_{inventory} + C_{facilities} + C_{energy}$$

$$C_{assembly} = \frac{assembly\ content \cdot assembly\ labor\ cost}{assembly\ productivity \cdot assembly\ yield}$$

$$C_{purchased\ parts} = \frac{total\ purchased\ parts\ cost}{sourcing\ efficiency \cdot purchased\ parts\ yield}$$

$$C_{molded\ parts} = \frac{C_{plastic} + C_{molding}}{molded\ part\ yield}$$

$$C_{plastic} = plastic\ use \cdot polypropylene\ cost \cdot (1 - plastic\ regrind\ rate)$$

	Assembly Content (BDI hours)	Total Purchased Parts (US$)	Sheet Metal Use (eqv. kg mild steel)	Sh. Metal Processing Time (hours × 1,000)	Sh. Metal Press Req'ts (kN-hours)	Sh. Metal Tooling Fabrication Time (k-hours)	Plastic Use (equiv. kg polypropylene)	Molding Processing Time (hours × 1,000)	Molding Machine Req'ts (kN-hours)	Total Plastic Mass (kg)	Mold Fabrication Time (k-hours)	Tooling Lead Time (weeks)	No. of Parts	No. of Molded Parts	No. of Unique Parts	No. of Fasteners
Black and Decker																
DCM90	0.110	4.39	0.86	1.67	0.52	1.10	1.25	33.70	71.1	1.17	11.9	6.8	52	15	43	11
DCM900	0.094	3.64	0.41	2.08	0.59	1.35	1.11	29.50	66.6	1.00	14.5	15.8	46	12	42	4
Mr. Coffee																
International	0.110	3.83	1.29	2.50	0.96	2.15	0.75	22.30	52.6	0.67	7.8	8.2	71	7	50	23
SR-12	0.102	4.82	1.29	2.50	0.96	2.15	0.93	23.90	56.8	0.88	11.6	10.9	54	9	42	15
Accel	0.122	4.22	1.02	2.50	0.96	1.95	1.35	35.50	87.4	1.25	15.8	23.1	67	12	46	23
Expert	0.160	5.03	1.33	1.67	0.70	1.50	1.35	37.60	87.5	1.23	15.3	11.9	80	13	46	32
N.A.P.																
Norelco 663	0.079	3.68	0.98	2.08	0.89	1.05	1.39	23.90	69.7	1.10	12.3	16.6	41	11	33	9
Proctor Silex																
A6278	0.103	3.67	1.04	2.08	0.89	1.70	0.95	28.90	60.3	0.89	11.6	8.9	47	12	36	11
A8737	0.129	4.06	0.55	2.08	0.70	1.35	1.27	38.70	77.1	1.08	14.9	12.3	61	14	43	22
Regal																
Regal	0.094	3.52	0.49	2.08	1.00	0.80	0.87	25.50	63.8	0.75	12.4	14.4	42	10	31	14
Toastmaster																
Toastmaster	0.105	4.37	0.30	1.25	0.45	0.55	1.71	33.00	73.0	1.14	13.4	13.5	57	12	57	11
Braun																
KF400	0.074	4.03	1.15	1.67	0.63	1.85	0.97	30.10	59.3	0.77	17.6	19.3	49	16	43	5
KF650	0.092	3.91	0.22	1.25	0.45	0.85	1.95	36.50	78.7	1.17	17.7	24.7	42	14	37	4
Krups																
130	0.092	4.99	0.34	1.25	0.45	0.85	2.10	33.50	77.8	1.21	17.9	21.6	47	13	40	4
150	0.078	5.03	0.34	1.25	0.45	0.85	2.08	25.60	54.6	0.89	13.4	14.2	39	10	36	3
178	0.117	4.57	0.39	1.67	0.52	1.10	3.43	35.40	73.4	1.06	15.6	16.0	52	13	44	7
Rowenta																
FG22-0	0.090	3.28	0.30	1.25	0.45	0.80	1.03	27.30	53.8	0.86	15.1	10.4	46	13	36	5
FK26-S	0.101	3.85	0.27	0.83	0.26	0.55	1.33	34.80	77.9	1.14	21.2	24.0	60	16	43	2
Mean	0.103	4.16	0.70	1.76	0.66	1.25	1.43	30.87	69.0	1.01	14.4	15.1	53	12	42	11
Minimum	0.074	3.28	0.22	0.83	0.26	0.55	0.75	22.30	52.6	0.67	7.8	6.8	39	7	31	2
Maximum	0.160	5.03	1.33	2.50	1.00	2.15	3.43	38.70	87.5	1.25	21.2	24.7	80	16	57	32

FIGURE 5.23: COST DRIVERS FOR COFFEE MAKERS

(Source: Ulrich and Pearson 1998. Adapted and printed by permission of INFORMS, Linthicum, Maryland.)

$$C_{molding} = \left(\frac{molding\ processing\ time}{production\ rate} \right) \cdot$$

$$\left\{ base\ machine\ rate + (machine\ capacity\ rate) \atop x\,(molding\ machine\ requirements) + \frac{operator\ labor\ cost}{molding\ machines\ per\ operator} \right\}$$

$Base\ machine\ rate =$

$$\frac{r(1 + r)^n}{(1 + r)^n - 1} \cdot \frac{base\ molding\ machine\ cost}{days\ per\ year \cdot hours\ per\ day \cdot equipment\ utilization}$$

$Machine\ cap\ rate =$
$$\frac{r(1 + r)^n}{(1 + r)^n - 1} \cdot \frac{mold\ machine\ cap\ cost}{days\ per\ year \cdot hours\ per\ day \cdot equipment\ utilization}$$

where r is the cost of capital and n is the useful life of the machine.

$$C_{sheet\ metal\ parts} = \frac{C_{metal} + C_{stamping}}{stamped\ part\ yield}$$

$$C_{metal} = sheet\ metal\ use \cdot mild\ steel\ cost$$

$$C_{stamping} = \left(\frac{sheet\ metal\ processing\ time}{production\ rate}\right) \cdot$$
$$\left\{ \begin{array}{l} base\ press\ rate + (press\ cap\ rate) \cdot (sheet\ metal\ requirements) \\ + \dfrac{operator\ labor\ cost}{stamping\ machine\ per\ operator} \end{array} \right\}$$

$Base\ press\ rate = \dfrac{r(1 + r)^n}{(1 + r)^n - 1} \cdot$

$$\frac{base\ stamping\ machine\ cost}{days\ per\ year \cdot hours\ per\ day \cdot equipment\ utilization}$$

$Press\ rate = \dfrac{r(1 + r)^n}{(1 + r)^n - 1} \cdot \dfrac{stamping\ machine\ capacity\ cost}{days\ per\ year \cdot hours\ per\ day \cdot equipment\ utilization}$

where r is the cost of capital and n is the useful life of the machine.

$C_{tooling} =$
$$\frac{(mold\ fabrication\ time + sheet\ metal\ tooling\ fabrication\ time) \cdot toolmaking\ cost}{tool\ life}$$

$$C_{supervision} = \left(\frac{C_{assembly}}{assembly\ labor\ cost \cdot span\ of\ control}\right) \cdot supervisory\ labor\ cost$$

$$C_{inventory} = C_{variable} \cdot inventory\ holding\ cost \cdot \left(\frac{inventory\ level}{days\ of\ operation\ per\ year}\right)$$

where

$$C_{variable} = C_{assembly} + C_{purchased\ parts} + C_{molded\ parts} + C_{sheet\ metal\ parts} + C_{supervision} + C_{energy}$$

$$C_{facilities} = base\ facility\ size \cdot \frac{base\ yearly\ hours}{days\ operation\ per\ year \cdot hours\ per\ day}$$
$$\cdot space\ utilization\ factor \cdot facility\ cost \cdot \frac{1}{production\ rate}$$

where

$$Space\ utilization\ factor = \frac{3}{(assembly\ production \cdot assembly\ yield + equipment\ utilization \cdot mold\ yield + base\ inventory\ /\ inventory\ level)}$$

and base facility size is 5,000 m², base yearly hours is 4,000 hours/year, and base inventory is 60 days. The space utilization factor assumes that the required floor space for a given annual production is proportional to the average of the yield-adjusted equipment utilization and the inventory levels in the plant.

$$C_{energy} = total\ plastic\ mass \cong plastic\ processing\ energy \cong energy\ cost$$

where plastic processing energy has a value of 0.75 kwh/kg.

The manufacturing parameter values in Figure 5.24, unlike the cost drivers in Figure 5.23, are independent of product types. The values in Figure 5.23 and 5.24 are used in the above set of equations to obtain the estimated manufacturing cost by product type, shown in Figure 5.25. A sample set of calculations for the product DCM90 is described next. The constituents of the total cost are also shown for comparison.

Notice the difference between the estimated manufacturing cost among products in Figure 5.25. The least costly product has a cost of $5.92 and the most costly product has a cost of $9.28. The variation is ±25%

Parameter	Value
Assembly and operator labor cost, including benefits (U.S. $ hr.)	11.00
Supervisory labor cost, including benefits (U.S. $/hr)	16.50
Toolmaking cost, shop, and labor (U.S. $/hr)	38.00
Days operations per year	240
Hours per day	16
Facility cost (U.S. $m^2/yr)	25.00
Assembly productivity	1.20
Assembly yield	1.00
Sourcing efficiency	1.05
Purchased parts yield	1.00
Polypropylene cost (U.S. $/kg)	0.84
Mild steel cost (U.S. $/kg)	0.33
Molding machines per operator	3
Molded part yield	0.995
Stamping machines per operator	1
Stamped part yield	0.995
Equipment utilization	0.80
Span of control	10
Inventory level, including raw materials, WIP, and finished goods (expressed in equivalent days of finished goods)	30 days
Molding machine cost (U.S. $)	
Basic molding machine cost	21,383
Molding machine capacity cost	+ 59 per kN capacity
Stamping machine cost (U.S. $)	
Basic stamping machine cost	30,400
Stamping machine capacity cost	+ 73 per kN capacity
Inventory holding cost	20% per year
Cost of capital	10% per year
Plastic regrind rate	20%
U.S.Eful machine life	6 years
Energy cost	0.10 $/kwh
Production rate	1,0000,000 units/year
Tool life	1,000,000 units

FIGURE 5.24: MANUFACTURING PARAMETER VALUES (PRODUCT-INDEPENDENT)

(Source: Ulrich and Pearson 1998. Adapted and printed by permission of INFORMS, Linthicum, Maryland.)

Manufacturer and Model	Estimated Mfg. Cost
Black & Decker	
DCM90	7.53
DCM900	6.45
Mr. Coffee	
International	6.56
SR-12	7.74
Accel	7.84
Expert	9.11
N.A.P.	
Norelco 663	6.55
Proctor Silex	
A6278	6.57
A8737	7.48
Regal	
Regal	6.05
Toastmaster	
Toastmaster	7.61
Braun	
KF400	6.87
KF650	7.36
Krups	
130	8.55
150	8.16
178	9.28
Rowenta	
FG22-0	5.92
FK26-S	7.09
Mean	7.37
Minimum	5.92
Maximum	9.28

FIGURE 5.25: ESTIMATED MANUFACTURING COST

(Source: Ulrich and Pearson 1998. Adapted and printed by permission
of INFORMS, Linthicum, Maryland.)

from the mean cost. Ulrich and Pearson state that in their study they
equalize the number and type of features for all products and have assumed identical manufacturing practices for all products. It therefore follows that most of the above difference may be caused by differences in
feature design and are reflected in the values of manufacturing content of
each cost driver in Figure 5.23. There is a tremendous variation. For ex-

ample, sheet metal use varies $\pm 90\%$ from the mean, the number of molding machines used varies $\pm 27\%$, and the number of fasteners used varies $\pm 191\%$. A breakdown of manufacturing content by individual features of each product would have been more revealing. Nevertheless, the relationship between design variation and manufacturing cost variations is well established.

REFERENCES

Black, T., C. Fine, and E. Sachs (1990), "A Method for Systems Design Using Precedence Relationships," Working Paper 3208-90-MS, MIT Sloan School of Management, Cambridge, Mass.

Clark, K., and S. Wheelwright (1993), *Managing New Product and Process Development*, The Free Press, New York, N.Y.

Hendrix, E., C. Mecking, and H. Hendriks (1996), "Finding Robust System for Product Design Problems," *European Journal of Operational Research*, vol. 92, pp. 28–36.

Pahl, G., and W. Beitz (1988), *Engineering Design: A Systematic Approach*, Design Council, London.

Singhal, J., K. Singhal, and A. Weeks (1988), "Long Range Process Design and Compatibility Among Operations," *Management Science*, no. 4, pp. 619–632.

Smith, R., and S. Eppinger (1997), "A Predictive Model of Sequential Iteration in Engineering Design," *Management Science*, vol. 43, no. 8, pp. 1104–1120.

Steward, D. (1981), "The Design Structure System: A Method for Managing the Design of Complex Systems," *IEEE Transaction on Engineering Management*, EM-28, vol. 3, pp. 71–74.

Suh, N. (1990), *The Principles of Design*, Oxford University Press, New York.

Ulrich, K., D. Sartorius, S. Pearson, and M. Jakiela (1993), "Including the Value of Time in Design-for-Manufacturing Decision Making," *Management Science*, vol. 39, no. 4, pp. 429–447.

Ulrich, K., and S. Pearson (1998), "Assessing the Importance of Design Through Product Archaeology," *Management Science*, vol. 44, no. 3, pp. 352–369.

Wu, Z., W. Elmarahy, and H. Elmarahy (1988), "Evaluation of Cost-Tolerance Algorithms for Design Tolerance Analysis and Synthesis," *Manufacturing Review ASME*, vol. 1, no. 3, pp. 168–179.

Von Hippel, E. (1990), "Task Partitioning: An Innovation Process Variable," *Research Policy*, vol. 19, pp. 407–418.

Zangemeister, C. (1970), *Nutzweranalyse in der Systemtechnik*, Wiltenmanshe Buchhandlung, Munich.

6

Concurrent Mapping
of Product Features

6.1 RELEVANCE OF MAPPING

"Product feature" is a term commonly used by both marketers and engineers. However, it connotes different meanings to the two groups of people. Marketers emphasize product characteristics that are valued by customers and match their needs. Those include performance, reliability, durability, appearance, and price, among others. Design engineers, on the other hand, refer to product features in terms of the engineering components, such as cylinders, valves, crankshaft, gear box, transmission, and disc brake. Obviously there exists a relationship between the engineering components of the product and the customer-desired attributes of the product. In Chapter 2 we referred to this relationship as linkage L_I and discussed how such a linkage can be used to propel the system toward optimization. We did not delve into the specifics of such a linkage, however. A map (or a linkage) in this context is an expression of the relationship (abstract or simple) between the customer-desired attributes and engineering components of a product. In Chapter 4 we mostly dealt with customer-desired attributes. Recollect, in the context of conjoint analysis, that customers are asked to evaluate products based on their features. The features in both the pizza and food processor examples were well understood by customers and product designers. The features in the pizza case were crust, topping, cheese type, cheese amount, and price. The features of the food processor were motor, bowl, blades, and feed tube. In Chapter

5 we saw how design engineers determine the set of components and assemblies that are required for the product to perform certain physical functions. Little attempt was made to relate physical functions to customer-desired attributes such as reliability, durability, and appearance. The procedure for designing a test device, discussed in Chapter 5 (Figure 5.11), comes close to the notion of a L_I linkage. Notice, however, that engineering attributes and customer-desired attributes are lumped together as product attributes. The relative importance of these attributes is determined by the designer (i.e., the customer is not involved), and price and cost characteristics are not included.

It is clear that feature design requires the participation of several functional areas: manufacturing, marketing, and engineering design. However, if the involvement of the groups is not concurrent, most of the benefits of joint participation cannot be realized. Integration of design with other tasks such as manufacturing, quality, and marketing is known as concurrent engineering. We shall therefore study the links between concurrent feature mapping and concurrent engineering.

In this chapter we first describe the notion of mapping customers' voice to engineers' voice, using quality function deployment. We then discuss some mathematical modeling approaches that have been used to capture such mappings. Procedures for capturing the voice of customers that require real-time participation of customers in the product design process are described next. We show how information systems and artificial intelligence can be useful in implementing functional concurrency.

6.2 QUALITY FUNCTION DEPLOYMENT (QFD)

Quality function deployment (QFD) is a set of relationships among customer needs, product components, and production needs. The first.application of QFD can be traced to the Kobe shipyard of Mitsubishi Heavy Electrics in Japan in 1972. It was first used in the United States by Ford Motor Company and by Xerox, in 1986, and since then has found wide application in many countries of Asia, North America, and Europe. Substantial cost savings in design arising from the use of QFD have been claimed (Hauser 1993).

QFD can be used in four phases of manufacturing: product planning, parts deployment, process planning, and production planning. One or more charts may be used to describe relationships in each phase. Hauser and Clausing (1988) provide a special construct for such charts, which they call the house of quality (HOQ), and suggest using four houses for the four phases. Since the same concepts are used in constructing all four HOQs, we describe only the HOQ for product planning, in detail, and then discuss ways of linking the four charts together.

As shown in Figure 6.1, a house of quality comprises several linked rectangles and a triangle. The rectangles, or "rooms," are well-defined set of relevant factors such as customer characteristics, design attributes, and costs. A "roof," in the form of a triangle, is used to capture the interactions (if any) among the characteristics describing a room. Customer characteristics, listed in rows in room 1, can be obtained by asking the question, "What do the customers need?" Hauser (1993) describes customer needs in two categories: primary and secondary (there could conceivably be a third category of tertiary needs). A primary need such as ease of use can be broken down into secondary needs such as being easy to set up and easy to

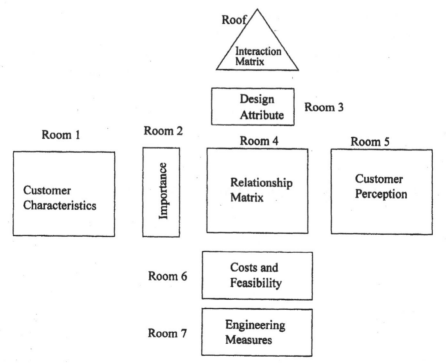

FIGURE 6.1: LAYOUT OF THE HOUSE OF QUALITY

operate. Since all needs listed in room 1 may not be equally important to a customer, the customers are asked to prioritize their needs. These priorities are recorded, using an appropriate scale, as importance weights in room 2 of the HOQ.

The design attributes are a set of parameters that can be used to describe the engineering performance of a product. This set of parameters must be measurable (in quantitative terms) and well understood by design engineers. They therefore form the starting point of any engineering design effort. These performance parameters must be chosen carefully so that they satisfy the customer needs. For example, performance parameters corresponding to the customer need for ease of operation may include time to perform the task, degree of coordination required in task performance, online support available in performing the task, and simplicity of task. Similarly, performance parameters corresponding to the customer need for a good ride may include shock isolation, anti-roll performance, and dampening performance. Design attributes are listed in columns in room 3. It should be clear that while a single attribute may partially satisfy several customer needs, it may require many more design attributes to fully satisfy all of the customer's demands.

The extent to which a design attribute can satisfy a customer need is recorded at the intersection of the customer-need row and the design-attribute column of the relationship matrix, in room 4. The symbols used for recording the strength of relationship between a customer need and a design attribute are shown in Figure 6.2. Of the two scoring alternatives, the alternative A (scores: 9, 3, 1, 0) is used more often. A symbolic relationship becomes more appropriate for qualitative reasoning in the early

		Quantitative Score	
Strength of Relationship	**Symbols**	**A**	**B**
Very Strong	●	9	9
Strong	○	3	5
Weak	△	1	1
No Relationship	[blank]	0	0

FIGURE 6.2: RELATIONSHIP STRENGTH

stages of HOQ construction and in presentations to upper-level manage-ment. Recording strength of relationship, qualitative or quantitative, is clearly best done by a multifunctional team.

The marginal cost of a design attribute is entered in room 6. Other en-gineering measures are recorded in room 7. For example, the rating of a product in terms of each of the design attributes may be high, medium, or low, depending on the product. Thus in this example three rows may be used—one for high, one for medium, and one for low—in each column (design attribute) corresponding to a competitive product. Customer per-ceptions of competing products in regard to each of the customer needs are similarly recorded in room 5.

The roof of the house is used to record interactions among design at-tributes. For example, the design attributes of task time and coordination required in task may not be independent. The higher the level of coordi-nation required, the higher the task time is expected to be. Such interac-tions are shown at the intersection of any two design attribute columns in the roof triangle.

We describe two examples. The first example is that of a hypothetical writing instrument (Wasserman 1993), and the second example is adapted from a real application on a patient care unit of Puritan-Bennett (Hauser 1993).

Writing Instrument

Four primary customer needs are considered: easy to hold, does not smear, point lasts, and does not roll. The important weights of the cus-tomer needs are assessed to be 15, 25, 45, and 15, respectively. Five major design attributes of the instrument considered are length of pencil, time between sharpening, lead dust generated, hexagonality, and minimum erasure residue. The HOQ chart for this instrument is shown in Figure 6.3. Observe that there is a very strong interaction between the design attrib-utes lead dust generated and minimal erasure residue. There is also a strong relationship of time between sharpening with lead dust generated and minimal erasure residue. It is clear from the relationship matrix that lead dust generated, hexagonality, and minimal erasure residue all have strong relationships with customer needs, while the remaining two design

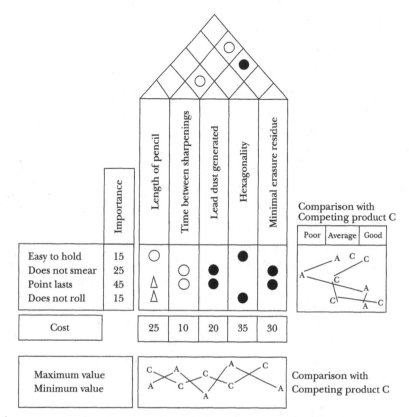

FIGURE 6.3: HOUSE OF QUALITY (WRITING INSTRUMENT)

(Source: Wasserman 1993. Adapted and printed with permission
from Kluwer Academic Publishers.)

attributes have strong and weak relationships, respectively, with customer
needs. It is also clear that customers generally perceive the competing
product (C) as superior to the company's product (A). The design engineers, however, do not consider product C to dominate product A in all
five design attributes. It can be seen that there are two design attributes,
lead dust generated and minimal erasure residue, where C outperfoms A;
these attributes are very strongly related to *popular* customer needs and
outweighs the superiority of A in the design attributes time between
sharpenings and hexagonality. It is also clear that the design engineers
have been motivated by the cost of design attributes in conceding ground
to C in minimal erasure residue.

Spirometer

Hauser (1993) describes the use of HOQ in designing a new lung-capacity measurement machine, called a spirometer, at Puritan-Bennett Company. The redesigned equipment provided only the basic functions that most general practitioners wanted. Since it was designed in a modular format, customers could buy the basic unit at $1,590, or they could upgrade it with other modules for a total cost of $4,088. The pressure to redesign came from a competitor who had introduced a base unit at a cost of $1,995. In Figure 6.4 we show an abridged version of the relationship matrix, with 13 customer needs and 13 design attributes. Hauser's full matrix has 26 customer needs and 56 design attributes.

It should be clear that although HOQ charts can act as a medium for channeling discussion in multifunctional groups, they are by no means substitutes for optimization of product design. As a graphical technique, these charts have the power of visualization, but they are also extremely cumbersome given the hundreds (if not thousands) of components in a realistic product design exercise. The economics of a product, such as potential sales and marginal costs of processes and raw materials, are not captured in an HOQ. The HOQ information is good only for engineering a *single* unit of a product. Consider, for example, a situation where a large variety is desired, with different sales potential, different unit costs, and different levels of supplier involvement. HOQ provides no guidance in deciding which subset of products would be optimal for the company. In addition, there are issues such as how to organize a multifunctional team (to chart the relationship matrix) so as not to let inherent differences within the team drag out the project. We shall discuss this further in the context of customer interface.

6.3 MATHEMATICAL MODELING OF QUALITY FUNCTION DEPLOYMENT

As mentioned earlier, HOQ does not provide a procedure per se for deciding the optimal levels of design attributes to be engineered into a product. Mathematical modeling approaches can be very useful in such decisions. We discuss them, starting from a very simple scenario.

	No. of hands required to hold	ATS waveform	Accuracy (%)	Time to enter data	Time to clean	Sensor weight	No. of records stored	Time to retrieve record	Machine order to shipment time	Supply order lead time	Instrument failure rate	Price of product	Price of disposables
Affordability												●	
Portability	●					○							
Reliability											●		
Accurate reading		●	○										
Easy to operate				○			○						
Fast to use				○			○						
Easy to calibrate				○									
Easy to clean					●								
Low-cost supplies													●
Availability of machine									●	●			
Availability of supplies									●	●			
Cost of repairs and service										○			
Effective data storage/retrieval							●	●					

FIGURE 6.4: RELATIONSHIP MATRIX OF SPIROMETER

(Source: Hauser 1993. Adapted and printed with the permission of Sloan Management Review Association.)

Let I $(i \in I)$ be a set of customer needs and J $(j \in J)$ a set of design attributes, corresponding to Figure 6.1. For the base model we assume that there are no interactions among the designed attributes, so the roof matrix is not required in the HOQ.
Let

w_i = importance weight of the ith customer need

R_{ij} = strength of relationship between customer need i and design attribute j, R_{ij} = 9, 3, 1, or 0

r_{ij} = normalized value of R_{ij}, $r_{ij} = R_{ij}/\sum\limits_{j \in J} R_{ij}$, assuming no interactions among $j \in J$

x_j = value of design attribute relative to that in a world-class product, with x_i = 1 indicating the jth attribute to be at a world-class level

c_j = cost of providing design attribute j at a world-class level (x_j = 1),

t_j = time required to engineer attribute j at a world-class level

T = time to market introduction of the product

K = targeted cost of the product

The objective of product design is assumed to be maximization of customer satisfaction, subject to cost and time constraints. For given values of x_j, customer satisfaction of the ith need can be written as

$$\text{Satisfaction} = \sum_{j \in J} w_i\, r_{ij}\, x_j$$

Therefore,

$$\text{total satisfaction} = \sum_{i \in I} \sum_{j \in J} w_i\, r_{ij}\, x_j$$

The mathematical model can therefore be expressed as

$$\text{Maximize} \sum_{j \in J} W_j\, x_j \tag{6.1}$$

subject to

$$\sum_{j \in J} c_j\, x_j \le K \tag{6.2}$$

$$\sum_{j \in J} t_j\, x_j \le T \tag{6.3}$$

$$x_j < 1 \tag{6.4}$$

where

$$W_j = \sum_{i \in I} w_i \, r_{ij}$$

$$\sum_{j \in J} r_{ij} = 1$$

$$\sum_{i \in I} w_i = 1$$

$$\sum_{j \in J} W_j = 1$$

The weight W_j can be thought of as the value of design attribute j to customers, or the technical rating of attribute j. For the example in Figure 6.3, the R_{ij} values are shown in Table 6.1. It can be verified that the values of W_j (for given values of W_i and r_{ij}) will be as shown in Table 6.2.

The mathematical model corresponding to equations 6.1 to 6.4 is a linear program and can be solved for x_j using standard software. The linear program formulation corresponding to Table 6.2, with $K = 19$ and $T = 50$, can be written as

$$\textit{Maximize} \, .07x_1 + .10x_2 + .29x_3 + .25x_4 + .29x_5 \tag{6.5}$$

so that

		R_{ij}				
		$j =$				
		1	2	3	4	5
$i =$	1	3	0	0	9	0
	2	0	3	9	0	9
	3	1	3	9	0	9
	4	1	0	0	9	0

Table 6.1: Values of R_{ij}

w_i				r_{ij}		
				$j =$		
		1	2	3	4	5
$i =$	1	.15	.25	0	0	.75	0
	2	.25	0	.14	.43	0	.43
	3	.45	.04	.14	.41	0	.41
	4	.15	.10	0	0	.90	0
	W_j		.07	.10	.29	.25	.29
	Cj		1	2	15	7	12
	t_j		3	5	4	10	5

Table 6.2: Values of r_{ij}, W_j, c_j, and t_j

$$x_1 + 2x_2 + 15x_3 + 7x_4 + 12x_5 \le 19 \tag{6.6}$$

$$3x_1 + 5x_2 + 4x_3 + 10x_4 + 5x_5 \le 50 \tag{6.7}$$

$$x_1, x_2, x_3, x_4, x_5 \le 1$$

The linear program software produces the optimal solution as $x_1 = 1.0$, $x_2 = 1.0$, $x_3 = 0$, $x_4 = 1.0$, and $x_5 = 0.75$. If the number of constraints is not high, we may use simple economic arguments to generate interesting insights. For example, if cost is the only constraint, we can easily determine the equal value points (say, 1 unit value) for the five attributes by dividing the coefficient of x_j in equation 6.6 by the corresponding coefficient of x_j in equation 6.5; the ratio would be 14.3, 20, 51.7, 28, and 41.4. It is clear that attributes with low resource use per unit value will have preference. Thus attribute performance ordering will be 1, 2, 4, 5, 3. Since the maximum value of x_j is 1.0, the optimal solution can be verified to be $x_1 = 1$, $x_2 = 1$, $x_4 = 1$, $x_5 = 0.75$, and $x_3 = 0$. For the case of a *single* resource constraint, we can thus generalize that if n of possible m ($n \le m$) design attributes are engineered in a product, at least $n-1$ of them will be built at the world-class level. Including constraint 6.7, the equal-value vectors in two dimensions can be computed to be (14.3, 42.9), (20, 5), (51.7, 13.8), (28, 40), and (41.4, 17.2). The equal-value attribute vectors can be plotted in a two-

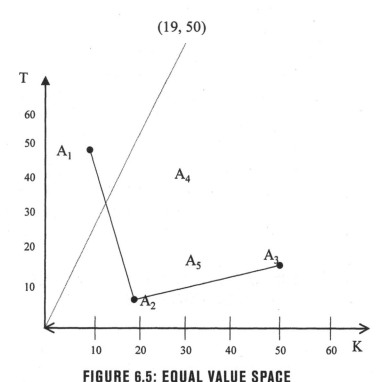

FIGURE 6.5: EQUAL VALUE SPACE

dimensional resource space, as shown in Figure 6.5. Observe that initially attributes A_4 and A_5 will not be considered in the solution, as they do not lie on the minimum-resource convex hull. The resource vector (21, 50) intersects the line joining A_1 and A_2, indicating that these two attributes will be in the optimal solution. However, since $x_j \leq 1$, setting $x_1 = 1$, $x_2 = 1$ does not saturate the resource vector (21, 50). Next, we exclude attributes A_1 and A_2 (i.e., $x_1 = 1$, $x_2 = 1$) and adjust the resource vector to (16, 42). Repeating the procedure as above, we can verify that A_4 and then A_5 will enter the solution in that order. Thus, $x_4 = 1$ and $x_5 = 0.75$. For the optimal solution ($x_1 = 1$, $x_2 = 1$, $x_4 = 1$, $x_5 = .75$), cost = \$19 and time to market = 21.75 units. That is, the cost would equal to the target of \$19, and it would be launched 28.25 periods ahead of time. The total customer satisfaction will be equal to 0.63. Observe that attribute A_5 dominates attribute A_3, as both produce the same customer satisfaction, 0.29 per unit, but A_5's resource utilization better matches resource availability. It can also be verified that the dual value of the cost target constraint is 0.33. It implies that customer satisfaction will increase by 0.33 units for every unit increase in the targeted cost. It is, of course, assumed that the price charged to the customer remains unchanged and that customers are indifferent to price increases.

The approach outlined above is aimed at designing a single product for all customers. It should be appreciated that the weight w_i is only the average of the importance of needs expressed by customers. We have seen in Chapter 4 that customers are by no means homogeneous, and as established in the context of conjoint analysis, the optimal solution may be to offer customers, several variations of the product instead of a standard product for all. This is especially true when competing products are available in the market. It therefore follows that QFD should be integrated with conjoint analysis. We shall examine the mechanics of such an integration in Chapter 7.

6.4 MODELING DESIGN ATTRIBUTES INTERACTIONS

The interactions among design attributes are captured in the roof matrix of an HOQ, as in Figure 6.3. In the mathematical model discussed, it was assumed that there were no such interactions. We will describe two possible ways of modifying the model to incorporate such interactions.

Consider again the example in Figure 6.3. The R_{ij} values and the interaction values are shown in Figure 6.6. Since columns 1 and 4 have no in-

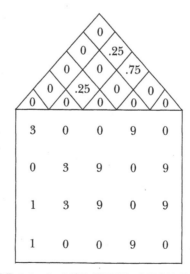

FIGURE 6.6: R_{ij} AND INTERACTION VALUES

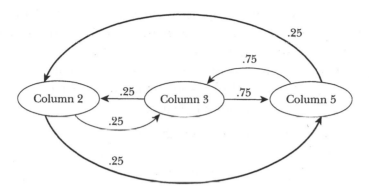

FIGURE 6.7: TRACE OF INTERACTIONS, ROW 2

teractions, the normalized values of R_{ij} (r_{ij}) in row 1 (Figure 6.6) will be $\frac{3}{12}$ and $\frac{9}{12}$, as before (Table 6.2). In row 2, however, we have substantial interactions among columns 2, 3, and 5. Those interactions are traced in Figure 6.7. Thus the revised values in row 2 and columns 2, 3, and 5 can be written as

$$Column\ 2 \rightarrow 3 + .25\ (col\ 3) + .25\ (col\ 5)$$

$$= 3 + .25\ (9) + .25\ (9) = 7.5$$

$$Column\ 3 \rightarrow 9 + .75\ (col\ 5) + .5\ (col\ 2)$$

$$= 9 + .75\ (9) + .5\ (3) = 17.25$$

$$Column\ 5 \rightarrow 9 + .75\ (col\ 3) + .25\ (col\ 2) = 17.25$$

Hence the normalized values would be

$$r_{22} = \frac{7.5}{42} \approx 0.18$$

$$r_{23} = 0.41$$

$$r_{25} = 0.41$$

In a similar way, interaction values can be incorporated while normalizing R_{ij} in row 3. For row 4, R_{ij} will remain unchanged due to lack of interactions. A linear programming model, similar to equations 6.5 to 6.7, can be formulated with revised r_{ij} values.

Belhe and Kusiak (1996) propose an alternative way of incorporating design attribute interactions. They suggest normalizing the row value, $\sum_j R_{ij}x_j$, instead of normalizing the R_{ij} in each cell. They also suggest explicit incorporation of interaction terms in the form of pairs of variables, as is the practice in statistical regression analysis. Consider the problem solved by Belhe and Kusiak, shown in Figure 6.8.

The row values, corresponding to Figure 6.8, can be written as

$$y_1' = 3x_1 + x_2 - x_4 + 9x_5 - x_1x_2 - x_1x_5 + 3x_2x_4$$

$$y_2' = 9x_1 + 3x_3 + x_5 - x_1x_5 + 3x_3x_5$$

$$y_3' = 9x_1 + x_2 + 9x_3 + 3x_5 - x_1x_2 - x_1x_5 + 3x_3x_5$$

$$y_4' = 3x_1 + 9x_2 + 3x_3 - x_4 - x_1x_2 + 3x_2x_4$$

Since $x_j \leq 1$, the authors show that the maximum value of y_i' is obtained when $x_j = 1$ for all j. Thus

$$Max\ y_1' = 13$$

$$Max\ y_2' = 13$$

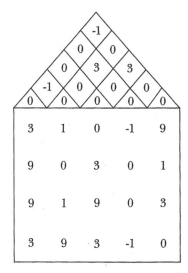

FIGURE 6.8: DESIGN ATTRIBUTE INTERACTIONS

(Source: Belhe and Kusiak 1995.)

$$Max\ y'_3 = 17$$

$$Max\ y'_4 = 16$$

The normalized values of y'_1, in the range 0 to 1, can therefore be written as

$$y_1 = \frac{1}{13}(3x_1 + x_2 - x_4 + 9x_5 - x_1x_2 - x_1x_5 + 3x_2x_4) \qquad (6.8)$$

$$y_2 = \frac{1}{13}(9x_1 + 3x_3 + x_5 - x_1x_5 + 3x_3x_5) \qquad (6.9)$$

$$y_3 = \frac{1}{17}(9x_1 + x_2 + 9x_3 + 3x_5 - x_1x_2 - x_1x_5 + 3x_3x_5) \qquad (6.10)$$

$$y_4 = \frac{1}{16}(3x_1 + 9x_2 + 3x_3 - x_4 - x_1x_2 + 3x_2x_4) \qquad (6.11)$$

Assuming customer-need importance weights to be 0.12, 0.44, 0.25, and 0.19, the formulation for maximizing customer satisfaction can be expressed as

$$Maximize \quad 0.12y_1 + 0.44y_2 + 0.25y_3 + 0.19y_4 \qquad (6.12)$$

Substituting for y_1 to y_4 from equation 6.8 to equation 6.11, respectively, the objective function 6.12 can be rewritten as

$$Maximize \quad \begin{aligned} &0.49x_1 + 0.127x_2 + 0.268x_3 - 0.021x_4 + \\ &\quad 0.172x_5 - 0.036x_1x_2 - 0.058x_1x_5 - \\ &\quad\quad 0.066x_2x_4 + 0.15x_2x_5 \end{aligned} \qquad (6.13)$$

$$subject\ to\ x_j \le 1,\ \forall_j$$

Belhe and Kusiak solve equation 6.13 as a geometric program, and the optimal values of design attributes are $x_1 = 0.75$, $x_2 = 0.93$, $x_3 = 0.832$, $x_4 = 0.1$, and $x_5 = 0.95$. Note that the optimal solution values are high, as we do not have constraints related to cost targets and time-to-market, as in constraints 6.6 and 6.7, respectively. Adding such constraints would make the problem extremely cumbersome to solve.

6.5 PATTERN RECOGNITION MODEL OF MAPPING

The most crucial element of any HOQ chart is the relationship matrix, where the strength of the relationship between customer needs and design attributes is recorded. This information is also hardest to obtain, as it requires a multifunctional team performing in a conflict-free environment. An alternative approach, proposed by Balakrishnan, Chakravarty, and Ghose (1997), bypasses multifunctional teams and attempts to discern the levels of design attributes (x_j) in the mathematical model corresponding to equations 6.1 to 6.4 directly from the features of existing products. They use an artificial-intelligence model, called neural net, for modeling the mapping of customer needs to design attributes.

The generic form of a neural net model is shown in Figure 6.9. It has an input layer, an output layer, and one or more intermediate layers. The nodes in the input layer represent the source variables (usually one variable per node) for mapping. The nodes in the output layer represent the target variables for mapping. The input nodes are connected to the output nodes through the nodes in the intermediate layer, called the hidden layer. The hidden layer may comprise several sublayers. The hidden layer is used, essentially, to increase the degree of freedom in mapping. In-

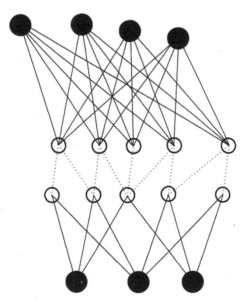

FIGURE 6.9: NEURAL NET MODEL

creasing the number of sublayers has been shown to increase mapping accuracy in many applications. The arcs connecting the nodes, belonging to different layers, have weights attached to them, and those weights are analogous to the strength of relationship in the matrix of an HOQ. They are determined for a specific problem by "training" the neural net, using the known values of input variables. The predicted values of the output variables are compared with the known values of the output variables. The arc weights are then adjusted until the errors between the predicted and known values of the output variables are minimized. In the trained net, the set of arc weights can then be used to predict output values corresponding to any set of arbitrary input values.

Balakrishnan, Chakravarty, and Ghose used this approach to study the relationship between marketing and design variables in the automobile industry. They defined customer-desired attributes of a car as those product features (e.g., turning circle, miles per gallon, and acceleration) that customers can assess by direct observation, or from their familiarity with the features, before making a buying decision. Design attributes, according to this dichotomy, are those features that are less visible or for which customers cannot directly impute values (e.g., cooling system capacity, oil capacity, and maximum output). Accordingly, the authors list seven design attributes:

A1 Cylinder capacity (cc)
A2 Compression ratio
A3 Oil capacity (liters)
A4 Cooling system capacity (liters)
A5 Power-to-weight ratio (kw/kg)
A6 Curb weight (kg)
A7 Maximum power output (HP)

The customer-desired features listed are

A8 Wheelbase (cm)
A9 Track (front tires) (cm)
A10 Ground clearance (cm)
A11 Turning circle (meters)
A12 Trunk space (cc)
A13 Maximum miles per hour (mph)
A14 Miles per gallon (mpg)

A set of 51 cars (U.S., European, and Japanese) was used to train the neural net, with attributes $A8$ to $A14$ as input variables and $A1$ to $A7$ as output variables. They found that prediction efficiency improved remarkably when the 51 cars were divided into four clusters (called envelopes) with individual neural nets trained for each cluster. The 51 cars are listed in four clusters in Table 6.3. Four cars, one from each cluster, were held out from the training set and were used to test the predictive accuracy of the neural net. The comparison of the predicted and actual output values of the four held-out cars (Plymouth Grand Fury, Dodge Daytona coupe, Honda Concerto, and Ford Probe coupe) is shown in Table 6.4.

Thus the weights associated with the arcs in a trained neural net capture the pattern of transformation of customer needs to designed attributes. Such a trained neural net can therefore be used to determine the design attributes of a new car for which the customer needs are known. Note that the market conditions revealed by importance weights of customer needs and/or product variety have not been explicitly captured in the above neural net model. Neither has any attention been paid to maximization of an objective function or satisfying cost, time, or some other forms of constraint. In Chapter 7 we shall discuss how to model such a decision process within the framework of a model for attribute transformation.

6.6 MANAGING THE MAPPING PROCESS

We study two aspects of mapping coordination: interacting with customers and information flows within and between functions.

Customer Interface

To complete the QFD (HOQ) chart, interaction with customers is required at several levels. It is clear from Figure 6.3 that customers need to be solicited to obtain three types of information in the QFD chart. These are identification of customer needs related to a product, structuring the needs, and determining the importance of the needs. Griffin and Hauser (1993) discuss elaborate experiments they have conducted to capture the above information.

Cluster 1 (10 cars)	**Cluster 2 (13 cars)**
Buick Century sedan	Chrysler GTS sedan
Buick Regal coupe	Chrysler LeBaron coupe
Cadillac Seville sedan	Dodge Aries America sedan
Chevrolet Celebrity sedan	Dodge Dynasty sedan
Dodge Diplomat sedan	Dodge Lancer sedan
Ford LTD Crown Victoria	Ford Mustang coupe
Mercury Cougar coupe	Ford Taurus sedan
Mercury Grand Marquis	Peugeot 505
Pontiac Bonneville	Plymouth Acclaim sedan
Pontiac Grand Prix coupe	Plymouth Horizon sedan
	Plymouth Reliant sedan
	Plymouth Sundance coupe
	Volvo 740
Plymouth Grand Fury*	Dodge Daytona coupe*

Cluster 3 (18 cars)	**Cluster 4 (10 cars)**
Ford Escort	Buick Skyhawk sedan
Ford Fiesta	Buick Skylark coupe
Ford Orion	Chevrolet Cavalier sedan
Honda Accord	Chevrolet Corsica sedan
Honda Civic sedan	Ford Tempo sedan
Hyundai Excel	Geo Prizm sedan
Mazda 626 sedan	Ginetta G32 coupe
Mazda 929 sedan	Honda Legend
Mercedez-Benz 190	Honda Prelude coupe
Mercedez-Benz 300	Pontiac Sunbird coupe
Mitsubishi Galant	
Nissan Stanza	
Nissan Sunny	
Plymouth Laser coupe	
Saab 900 coupe	
Toyota Camry	
Volkswagen Jetta	
Volkswagen coupe	
Honda Concerto*	Ford Probe coupe*

*Held-out cars

Table 6.3 Cluster of Cars

(Source: Balakrishnan, Chakravarty, and Ghose 1997. Reprinted with permission from Elsevier Science.)

	Attribute	Actual	Prediction	% Off
	Plymouth Grand Fury			
	Capacity	5,210.00	5,12.21	1.61
	Compression ratio	9.00	9.08	0.92
	Oil capacity	3.80	3.88	2.20
Cluster 1	Cooling capacity	14.70	13.32	9.41
	Power-to-weight ratio	15.50	16.40	5.79
	Curb weight	1,630.00	1,603.04	1.65
	Power output	140.00	136.15	2.75
	Dodge Daytona coupe			
	Capacity	2,501.00	2,434.29	2.67
	Compression ratio	9.00	9.24	2.61
	Oil capacity	3.80	4.19	10.31
Cluster 2	Cooling capacity	8.50	8.83	3.89
	Power-to-weight ratio	16.90	17.16	1.56
	Curb weight	1,250.00	1,273.88	1.91
	Power output	100.00	97.51	2.49
	Honda Concerto			
	Capacity	1,493.00	1,673.97	12.12
	Compression ratio	9.20	8.95	2.72
	Oil capacity	3.50	3.75	7.14
Cluster 3	Cooling capacity	5.00	5.49	9.87
	Power-to-weight ratio	14.20	13.79	2.85
	Curb weight	950.00	999.55	5.21
	Power output	90.00	89.80	0.22
	Ford Probe coupe			
	Capacity	2,183.98	1,772.50	18.84
	Compression ratio	8.60	9.24	7.42
	Oil capacity	4.60	3.92	14.83
Cluster 4	Cooling capacity	7.50	7.10	5.28
	Power-to-weight ratio	15.00	12.84	14.40
	Curb weight	1,230.00	1,028.25	16.40
	Power output	110.00	112.00	1.81

Table 6.4: Design Attribute Prediction

(Source: Balakrishnan, Chakravarty, and Ghose 1997. Reprinted with permission from Elsevier Science.)

To identify customer needs, Griffin and Hauser prefer use of one-on-one interviews instead of focus group interviews. The primary issue in one-on-one interviews is figuring out how many customers it is necessary to interview to be reasonably sure of covering all customer needs related to the product. Assume that a randomly selected customer identifies customer need i with probability p_i. Then the probability that none of the n customers in a sample identifies need i will be

$$(1 - p_i)^n$$

Therefore, the probability that need i is identified by at least one customer in the sample can be written as

$$q_i = 1 - (1 - p_i)^n \tag{6.14}$$

However, since the values of p_i are not known, Griffin and Hauser assume $g(p_i)$, the pdf of p_i, to have a beta distribution over customer needs. Using q_i, the expected value of the proportion of customer needs identified in a sample of n customers can be written as

$$Q_n = \frac{\Gamma(n + \beta)\Gamma(\alpha + \beta)}{\Gamma(n + \alpha + \beta)\Gamma(\beta)} \tag{6.15}$$

and

$$g(p) = p^{\alpha-1}(1 - p)^{\beta-1}/B(\alpha, \beta)$$

$$B(a, \beta) = \Gamma(\alpha)\Gamma(\beta)/\Gamma(\alpha, \beta)$$

For their data set Griffin and Hauser estimated α and β to be 1.45 and 7.64, respectively.

In a one-on-one interview the customer is not asked to differentiate the needs as primary, secondary, or tertiary. Differentiating customer needs in this way is called hierarchical structuring of needs. This occurs in the second phase, where customers are asked to place the needs from the master list into different groups so that there is a close affinity among the needs in each group.

Finally, to obtain the importance of customer needs, customers are asked to rate different concepts of the product. For example, if there are five primary customer needs, five product concepts will be designed. In product concept k, the kth customer need will dominate, by far, over the other $k - 1$ needs. These ratings are then converted to importance weights. The authors, however, caution that how customers are asked to rate the product concepts may have a significant effect on their reliability. For example, if customers are asked to rate the product concept they choose, independent of other concepts, the revealed importance values do not correlate with the preference for the concept. However, if customers rate the concepts relative to other concepts, the said correlation improves markedly.

In an ideal form of participation, customers would be involved as co-designers. To make this happen, of course, there needs to be adequate infrastructural support. Reich et al. (1995) formulate infrastructural support in terms of (1) recording of histrical data on participation activities that include design decisions, project' progress, and outcomes, (2) provision of improved educational material on technical issues, for future participants, (3) articulation of informal knowledge possessed by users (customers) and designers, and (4) asynchronous communication among participants. The authors point out how the above infrastructural support may help to broaden the scope of the form of participation, the role and quality of contribution of participants, and the duration of participation.

The level of direct participation can be increased through technology in the form of high-speed computers with vast storage capacities, network linkages, and machine learning capabilities. They permit rapid analysis of diverse pieces of information and easy context-oriented search of databases. A participant's role can improve significantly if he/she has an opportunity to study what the previous participants' roles were. A recording of previous sessions can be used with great effectiveness for this purpose. Similarly, accumulation of educational material, along with the recording of previous participants, can vastly improve the quality of participation. Finally, to support participation in a long-term project, the records of past participation and educational material need to be organized so that they can be accessed and used over a long period. An appropriately designed data warehouse would be an effective way of realizing this. For a long-term project, it may not be possible to synchronize communication among participants in different time periods. Thus there needs to be a provision for providing asynchronous communication in a longer-term project.

Culturally Correct QFD

Hales (1995) suggests that the inherent incompatibilities between QFD characteristics and the traits of American culture have hindered successful implementation of QFD projects on a large scale in the United States. American culture encourages individuality motivated by imagination and emphasizes action with expectations of immediate rewards. QFD, on the other hand, requires a team-oriented systematic approach built upon extensive communication and open-mindedness. Because of these differences, it becomes painful for team members to be involved with a QFD project for too long. Hales suggests modifying the approach to QFD implementation by minimizing the need to work in teams. Consider the problem of finding answers that would explain the relationships governing a cell in a relationship matrix. The QFD approach expects it to be done in a team meeting where everyone focuses on the particular cell. Very often hours of meeting time may fail to discover any significant relationship at all. A different approach would be to ask each participant to prepare a list of design attributes (and their relative importance) that contribute to a customer need, corresponding to a row in the HOQ. This information (from all participants) can then be compiled for the overall cell value.

Let

$$x_{ijn} = \text{rating (percentage) of the } j\text{th design attribute (column)}$$
$$\text{contributing to the } i\text{th customer need (row),}$$
$$\text{by the } n\text{th member of the HOQ team}$$

Then

$$\sum_{j} x_{ijn} = 100, \text{ for all } i \text{ and } n$$

The overall entry in the cell corresponding to the ith row and jth column can be computed as

$$R_{ij} = \sum_{n=1}^{N} x_{ijn} \bigg/ \sum_{j} \sum_{n=1}^{N} x_{ijn}$$

where N is the number of participants (members) in the team. Since

$$\sum_j x_{ijn} = 100$$

R_{ij} can be rewritten as

$$R_{ij} = \sum_{n=1}^{N} x_{ijn}/(100 \cdot N)$$

Therefore, in terms of percentages, we have

$$R_{ij}\,(percentage) = \sum_{n=1}^{N} x_{ijn}/N$$

In a follow-up team meeting, only the cells with reasonably high values of R_{ij} will be reviewed, and the values of R_{ij} in these cells will be modified if necessary. A large number of cells with low values of R_{ij} will thus be eliminated from further discussion, saving valuable time.

To ensure that individual team members' contributions are meaningful, care should be taken to provide learning opportunities to participants at every step of the QFD. This could be in the form of database access, manuals, and decision-making techniques. The team members should also have easy access to all functional areas they need information from in order to complete their tasks. Hales suggests that each team meeting should end with a set of questions that the team members are asked to resolve by the time of the next meeting. They do so individually, using whatever resources they deem necessary. At the beginning of any meeting, all the questions from the previous meeting are reviewed and the contributions of the team members are discussed, with clear identification of good solutions. This provides valuable feedback and team recognition of individual contributions.

Multifunction Coordination

Effectiveness of multifunctional teams is crucial to success of QFD in the context of structuring various relationship matrices. One way of assessing the degree of coordination is to measure the amount of communication that takes place per week among functional units in an organization. Griffin and Hauser (1992) compared the intra-organization communication flow between an organization that had implemented QFD, and one that had a more traditional system, which they call a "phase review system." The results of their survey are shown in Tables 6.5 and 6.6 and in Figures 6.10 and 6.11. The numbers in Tables 6.5 and 6.6 signify the extent of weekly communication between functions, considered in pairs. There was, for example, about twice as much communication between engineering and OEM management than between engineering and manufacturing. It can be seen clearly that in the QFD mode there is much more communication within the organization compared to the phase review mode. For example, the communication with suppliers is extensive in phase review, whereas there is hardly any communication with suppliers in the QFD mode. This comes through much more succinctly in Figures 6.10 and 6.11. Although (as in Figure 6.10) overall communication is 8.5, only 20% of this (less than 2) is external communication in the QFD. Similarly, in the within-function category, all communication in QFD mode is internal. In the phase review mode, the percentage external communication is high, relatively speaking.

It therefore follows that the teams are much more integrated and cooperative in a QFD environment. But at the same time they appear to be more self-contained (or inward-looking, as Griffin and Hauser put it). Increase in internal communication is what we would expect for QFD to be effective. But the sharp drop in external communication could be of concern, especially if the project could benefit from information available outside the QFD team's domain. The flip side of this, where a QFD team does

	Engineering	Manufacturing	Marketing	OEM Management	Supplier
Engineering	x	8	11	15	0
Manufacturing			6	3	0
Marketing				2	0
OEM Management					0

Table 6.5: Interfunctional Communication Frequency in QFD

	Engineering	Manufacturing	Marketing	OEM Management	Supplier
Engineering	x	6	3	6	8
Manufacturing			5	1	5
Marketing				5	6
OEM Management					4

**Table 6.6: Interfunctional Communication Frequency
in Phase Review**

not show a significant increase in internal communication, is a much more serious situation. This would indicate that the QFD team is not functioning as it should, pointing out serious management flaws in planning and/or implementing the feature mapping process.

Extended QFD

As mentioned earlier, QFD comprises four phases—product planning, parts deployment, process planning, and production planning—which employ four major HOQ charts. The inputs and outputs in the first chart, as shown in Figure 6.1, are customer needs and design attributes, respectively. The inputs and outputs of the three other charts can be constructed in a similar way and linked together as explained below.

In the product planning HOQ, the optimal values of design attributes set the goals of product design. To obtain a product, however, we need the right parts, such as (in the case of a car door) a frame, sheet metal, weatherstripping, and hinges (Hauser and Clausing 1988). This information is obtained from the part deployment HOQ. Next, to manufacture the parts, the right set of processes, such as molding, turning, milling, heat treatment, and the like, needs to be defined. This would be the output of the process planning HOQ. Finally, a production plan is required to ensure customer fulfillment and resource sharing. This would be the output of the production planning HOQ. In any HOQ the "whats" (input), such as customer needs, are related to the "hows" (output), such as design attri-

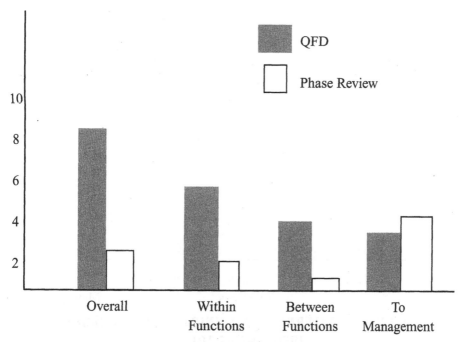

FIGURE 6.10: INTERNAL COMMUNICATION FREQUENCY

butes. It is clear that the "how" of an HOQ chart would become the "what" of a subsequent HOQ chart.

6.7 CONCURRENCY IN MAPPING

The phased nature of the four HOQs suggests a sequential process. That is, a phase 2 HOQ is initiated only after a phase 1 HOQ is complete. This may pose several problems. First, the process may be suboptimized because there is no feedback from one phase of QFD to its prior phase, and decisions in a phase are independent of the implications of these decisions in a subsequent phase. Second, a handoff of "how" (information) between phases may lead to substantial loss of information. This problem may be severe, as different teams may be involved in two different phases of QFD. Third, there may be severe delays in the system. This may be caused by handoff problems or if a phase n HOQ must be modified because the

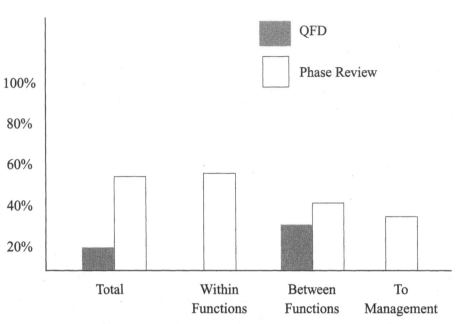

FIGURE 6.11: EXTERNAL COMMUNICATION FROM GROUPS (PERCENT OF TOTAL)

phase $n + 1$ HOQ cannot be made compatible with phase n. Finally, creativity may be hampered by the sequential nature of the four phases.

Concurrency is a way of providing parallelism to avoid the problems of sequential HOQs. Concurrency, obviously, needs to be provided between the "whats" and "hows" in an HOQ as well as among the phases of QFD. We briefly discussed the role of communication is providing concurrency in a multifunctional team context. We also alluded to the role of technology such as computers, networking, and artificial intelligence in making the customer interface more efficient. What is required is an overall architecture for providing visibility in a team, and between teams, in a real-time mode.

Reich (1995) proposes a computer-supported modeling approach that can be effectively used for providing concurrency extending from customer interface to integrating multiple HOQs. It is a graphical modeling approach that employs HOQ-like charts at its core. The graphical models can be easily structured, analyzed, combined (and/or split), and can provide attachments and links to other services.

The graphical modes are based on "atomic" concepts, which can be structured as objects or links in the graph. The objects (concepts) are similar to "whats" and "hows" in an HOQ and the links are analogous to the relationships in the relation matrix of the HOQ. The role of a concept is also built into a graph. For example, a product may be represented by three different objects. The first object may be used to depict product structure, by attaching the object to the product's bill of material (BOM). A second object may be attached to a file describing the product's functionality, and a third object may be attached to a file describing the manufacturing processes required to produce it. Since design is a shared understanding of a product and processes, arrived at through trade-offs and constraint negotiations, designers must be able to exchange information in different representational forms. This may require use of different media, manual as well as electronic.

Reich (1995) emphasizes that such a graphical model can provide benefits of concurrency by (1) providing facilities for storing, organizing, and retrieving graph-based information models, (2) permitting attachment of computational services to the graphical models, (3) providing support for adding expressive power to available techniques, (4) including graphical user interfaces (GUIs) that support easy entry and manipulation of data to support different users' preferences and/or different QFD tools, and (5) providing a forum for fostering creativity. The graphical model can be easily modified to include changes to layout and content. Portions of the model can be copied for use with new product development projects. Depending on the need, one can focus on parts of a data set while ignoring the rest, or one can quickly aggregate data from different sources. A graphical model can be used for detecting conflicts, for checking consistency, and for quick sensitivity tests on HOQ analysis.

Thus, using a formal model of information (graphical or otherwise), information used in HOQ in different phases of QFD can be made available to all, on demand. This is the so-called transparency property of the database. A multifunctional team can use this information with additional decision support tools, embedded in the model, to quickly arrive at a conclusion of a difficult problem. The adaptability of the model in terms of splitting, zooming in on a specific part of the model, and using patterns from historical data and decisions adds a valuable dimension to making the whole mapping process concurrent.

REFERENCES

Balakrishnan, N., A. Chakravarty, and S. Ghose (1997), "Role of Design Philosophy in Interfacing Manufacturing with Marketing," *European Journal of Operational Research,* vol. 103, pp. 453–469.

Belhe, U., and A. Kusiak (1996), "The House of Quality in Design Process," *International Journal of Production Research,* vol. 34, no. 8, pp. 2119–2131.

Griffin, A., and J. Hauser (1992), "Patterns of Communication Among Marketing, Engineering, and Manufacturing—a Comparison Between Two New Product Teams," *Management Science,* vol. 38, no. 3, pp. 360–373.

Griffin, A., and J. Hauser (1993), "The Voice of the Customer," *Marketing Science,* vol. 12, no. 1, pp. 1–25.

Hales, R. (1995), "Adaptive Quality Function Deployment to the U.S. Culture," *IIE Solutions,* October, pp. 15–18.

Hauser, J. (1993), "How Puritan-Bennett Used the House of Quality," *Sloan Management Review,* spring, pp. 61–68.

Hauser, J., and D. Clausing (1988), "The House of Quality," *Harvard Business Review,* May–June, pp. 63–73.

Reich, Y. (1995), "Computational Quality Function Deployment in Knowledge Intensive Engineering," Proceedings of KIC-1 (Knowledge-Intensive CAD), IFIP WG5.2, Helsinki, September.

Reich, Y., et al. (1995), "Varieties and Issues of Participation and Design," Technology Report, Software Engineering Institute, Carnegie-Mellon University.

Wasserman, G. (1993), "On How to Prioritize Design Requirements During the QFD Planning Process," *IIE Transactions,* May, pp. 59–65.

7

Product Platform and Variety

7.1 SCOPE OF VARIETY

In Chapter 4 we discussed the notion of positioning individual products that maximize the probability of customers buying that product. The idea was to determine points in customers' utility space that would be close to customers' ideal points. This probability can obviously be increased by offering multiple products that do not have identical positioning in the attribute space. One way of conceptualizing product variety is to think in terms of the area in the customers' attribute space enclosed by a company's multiple product offerings—the larger the space, the greater the variety. The major factors that limit variety are the cost of production and suppliers, and customer preferences.

As new business paradigms evolve, we are discovering that the above definition of variety may be too limiting. While production economics is one of the determinants of variety, the scope of variety must be broadened to include customer economics and system economics (Hax and Wilde 1999). Production economics requires that component variety be controlled, while the variety of finished product is maximized (Ho and Tang 1998). This can be restated as simultaneous maximization of product variety and component commonality (Anderson, Fornell, and Rust 1997). This, in turn, requires tapping of the cumulative experience, pursuit of economies of scale, and rationalization of products and processes.

The emphasis in customer economics, on the other hand, is on decreasing the customer's cost and thereby increasing his/her profit. Instead of selling products to a customer, the company's effort is targeted to fulfilling the customer's unique needs. These include the switching of the customer's current system over to the new product and helping the customer acquire other, related products that may not all be produced by the company. To do this, several other vendors may have to be brought into a close alliance with the company. Joint product development and outsourcing are natural outcomes of a customer-solution approach. Thus the definition of variety may not be limited to the main product. It may also incorporate product customization, to minimize the customer's switching cost, and creation of a portfolio of all related products for the total solution.

System economics broadens the view of variety further by including in the system all meaningful players who contribute to the creation of economic value. In the telecommunications industry, for example, some of the meaningful players would be backbone providers, access providers, equipment suppliers, and software vendors. Each group of participants in the network may adopt multiple options related to products and services. Combinations of these options generate a dynamic system with almost limitless variety. A company, therefore, needs to nurture relationships with other service providers on the network, and needs to be innovative in managing its operations in the context of such related companies. Hax and Wilde (1999) call such networked companies complementors.

Baldwin and Clark (1997) suggest modularity as a strategy for organizing complex products and processes. They define product modules as units that are designed independently but which function together as an integral whole. For this to happen, characteristics of modularity, such as product architecture, module interfaces, and standards, must be well defined and implemented (Baldwin and Clark 1999). An architecture defines what modules will be part of the system and what their functions will be. Interfaces describe how the modules will interact, including connectivity and communication, and standards describe how the module design and performance can be evaluated.

In the context of networked companies in a system, a company is like a module. The difference is that modules, architecture, and standards are not created by a grand design; they evolve and change, governed by market dynamics. The module interfaces, in this sense, determine which companies become complementors and which become competitors. The rules of module interfaces are not permanent, however, and they can be bypassed by an innovative new entrant to create an alternative interface.

7.2 ROLE OF PRODUCT VARIETY

Balakrishnan, Chakravarty, and Ghose (1997) have studied the role of variety in a multidimensional attribute space for an entire industry. Working with data from the automobile industry, they identify fourteen product attributes, discussed in Chapter 6. As noted in Chapter 6, they include both customer-desired attributes (A8 to A14) and engineering design attributes (A1 to A7). They collect attribute data on 51 cars, shown in Table 6.3. They then create a spatial representation of the cars using the procedure of multidimensional scaling (MDS). A plot of the two-dimensional MDS is shown in Figure 7.1. The corresponding directional cosines of the fourteen attributes are shown in Table 7.1.

Observe in Figure 7.1 that the 51 cars are positioned in four quadrants. If the two dimensions (X and Y coordinates in Figure 7.1) are rotated, the

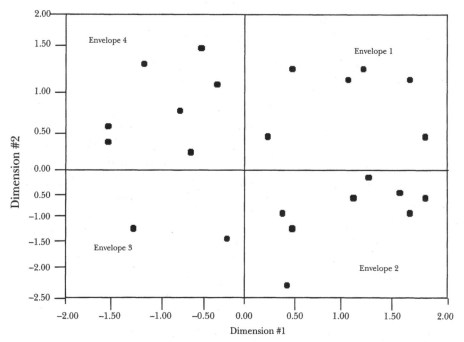

FIGURE 7.1: TWO-DIMENSIONAL SPATIAL REPRESENTATION

(Source: Balakrishnan, Chakravarty, and Ghose 1997.
Reprinted by permission of Elsevier Science.)

	Directional Cosine	
	Dimension 1	**Dimension 2**
A1	0.9824	0.1869
A2	−0.4123	−0.9110
A3	0.9183	0.3958
A4	0.9993	0.0363
A5	−0.5845	−0.8114
A6	0.9764	0.2159
A7	0.8754	0.4834
A8	0.9687	0.2483
A9	0.9943	0.1063
A10	−0.4244	0.9055
A11	0.9724	0.2333
A12	0.9953	−0.0965
A13	0.6501	0.7598
A14	−0.9835	−0.1808

Table 7.1: Directional Cosines and Angles

(Source: Balakrishnan, Chakravarty, and Ghose 1997. Reprinted
with permission from Elsevier Science.)

directional cosines and angles of the cars with respect to the two axes will
change. In fixing the two dimensions, an attempt is therefore made to
maximize the number of attributes that are at 0° (or 180°), or close to it,
with respect to one of the two dimensions. Thus if the direction cosines of
an attribute with dimension 1 are exactly ±1.0, it would imply that, on average, there is no difference in the level of that attribute in all 51 cars. The
fact that most of the cars do not lie along dimension 1 or dimension 2 tells
us that a substantial level of variety is present in the 51 cars as a whole. The
cars in Figure 7.1 appear to split into two groups across the northeast diagonal. This implies that customer preferences and engineering design
are such that although there is substantial overall variety, the variety within
each of the two groups is not very high. The absence of cars in the south-
west corner tells us that variety can be increased further by positioning cars
in that corner. It may be the case, however, that it is not technologically
and/or commercially feasible to have cars there.

Observe that in the above study the focus is on the variety in the auto industry as a whole, and it says nothing about the variety in product lines of individual companies. Where a company positions its cars in the two-dimensional space in Figure 7.1 depends on its technology, markets, and manufacturing costs. The coordinated design of products balances marketing considerations against manufacturing cost. Kekre and Srinivasan (1990) study empirical measures of variables that determine the breadth of product lines.

7.3 AGGREGATE DESIGN BENCHMARK

The concept of a generic product in an industry (independent of companies) is a notion that can be used to quickly establish the approximate levels (values) of product features for a given set of customer preferences. A company may use a generic product as a vehicle for populating product features in its platforms, corresponding to customer attributes it plans to satisfy. In this sense, a generic product acts as an aggregate benchmark for mapping customer-desired attributes to product features. To create such a benchmark, performances of several product feature bundles are studied, and a performance frontier is determined for a given set of features. The notion is Darwinian in that the products that survive and do well in the marketplace are qualified as good products from a commercial point of view.

Chakravarty and Ghose (1993) have developed this idea further with the help of a technique called data envelopment analysis. Using automobiles as units of study, they build a model to capture patterns of relationship between sets of customer-desired attributes and product features. They define the value of a generic product feature as a convex combination (weighted sum) of the values of the same feature in the cars being studied. The model for the performance frontier of the generic product is described below. Let

$I(i \in I)$ be the set of customer-desired attributes

$J(j \in J)$ be the set of engineering attributes

$N(n \in N)$ be the set of cars studied (observations)

K_{in} = value of the ith customer attribute in car n

F_{jn} = value of the jth engineering attribute in car n

$\lambda_n \, (< 1)$ = a weight associated with the nth car, such that $\sum_{n \in N} \lambda_n = 1$

Since a convex combination is a weighted sum of the set N of cars, the values of customer attributes and engineering features, for the generic product, can be written as

$$\text{Value of the ith customer attribute} = \sum_n \lambda_n K_{in}, \text{ for all } i \in I$$

$$\text{Value of the jth engineering feature} = \sum_n \lambda_n F_{jn}, \text{ for all } j \in J$$

Observe that the values of the attributes of the generic product are unknown, as the values of λ_n are unknown. But whatever these values happen to be, they are expected to be on the efficient frontier in the space defined by customer-desired-attributes and engineering attributes. By definition, therefore, the attributes of the new product cannot be superior to the benchmark. Thus if the ith customer attribute is miles per gallon, which the customer wishes to maximize, the benchmark constraint would be expressed as

$$K_{im} \leq \sum_{n \in N} \lambda_n K_{in}$$

where m denotes the new product. The above constraint implies that the value of K_{im} can at best be equal to the benchmark value; it can never exceed it. The product designer, therefore, attempts to maximize K_{im} without violating the above constraint. If, on the other hand, the attribute is an undesirable one, such as noise level which the customer wishes to minimize, the constraint would be expressed as

$$K_{im} \geq \sum_n \lambda_n K_{in}$$

Consider now an engineering attribute such as the number of cylinders in a car. Suppose the objective is to minimize cost. Since cost can be reduced by decreasing the number of cylinders, the model will attempt to set the number of cylinders as low as possible. But because of the benchmark constraint, we may not reduce the number of cylinders to below the benchmark value. Thus

$$F_{jm} \geq \sum_n \lambda_n F_{jn}$$

It is clear that the above constraints can be satisfied for any value of K_{im} and F_{jn} by appropriately choosing the values of λ_n. However, since the same set of λ_n must also satisfy the constraint for all other attributes ($i \in I$), λ_n must be chosen carefully to generate feasible solutions. The optimal solution can then be chosen from the set of feasible solutions to minimize cost or to maximize profit.

Hence the cost minimization model can be stated as

$$Minimize \sum_{j \in J} c_{jm} F_{jm} \qquad (7.1)$$

subject to

$$F_{jm} \geq \sum_{n \in N} \lambda_n F_{jn}, \text{ if cost increases with } F_{jn} \qquad (7.2)$$

$$F_{jm} \leq \sum_{n \in N} \lambda_n F_{jn}, \text{ if cost decreases with } F_{jn} \qquad (7.3)$$

$$K_{im} \leq \sum_{n \in N} \lambda_n K_{in}, \text{ if customer wishes to increase } K_{im} \qquad (7.4)$$

$$K_{im} \geq \sum_{n \in N} \lambda_n K_{in}, \text{ if customer wishes to decrease } K_{im} \qquad (7.5)$$

$$\sum_{n \in N} \lambda_n = 1 \qquad (7.6)$$

where c_{jm} is the cost per unit of engineering attribute j. Clearly, c_{jm} will have to be known, or determined from a separate study.

In the Chakravarty and Ghose study, the customer-desired attribute set comprised performance attributes horsepower (HP), maximum miles per hour (MPH), and maximum miles per gallon (MPG). The set of manufacturing attributes comprised cylinder capacity (CAP), oil system capacity (OILCAP), coolant capacity (COOLCAP), and transmission drive ratio (TR).

The first phase of the model involved running a multiple-regression model with the cost (or price) of the car as the dependent variable and the four manufacturing attributes (CAP, OILCAP, COOLCAP, and TR) as independent variables. The regression equation obtained was of the form

$$Cost\,(or\,price) = 14{,}617 + 4.9\,CAP + 439\,OILCAP \\ + 309\,COOLCAP + 3{,}422\,TR$$

In the second phase, the regression equation becomes the objective function (to be minimized).

The values of four engineering attributes, three customer-desired attributes, and costs are shown for 29 cars in Table 7.2. The detailed formulation of the model for a new car for desired customer attribute values of 83, 106, and 39.20, respectively, is shown in Figure 7.2. The model generated the following engineering attribute values:

$$CAP = 1{,}374$$

$$OILCAP = 3.39$$

$$COOLCAP = 5.47$$

$$TR = 3.29\ \text{to}\ 1$$

The cost of the new car is $6,552.

In certain cases the model may have to be modified before it can be applied. For example, there is no guarantee that the model's solution will generate a car design that would be technologically feasible. The assumption of linearity between K_{in} and F_{in}, even in the short interval between two adjacent cars, may not hold, and a more complex approximation may be required.

7.4 PRODUCT PLATFORM CONCEPTS

A product platform is a way of managing component commonality. To understand a product platform, assume that all possible variations of product

	Manufacturing Variables					Marketing Variables		
	MFG. COST	CAPACITY	OILCAP	COOLCAP	TR	HP	MPH	MPG
Buick Century sedan	10,688.94	2,471.00	2.80	9.20	2.84	98.00	99.50	26.10
Buick Regal coupe	12,224.04	2,838.00	3.80	10.70	3.51	130.00	115.00	26.10
Buick Skylark coupe	9,892.35	2,471.00	3.80	7.40	2.84	110.00	112.00	29.40
Cadillac Seville sedan	25,287.50	4,467.00	4.70	11.40	2.97	156.00	115.00	21.40
Chevrolet Cavalier sedan	7,907.40	1,991.00	3.80	7.50	3.45	90.00	102.50	29.40
Chevrolet Celebrity sedan	9,885.70	2,471.00	2.80	9.20	2.84	98.00	99.00	26.10
Chevrolet Corsica sedan	8,886.65	1,991.00	3.80	8.20	3.83	90.00	106.00	29.40
Dodge Aries sedan (American)	6,911.45	2,213.00	3.80	8.50	3.02	93.00	99.00	29.40
Ford Escort	6,406.88	1,117.00	3.25	6.70	3.84	49.00	90.00	46.10
Ford Taurus sedan	10,129.08	2,498.00	3.80	8.00	3.26	90.00	96.00	23.50
Ford Tempo sedan	8,286.30	2,300.00	3.80	7.60	3.40	98.00	102.00	33.69
Geo Prizm	8,500.80	1,587.00	3.70	6.00	3.72	102.00	115.00	29.40
Honda Accord	9,886.80	1,830.00	3.50	6.00	4.07	109.00	109.00	41.30
Honda Civic sedan	8,271.00	1,343.00	3.50	4.40	4.06	74.00	104.40	46.10
Hyundai Excel	6,323.59	1,299.00	4.00	7.30	3.17	66.00	96.00	41.30
Hyundai Sonata	8,240.00	1,796.00	4.00	6.10	4.59	98.00	106.00	36.80
Mazda 323 sedan	7,635.08	1,323.00	3.40	5.00	4.10	66.00	99.50	44.40
Mazda 626 sedan	11,523.14	1,597.00	3.40	6.00	3.85	72.00	99.50	33.60
Mazda 929 sedan	18,193.60	1,998.00	4.10	7.50	3.73	80.50	102.50	33.60
Mitsubishi Galant	9,544.77	1,755.00	3.90	6.10	4.02	93.00	104.00	39.90
Plymouth Acclaim sedan	8,828.80	2,501.00	3.80	8.50	2.51	99.00	103.00	23.50
Plymouth Horizon sedan	6,001.45	2,213.00	3.80	8.50	2.55	93.00	103.00	29.40
Plymouth Reliant sedan	6,911.45	2,213.00	3.80	8.50	2.51	93.00	99.00	29.40
Plymouth Sundance coupe	7,821.45	2,213.00	3.80	8.50	2.51	93.00	103.00	33.60
Saab 900	14,712.60	1,985.00	3.50	10.00	3.89	108.50	106.00	31.40
Toyota Corolla	7,910.28	1,296.00	3.20	4.90	3.72	72.00	99.00	47.00
Toyota Tercel	7,769.58	1,456.00	3.20	4.90	3.09	109.00	115.00	33.60
Volkswagen Jetta	8,720.80	1,272.00	3.50	6.30	4.27	54.00	93.00	42.80
Volvo 740	16,587.55	2,316.00	3.80	9.50	3.31	115.00	111.00	35.60

Table 7.2: Automobile Data

(Source: Chakravarty and Ghose 1993. Reprinted with permission from P.O.M.S.)

$$\text{MIN } 4.90\ Q_1 + 439\ Q_2 + 309\ Q_3 + 3422\ Q_4$$

Subject to

2. $2{,}471.00\ \alpha_1 + 2{,}838.00\ \alpha_2 + 2{,}471.00\ \alpha_3 + 4{,}467.00\ \alpha_4 + 1{,}991.00\ \alpha_5 + 2{,}471.00\ \alpha_6 + 1{,}991.00\ \alpha_7 + 2{,}213.00\ \alpha_8 + 1{,}117.00\ \alpha_9 + 2{,}498.00\ \alpha_{10} + 2{,}300.00\ \alpha_{11} + 1{,}587.00\ \alpha_{12} + 1{,}830.00\ \alpha_{13} + 1{,}343.00\ \alpha_{14} + 1{,}299.00\ \alpha_{15} + 1{,}796.00\ \alpha_{16} + 1{,}3230.00\ \alpha_{17} + 1{,}597.00\ \alpha_{18} + 1{,}998.00\ \alpha_{19} + 1{,}755.00\ \alpha_{20} + 2{,}501.00\ \alpha_{21} + 2{,}213.00\ \alpha_{22} + 2{,}213.00\ \alpha_{23} + 2{,}213.00\ \alpha_{24} + 1{,}985.00\ \alpha_{25} + 1{,}296.00\ \alpha_{26} + 1{,}456.00\ \alpha_{27} + 1{,}272.00\ \alpha_{28} + 2{,}316.00\ \alpha_{29} - Q_1 \le 0$

3. $2.80\ \alpha_1 + 3.80\ \alpha_2 + 3.80\ \alpha_3 + 4.70\ \alpha_4 + 3.80\ \alpha_5 + 2.80\ \alpha_6 + 3.80\ \alpha_7 + 3.80\ \alpha_8 + 3.25\ \alpha_9 + 3.80\ \alpha_{10} + 3.80\ \alpha_{11} + 3.70\ \alpha_{12} + 3.50\ \alpha_{13} + 3.50\ \alpha_{14} + 4.00\ \alpha_{15} + 4.00\ \alpha_{16} + 3.40\ \alpha_{17} + 3.40\ \alpha_{18} + 4.10\ \alpha_{19} + 3.90\ \alpha_{20} + 3.80\ \alpha_{21} + 3.80\ \alpha_{22} + 3.80\ \alpha_{23} + 3.80\ \alpha_{24} + 3.50\ \alpha_{25} + 3.20\ \alpha_{26} + 3.20\ \alpha_{27} + 3.50\ \alpha_{28} + 3.80\ \alpha_{29} - Q_2 \le 0$

4. $9.20\ \alpha_1 + 10.70\ \alpha_2 + 7.40\ \alpha_3 + 11.40\ \alpha_4 + 7.50\ \alpha_5 + 9.20\ \alpha_6 + 8.20\ \alpha_7 + 8.50\ \alpha_8 + 6.70\ \alpha_9 + 8.00\ \alpha_{10} + 7.60\ \alpha_{11} + 6.00\ \alpha_{12} + 6.00\ \alpha_{13} + 4.40\ \alpha_{14} + 7.30\ \alpha_{15} + 6.10\ \alpha_{16} + 5.00\ \alpha_{17} + 6.00\ \alpha_{18} + 7.50\ \alpha_{19} + 6.10\ \alpha_{20} + 8.50\ \alpha_{21} + 8.50\ \alpha_{22} + 8.50\ \alpha_{23} + 8.50\ \alpha_{24} + 10.00\ \alpha_{25} + 4.90\ \alpha_{26} + 4.90\ \alpha_{27} + 6.30\ \alpha_{28} + 9.50\ \alpha_{29} - Q_3 \le 0$

5. $2.84\ \alpha_1 + 3.61\ \alpha_2 + 2.84\ \alpha_3 + 2.97\ \alpha_4 + 3.45\ \alpha_5 + 2.84\ \alpha_6 + 3.83\ \alpha_7 + 3.02\ \alpha_8 + 3.84\ \alpha_9 + 3.26\ \alpha_{10} + 3.40\ \alpha_{11} + 3.72\ \alpha_{12} + 4.07\ \alpha_{13} + 4.06\ \alpha_{14} + 3.17\ \alpha_{15} + 4.59\ \alpha_{16} + 4.10\ \alpha_{17} + 3.85\ \alpha_{18} + 3.73\ \alpha_{19} + 4.02\ \alpha_{20} + 2.51\ \alpha_{21} + 2.55\ \alpha_{22} + 2.51\ \alpha_{23} + 2.51\ \alpha_{24} + 3.89\ \alpha_{25} + 3.72\ \alpha_{26} + 3.09\ \alpha_{27} + 4.27\ \alpha_{28} + 3.31\ \alpha_{29} - Q_4 \le 0$

6. $98.00\ \alpha_1 + 130.00\ \alpha_2 + 110.00\ \alpha_3 + 156.00\ \alpha_4 + 90.00\ \alpha_5 + 98.00\ \alpha_6 + 90.00\ \alpha_7 + 93.00\ \alpha_8 + 49.00\ \alpha_9 + 90.00\ \alpha_{10} + 98.00\ \alpha_{11} + 102.00\ \alpha_{12} + 109.00\ \alpha_{13} + 74.00\ \alpha_{14} + 66.00\ \alpha_{15} + 98.00\ \alpha_{16} + 66.00\ \alpha_{17} + 72.00\ \alpha_{18} + 80.50\ \alpha_{19} + 93.00\ \alpha_{20} + 99.00\ \alpha_{21} + 93.00\ \alpha_{22} + 93.00\ \alpha_{23} + 93.00\ \alpha_{24} + 108.50\ \alpha_{25} + 72.00\ \alpha_{26} + 109.00\ \alpha_{27} + 54.00\ \alpha_{28} + 115.00\ \alpha_{29} \ge 83$

7. $99.50\ \alpha_1 + 115.00\ \alpha_2 + 112.00\ \alpha_3 + 115.00\ \alpha_4 + 102.50\ \alpha_5 + 99.00\ \alpha_6 + 106.00\ \alpha_7 + 99.00\ \alpha_8 + 90.00\ \alpha_9 + 96.00\ \alpha_{10} + 102.00\ \alpha_{11} + 115.00\ \alpha_{12} + 109.00\ \alpha_{13} + 104.40\ \alpha_{14} + 96.00\ \alpha_{15} + 106.00\ \alpha_{16} + 99.50\ \alpha_{17} + 99.50\ \alpha_{18} + 102.50\ \alpha_{19} + 104.00\ \alpha_{20} + 103.00\ \alpha_{21} + 103.00\ \alpha_{22} + 99.00\ \alpha_{23} + 103.00\ \alpha_{24} + 106.00\ \alpha_{25} + 99.00\ \alpha_{26} + 115.00\ \alpha_{27} + 93.00\ \alpha_{28} + 111.00\ \alpha_{29} \ge 106$

8. $26.10\ \alpha_1 + 26.10\ \alpha_2 + 29.40\ \alpha_3 + 21.40\ \alpha_4 + 29.40\ \alpha_5 + 26.10\ \alpha_6 + 29.40\ \alpha_7 + 29.40\ \alpha_8 + 46.10\ \alpha_9 + 23.50\ \alpha_{10} + 33.69\ \alpha_{11} + 29.40\ \alpha_{12} + 41.30\ \alpha_{13} + 46.10\ \alpha_{14} + 41.30\ \alpha_{15} + 36.80\ \alpha_{16} + 44.40\ \alpha_{17} + 33.60\ \alpha_{18} + 33.60\ \alpha_{19} + 39.90\ \alpha_{20} + 23.50\ \alpha_{21} + 29.40\ \alpha_{22} + 29.40\ \alpha_{23} + 33.60\ \alpha_{24} + 31.40\ \alpha_{25} + 47.00\ \alpha_{26} + 33.60\ \alpha_{27} + 42.80\ \alpha_{28} + 35.60\ \alpha_{29} \ge 39.20$

9. $\alpha_1 + \alpha_2 + \alpha_3 + \alpha_4 + \alpha_5 + \alpha_6 + \alpha_7 + \alpha_8 + \alpha_9 + \alpha_{10} + \alpha_{11} + \alpha_{12} + \alpha_{13} + \alpha_{14} + \alpha_{15} + \alpha_{16} + \alpha_{17} + \alpha_{18} + \alpha_{19} + \alpha_{20} + \alpha_{21} + \alpha_{22} + \alpha_{23} + \alpha_{24} + \alpha_{25} + \alpha_{26} + \alpha_{27} + \alpha_{28} + \alpha_{29} = 1$

FIGURE 7.2: MODEL FORMULATION

(Source: Chakravarty and Ghose 1993. Reprinted by permission of Production and Operations Management Society.)

offering are known. These products (variations of the original product) can be divided into focused groups with a high degree of commonality (components, processes, etc.) within each group. The common components and processes in a group are said to belong to a platform, so that engineering and managerial infrastructure can be targeted at supporting such platforms individually (Ulrich and Eppinger 1995). Since all possible product variations are seldom known a priori, platforms must be created with less than perfect information. This induces a degree of creativity in designing a platform, since product variations (called derivative products) would be limited by platform design. Observe, however, the simplicity with which components can be managed, and processes controlled, using product platforms.

Robertson and Ulrich (1998) have shown how elements of HOQ (house of quality), discussed in Chapter 6, can be used in forming product platforms in the auto industry. They considered three variations of car targeted to three customer segments: a sporty coupe for singles and young couples, a sedan for younger families and older couples, and a station wagon for families with children. Two important questions required resolution—how different the features of the cars needed to be to appeal to the respective market segments, and whether a platform approach that would maximize the number of common components could be used. The authors divided the components in the instrument panel of the car into nine broad functional groups, which they called "chunks." These chunks, along with development costs and the number of unique components, are shown in Table 7.3 Analogous to customer needs in an HOQ, the authors studied attributes that differentiate a coupe from a sedan and a station wagon. Seven differentiating attributes and their corresponding importance to customers are shown in Table 7.4.

Instrument Panel Chunks	Number of Unique Components	Development Cost ($ millions)	Tooling Cost ($ millions)
HVAC system	80	7.8	16.5
Dash cover and structure	100	7.8	13.5
Electrical equipment	180	6.0	4.3
Cross-car beam	24	4.0	4.0
Steering system and airbags	52	4.0	0.2
Instruments and gauges	29	1.8	0.4
Molding and trim	20	0.8	0.4
Insulation	4	0.3	0.2
Audio and radio	8	0.2	0

Table 7.3 Components and Chunks

(Source: Robertson and Ulrich 1998. Reprinted with permission from Sloan Management Review Association.)

Differentiating Attributes	Metric	Importance to Customer
Curvature of window glass	Curved/straight	3
Styling of instrument panel	Aesthetic/functional	3
Relationship between driver and instrument panel	High/low position	3
Front-end styling	Short/long nose	3
Colors and textures	Dark/light	2
Suspension stiffness	Stiff/soft	2
Interior noise	Decibels	1

Table 7.4: Differentiating Attributes and Customer Importance

(Source: Robertson and Ulrich 1998. Reprinted with permission from Sloan Management Review Association.)

The relationship between differentiating attributes and chunks, analogous to HOQ, is shown in Figure 7.3. The chunks are listed in decreasing order (left to right) of cost. The differentiating attributes are listed in decreasing order (top to bottom) of importance to customer. It is clear from Figure 7.3 that the upper left portion of the matrix—the rectangle marked by points *A*, *B*, *C*, and *D*—has a special significance. The chunks—HVAC, dashboard, electrical equipment, and cross-car beam—whose corresponding cells in the rectangle are filled (strong interdependency) are the critical elements on which platform planning can be focused. Inexpensive chunks that are outside the rectangle are candidates for standardization. A differentiating attribute such as front-end styling, which is not strongly related to the expensive chunks, can be varied arbitrarily according to market demand, as the cost of such variation is not expected to be high. The most expensive chunk in Figure 7.3 (HVAC system) is chosen as the first chunk to be redesigned. The components that are highly visible to customers are differentiated in terms of the coupe, sedan, and station wagon. The components hidden from customers are standardized. As shown in Table 7.5, the number of unique components in the redesigned HVAC is reduced to 53 from 80 (see Table 7.3). Consequently, component development cost drops to $4.4 million (from $7.8 million), and tooling cost drops to $9.5 million (from $16.5 million). Observe that no attempt is made to decrease the number of unique components in the chunks instruments and gauges, molding and trim, insulation, and audio and radio,

Differentiating Attributes	Chunks								
	HVAC system A	Dashboard cover and structure	Electrical equipment	Cross-car beam B	Steering system and airbag	Instruments and gauges	Molding and trim	Insulation	Audio and radio
Curvature of window glass	■	■		■	□		■	□	
Styling of instrument panel		■	□			■	■		
Relationship of driver to panel	■	■			■	□			
Suspension stiffness	D	■		C				□	
Front-end styling				□	■				
Interior noise		■						■	
Colors/Textures							■		

■ Strong interdependency
□ Weak interdependency

FIGURE 7.3: RELATIONSHIP DIAGRAM

(Source: Robertson and Ulrich 1998. Reprinted with permission
from Sloan Management Review Association.)

Instrument Panel Chunks	Number of Unique Components	Development Cost ($ millions)	Tooling Cost ($ millions)
HVAC system	53	4.4	9.5
Dash cover and structure	100	7.8	13.5
Electrical equipment	145	4.5	2.2
Cross-car beam	13	2.2	2.0
Steering system and airbags	47	3.0	0.1
Instruments and gauges	29	1.8	0.4
Molding and trim	20	0.8	0.4
Insulation	4	0.3	0.1
Audio and radio	8	0.2	0

Table 7.5: Commonality Plan

(Source: Robertson and Ulrich 1998. Reprinted with permission
from Sloan Management Review Association.)

as they lie outside the rectangle *ABCD* and therefore are not expensive. There is only a marginal reduction in the number of unique components in steering system and airbag.

In the auto example discussed above, the target for platform planning is the number of unique components. There are, of course, other dimensions of commonality in platform planning. These can be categorized as resource sharing and knowledge sharing. Unique components should be designed so as to share as much manufacturing and assembly equipment as possible. Similarly, use of common knowledge components should be maximized among platform components. Such knowledge may be related to design know-how, new technology, and manufacturing processes.

Companies that have used platforms evolve their own procedures for creating them. A telecommunications company with a large number of product offerings has developed several platforms to sharpen its competitiveness. They have formulated an elaborate five-step procedure for platform development. Setting goals for the platform is the first step. It comprises identification of business, product design, and marketing perspectives; conjoint analysis and target costing; and determination of the product's functional requirements and possible market segments. In step two the platform's scope and strategy are established. This involves identifying possible product families and exploring product architectures, common parts, and common assembly processes. In step three they identify opportunities and alternative platform concepts. Opportunities are defined in terms of market shares and revenues for different market segments, and alternative platform concepts are defined in terms of platform architectures that determine the interfaces among core subsystems on the platform. Step four is devoted to developing the platform responsible for the design, building, and testing of subsystems. It also requires ensuring that common interfaces are realized, limiting part proliferation, and controlling optional features. The final step involves development of actual products. It requires that meticulous attention be paid to the actual building and testing of the first product in terms of unique components and subsystems. Product modifications, design improvements, and design of new components are also part of this step.

An industrial product known as an MPG cabinet, which a group of products used for cooling purposes, was chosen for one such platform study. Product sales were showing two definite patterns: sales of full-featured (high-priced) products were dropping sharply, while sales of limited-feature (low-priced) products were increasing, especially in the Asian market. A subsequent manufacturability analysis revealed that there were too many similar components, and only a few of these were common. There were too many fasteners, and the stipulated painting and finishing specifications were set too high. The two important market segments were (1) internal, for the company's customer product lines (full-featured) and (2) OEM, for low-priced limited-feature products. Positioning of some of the company's current products in product feature and market segment space is shown in Figure 7.4. It is clear that most products fell into two shaded areas, shown. But the diagram also reveals that many full-featured products were being sold in the low-price OEM (original equipment manufacturer) market, which clearly did nothing to strengthen the company's competitiveness.

Based on the above preliminary study, the company settled on three

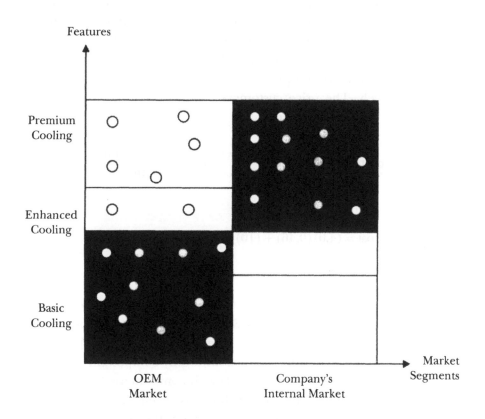

FIGURE 7.4: PRODUCT POSITIONING

broad goals: hold the company's internal market for MPG cabinets, penetrate deeper into the OEM market, and reduce unit cost by at least 30%. To attain its first two (market-related) goals, it was obvious that the company needed a better understanding of how customers valued the product features currently being offered. For this purpose the company employed a detailed conjoint analysis (discussed in Chapter 4) that established customer weighting of product features (part-worths). They then created a matrix (similar to HOQ, in Chapter 6) that established the contribution (weights) of components to each feature of the product. They combined the two sets of weights to determine what the customer would be willing to pay for the product, given component unit costs. This is called target costing. Consider the example in Table 7.6, which shows a fictitious product with three components and four features. It is clear in Table 7.6 that by multiplying a component's importance (contribution) to a feature by the corresponding (feature's) importance to the customer and adding the results for all features, we obtain the importance of individual components to customers. Thus the importance of component 1 to the customer will be equal to $(0.4)\ (0.8) + (0.1)\ (0.3) + (0.3)\ (0.1) + (0.2)\ (0.4) = 0.46$. Similarly, the importance of components 2 and 3 can be verified to be 0.31 and 0.23, respectively. The price customers would be willing to pay for the product can then be computed as

$$\text{Product cost} = \sum_{\text{components}} \begin{pmatrix} \text{component} \\ \text{unit cost} \end{pmatrix} \begin{pmatrix} \text{component importance} \\ \text{to customers} \end{pmatrix}$$

For the above example,

$$\text{Product cost} = (4.0)\,(0.46) + (10)\,(0.31) + (20)\,(0.23)$$

$$= \$26.10$$

It is clear that a product with all three components inserted will cost $70, far in excess of the target cost of $26.10. Obviously such a product will not perform well in the marketplace. There is a need to bundle the components in an innovative way: products with only component 1, products with components 1 and 3, and products with components 1 and 2. Procedures for determining such bundles will be discussed later.

To achieve the goal of a 30% cost reduction, the company conducted a detail study to increase component commonality. A bill of material for

	Component Importance to Features		
	Component 1	**Component 2**	**Component 3**
Feature 1	0.8	0.1	0.1
Feature 2	0.3	0.3	0.4
Feature 3	0.1	0.6	0.3
Feature 4	0.4	0.3	0.3

Unit Cost	**Feature Importance**
Component 1: $40	Feature 1: 0.4
Component 2: $10	Feature 2: 0.1
Component 3: $20	Feature 3: 0.3
	Feature 4: 0.2

Table 7.6: Target Costing

the MPG cabinet is shown in Figure 7.5. A redesign of some of the major subassemblies produced improvements, shown in Figures 7.6a and 7.6b.

The redesigned MPG cabinet had 47% fewer sheet metal parts and 60% fewer part types. The product development cycle time was cut to five months for the platform, with another one to three months to design, manufacture, and ship out individual products. In the old system, in contrast, it took nine to ten months from product concept to a working model for each individual product. Other benefits from platform architecture were fewer setups, reduced waste, and a reduction in work-in-process (WIP) and materials handling. It also enabled the workforce to be flexible in the face of demand variations, and it was easier to build to order.

Platform Dimensions

A product platform defines the range of capabilities (components, processes, etc.) that can be shared by a product family. Any *single* product may not be able to satisfy all its needs from this range of capabilities, but *all* products in the family would satisfy certain percentages (nonzero) of their needs from the platform. The obvious question is why the range of capabilities must be defined a priori, and why the capabilities cannot be added as needed. The simple answer is economy of scope. In developing

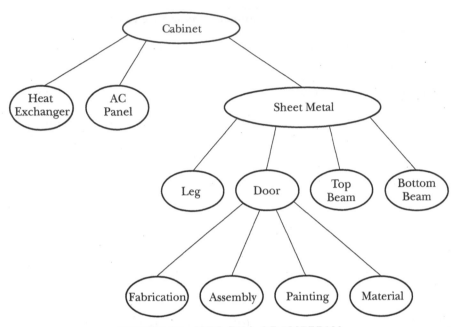

FIGURE 7.5: MPG BILL OF MATERIAL

	Old Design	New Design
Heat Exchanger (HX)	• Four Different HXs • Different beams and legs with and without HX	• One kind of HX • Same beam and legs with and without HX
Electronic Chamber Doors	• Six Different swing-out doors • One lift-up door	• Three different swing-out doors • Optional lifit-up door

FIGURE 7.6a: QUALITATIVE DESIGN IMPROVEMENTS

	Old	New
Protection Block Types	> 30	< 12
AC Panel	26	< 20
Legs	11	5

FIGURE 7.6b: QUANTITATIVE DESIGN IMPROVEMENTS

capabilities that may be interdependent, a significant potential for knowledge or resource sharing exists. Individual products in a family require a few unique components, with others that are shared. The common components, if known a priori, can share the cost of development. If they are acquired instead, suppliers can be chosen so that the smallest possible number is involved. The architecture of a platform is shown in Figure 7.7.

There are two major dimensions of product variety a platform can support: product breadth (or width) and product depth. Product breadth refers to the bundling of attributes such as miles per gallon and miles per hour. Consumer products such as cars, refrigerators, and food processors are usually designed for product breadth and have been discussed in Chapter 4. Product depth, on the other hand, refers to technology performance levels. Each level of technology may be used to create a different product. Krishnan, Singh, and Tirupati (1999) describe a product called a DAQ, used for acquiring data at the rate of several thousand samples per second (sampling rate). Sampling rate is the only customer-desired attribute, and it is determined by the performance level of technology available to the company. Thus if the technology level planned for a given platform is τ, the company can offer any number of DAQs with sampling rates in the range 0 to τ.

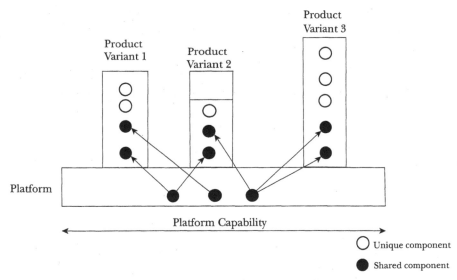

FIGURE 7.7: PLATFORM ARCHITECTURE

Mathematical models have been developed to help decide the optimal product families (DeGroot 1994). In the product-breadth models, the number and types of components and the suppliers are also decision variables. In product-depth models, product technology levels are chosen from a given set of discrete technology level options. We discuss some of these models in the next section, starting with product-breadth models for consumer products.

Industry Applications

To determine optimal product breadth, the Sunbeam Corporation (Page and Rosenbaum 1987) used conjoint analysis and simulation to estimate market shares. Choices of component attributes in five possible product variations, from their study, are shown in Table 7.7. The simulation involved $(3^4 \cdot 2^5)$ product varieties (Sunbeam had three variants each for four of its nine attributes, and two variants for each of the other five attributes). Sunbeam chose three products (out of 2,592), which increased its overall market share by 10%.

Moore, Louviere, and Verma (1999) describe a similar simulation to select a single variety of electronic test equipment. They considered eight product attributes of the equipment: display panel, backlight, data input/output device, accuracy, speed, connection method, stress level,

	Component Variants		
Product	Component 1	Component 2	Component 3
1	1	1	1
2	1	1	2
3	1	2	1
4	1	2	2
5	1	3	1
6	1	3	2
7	2	1	1
8	2	1	2
9	2	2	1
10	2	2	2
11	2	3	1
12	2	3	2

Table 7.7: Component Variants

and test size. Conjoint analysis was used to estimate the part-worth of attributes, and the logit model was used for market share estimation. Six of the attributes had two variants each, and the other two had four variants each. Using the variable and fixed costs of each attribute variant and the unit price of each product variety (1,024 varieties considered), profit and market share were computed for each product variety in order to determine the best product to offer. They repeated the experiment to choose one item from each of two product groups, X and Z (1,024 varieties each of X and Z, simultaneously). The simulation could exploit the common fixed costs of attributes to generate a solution that was more like a product platform.

7.5 PLATFORM MODEL FOR PRODUCT BREADTH

Consider a single-function product such as a bottle opener. Assume that its unit manufacturing cost and unit price are c and p, respectively. Let the fixed cost be F, so that the break-even quantity is $F/(p-c)$. Assume that its total sales is S. It is clear that the product will not be manufactured if $S < F/(p-c)$. Now consider adding a second function to the product's capability, opening cans. While the unit manufacturing cost may not increase substantially because of this addition, the added function may increase its sales significantly. Therefore if sales increase to much more than the break-even quantity, it would be commercially feasible to manufacture this product. The two functions of the product represented by the two attributes are clearly independent of each other.

In situations where product functions are not independent, it may not be possible to build the attributes into the same product. Consider a refrigerator whose design must serve dual objectives: easy access to food stored inside and minimization of heat gain when the door is open (Suh 1990). These two functional attributes are in conflict, and the choices would be either a single product with compromised customer satisfaction or a two-product design: a refrigerator for the accessibility attribute and a freezer for the heat-gain attribute. The two products may use common components such as a condenser and a temperature control unit, but the design of the respective doors would have to be very different. Thus while adding features may increase total demand, the cost may also increase substantially, especially when the features are bundled to create multiple products. If components of the product are being outsourced, supplier choice must be considered concurrently with product feature choice. There is a fixed cost of building a relationship with a supplier, which Sun Microsystems estimated for the semiconductor industry as $500,000 in 1996 (Farlow, Schmidt, and Tsay 1996). It should also be borne in mind that a general-purpose supplier may be capable of supplying a wide range of components but possibly at a higher unit cost, whereas a focused supplier may be capable of supplying only a few components but at a lower unit cost.

Consider a product family with three components. Assume that component 1 comes in two variants, component 2 in three variants, and component 3 in two variants. As shown in Table 7.7, it is clear that 12 products can be created from different variants of the components. Let J be the set of all such products. In this example $|J| = 12$, which would be the required span of the platform supporting the product family. Let p_j be the unit price of the jth ($j \in J$) product. Assume that all component variants can be procured from a set V of vendors. In our example we assume $|V| = 3$. That is, three vendors are being considered. Let the cost of building a relationship with vendor v ($v \in V$) = δ_v.

Let $\hat{J} (\hat{J} \subset J)$ and $\hat{V} (\hat{V} \subset V)$ be the set of products and the set of vendors supported on the platform, respectively. Let c_{ju} be the unit cost of product j, with components acquired from the set U of vendors. Since all vendors may not supply all component variants, it may not be possible to purchase components for all 12 products from a specific set $U(U \subset V)$ of vendors. The component variants carried by the three vendors in V and the corresponding unit prices are shown in Table 7.8.

Analogous to the food processor and pizza examples discussed in

Component	Variant	Vendors		
		1	2	3
1	1	$3	—	$18
	2	$5	$4	$16
2	1	$12	$12	$25
	2	—	$13	$27
	3	—	—	$20
3	1	—	$6	$20
	2	—	$5	$19

Table 7.8: Unit Price of Component Variants

Chapter 4, we assume that there is a one-to-one mapping between a component variant and a product feature. This is also the case with most electronic products, including personal computers. Assume that the part-worths of product features are determined from a conjoint analysis, and these part-worths are used to derive a rating for each of the 12 products. Let the estimated rating of product j be R_j, and the rating error e_j, and let the rating of a competing product be denoted as R_c. Assuming a logistics distribution of error e_j, the probability that a customer would purchase product j can be expressed as

$$P_j = \frac{exp\,(R_j)}{exp\,(R_c) + \sum\limits_{j \in \hat{J}} exp\,(R_j)} \tag{7.7}$$

where $\hat{J}$ is the set of products offered to customers. Assuming customer preferences to be uniformly distributed over all customers, the expression for the market share of j can be written as

$$s_j = \frac{A_j}{A_c + \sum\limits_{j \in \hat{J}} A_j} \tag{7.8}$$

where

$$A_j = exp\,(R_j)$$

Let $\hat{V}$ be the set of vendors chosen.

The profit corresponding to a known $\hat{J}$ and $\hat{V}$ can be written as

$$\Pi\,(\hat{J},\,\hat{V}) = \sum\limits_{j \in \hat{J}} M s_j\,(p_j - c_j \hat{V}_{j\hat{v}}) - \sum\limits_{U \in \hat{V}} \delta_U$$

Substituting for s_j from equation 7.8, we have

$$\Pi\,(\hat{J},\,\hat{V}) = M \sum\limits_{j \in \hat{J}} \left\{ (p_j - c_j \hat{v}) \frac{A_j}{A_c + \sum\limits_{r \in \hat{J}} A_r} \right\}$$

Letting $x_{jV} = M(P_j - c_{jv})A_j$ and $y_j = A_j$, we can rewrite the profit as

$$
\Pi(\hat{J}, \hat{V}) = \frac{\sum\limits_{j \in J} x_{j\hat{v}}}{A_c + \sum\limits_{j \in J} y_j} = \frac{X_{\hat{J},\,\hat{v}}}{A_c + Y_{\hat{J}}}
\tag{7.9}
$$

It is clear that $\Pi(\hat{J}, \hat{V})$ is linear in $X_{\hat{J},\,\hat{v}}$ and convex in $Y_{\hat{J}}$. Consider now a two-dimensional space in $X_{\hat{J},\,\hat{v}}$ and $Y_{\hat{J}}$ for a given $\hat{V}$. Let the values of $X_{\hat{J},\,\hat{v}}$ and $Y_{\hat{J}}$ be plotted in this space for all $\hat{J}(\hat{J} \subset J)$. It is clear that the maximum of $\Pi(\hat{J}, \hat{V})$ will be obtained at an extreme point of the convex hull in the $X_{\hat{J},\,\hat{v}},\ Y_{\hat{J}}$ space. Chakravarty and Balakrishnan (1999) have shown that if products are ranked in descending order of the ratio $r_j = x_{j\hat{v}}/y_j$, then any choice of q products from the top of the list will generate an extreme point in the above space.

Thus the optimization problem simplifies to adding products, starting at the top of the list, to the platform if the total platform profit $\Pi(\hat{J}, \hat{V})$ keeps increasing. Thus optimal $\hat{J}$ is defined as

$$
\Pi(\hat{J} - 1, \hat{V}) \le \Pi(\hat{J}, \hat{V}) \ge \Pi(\hat{J} + 1, \hat{V})
$$

This procedure is repeated for all $\hat{V} \subset V$ and the maximum value, that is,

$$
\underset{\hat{V}\,\subset\,V}{\text{Maximum}} \left\{ \Pi(\hat{J}, \hat{V}) \right\}
$$

is identified.

Simplification

The ratio r_j can be rewritten as

$$
r_j = \frac{x_{j\hat{v}}}{y_j} = \frac{M(p_j - c_{j\hat{v}})A_j}{A_j} = M(p_j - c_{j\hat{v}})
$$

Since M is a constant, it can be excluded from the ranking parameter, which simplifies to just the contribution margin, $m_{j\hat{v}} = p_j - c_{j\hat{v}}$. Thus the platform will comprise n products with the highest profit margins, where

n is the optimal number of products. This is an important property of the platform and tells us that the relative preference of a product for inclusion on the platform is *independent* of the product's attractiveness (A_j). Whether the product is included on the platform, however, depends on the optimal value of n, which in turn depends on the attractiveness of all products being considered for the platform (and also on the competitor's products). Goldberg and Zhu (1989) discuss a similar problem where cost minimization is the objective.

Example

The platform design problem, corresponding to Tables 7.7, 7.8, and 7.9, can now be solved. The additional information required is the values of unit price (p_j), product attractiveness (A_j) corresponding to the 12 products in Table 7.7, and the attractiveness of buying competing products (A_c) or not buying at all (A_o). These values are shown in Table 7.9. The market size is 100,000 buyers, and the cost of building a vendor relationship is $100,000 per vendor.

With three vendors, it is clear that there are seven possible vendor groups ($\hat{V}$): (v_1), (v_2), (v_3), (v_1, v_2), (v_2, v_3), (v_1, v_3), and (v_1, v_2, v_3). The vendor group $\hat{V} = (v_1)$ is not feasible, as vendor 1 does not supply any variant of component 3.

Consider $\hat{V} = (v_2)$. Since vendor 2 does not supply variant 1 of component 1 or variant 3 of component 2, it can be verified from Table 7.7 that the only products that vendor 2 can supply completely are products 7, 8, 9, and 10. The company's procurement costs can be determined using Table 7.8 and are shown in Table 7.10.

Using price information from Table 7.9, the profit margins can be verified to be 10, 4, 14, and 6, respectively. Thus the products would be ranked in the order product 9, product 7, product 10, and product 8. Using this information, the values of $x_j (= p_j - c_{jv_2})$, $y_j (= A_j)$, cumulative x_j,

Product, j	1	2	3	4	5	6	7	8	9	10	11	12	$A_c + A_o$
Attractiveness, A_j	50	80	40	70	70	60	35	60	40	45	80	50	200
Price, p_j	30	35	25	40	40	36	32	25	37	28	35	27	

Table 7.9: Product Attractiveness and Price

Product 7		Product 8		Product 9		Product 10	
Component, Variant	Cost	Component, Variant	Cost	Component, Variant	Cost	Component, Variant	Cost
1,2	4	1,2	4	1,2	4	1,2	4
2,1	12	2,1	12	2,2	13	2.2	13
3,1	6	3,2	5	3,1	6	3,2	5
Total	22		21		23		22

Table 7.10: Procurement Cost

cumulative y_j, and profit Π can be computed as shown in Table 7.11. Since profit keeps increasing, all four products will be on the platform if v_2 is the only supplier.

For $\hat{V} = v_3$, it can be verified that the profit margin on each of the 12 products will be negative. Hence v_3 by itself will not be a viable solution.

Consider now $\hat{V} = (v_1, v_2)$. In this scenario a component variant will be bought from the less expensive of the two vendors if a choice is available. Such is the case with variant 2 of component 1, which will be bought from vendor 2. Variant 1 of component 2 can be bought from either of the two vendors at the same price. Since variant 3 of component 2 cannot be supplied by either of the two vendors, the only admissible products are 1, 2, 3, 4, 7, 8, 9, and 10. The costs of these products can be computed using Table 7.7, as discussed earlier. The profit of the ranked products will be computed as shown in Table 7.12. Observe in Table 7.12 that profit increases as products are added to the platform; it reaches a maximum value of \$818,947 and then starts to decrease. Hence products 10, 8, and 3 will not be added to the platform. Therefore, the optimal composition of the platform for $\hat{V} = (v_1, v_2)$, will be {1, 2, 4, 7, 9}.

Ranked Products	$A (= y)$	$A \cdot M (p-c) (= x)$	cumulative x	$A_0 + A_c +$ cumulative y	Π
9	40	56×10^6	56×10^6	240	233,333
7	35	35×10^6	91×10^6	275	350,909
10	5	27×10^6	118×10^6	320	368,750
8	60	24×10^6	142×10^6	380	373,684

Table 7.11: Computations with Ranked Products

Ranked Products	Unit Cost	Margin	y	x	cumulative x	$A_o + A_c +$ cumulative y	Π
4	21	19	70	133×10^6	133×10^6	270	492,592
2	20	15	80	120×10^6	253×10^6	350	722,857
9	23	14	40	45×10^6	309×10^6	390	792,308
7	22	10	35	35×10^6	344×10^6	425	809,411
1	21	9	50	45×10^6	389×10^6	475	818,947
10	22	6	45	27×10^6	416×10^6	520	800,000
8	21	4	60	24×10^6	440×10^6	580	758,621
3	22	3	40	12×10^6	452×10^6	620	729,032

Table 7.12: Computations for $\hat{V} = (v_1, v_2)$

For $\hat{V} = (v_1, v_3)$, the only two products with positive margins are products 2 and 6. The platform will include both these products at a profit of $144,737. For $\hat{V} = (v_2, v_3)$, six products will have positive margins. The ranked products (left to right) and platform profit Π are shown in Table 7.13. It is clear from Table 7.13 that the optimal set of products in the platform would be {4, 7, 8, 9, 10, 11}, and the corresponding profit would be $396,226.

For $\hat{V} = (v_1, v_2, v_3)$, the products will be ranked as 4, 2, 9, 5, 7, 1, 6, 12, 10, 11, 8, 3. Since all 12 products have positive profit margins, they will be considered for the platform. The profit calculation is shown in Table 7.14, and the optional platform is {1, 2, 4, 5, 7, 9}. The corresponding profit is $855,046.

The above findings are summarized in Table 7.15.

It is clear that the highest net profit will equal $618,947 and that the optimal platform will comprise products 1, 2, 4, 7 and 9. The vendors used will be v_1 and v_2. For the above products, the breakdown of market share, sales, and profit contribution are shown in Table 7.16.

We can also compute the vendor's revenues as follows. Assuming that variant 1 of component 2 is bought from vendor 1, the unit cost of each product can be split between the two vendors, using Tables 7.7 and 7.8.

Product	9	7	10	11	4	8
Π	233,333	360,909	368,750	395,000	395,745	396,226

Table 7.13: Product Profits for $\hat{V} = (v_1, v_2)$

Ranked Products	Unit Cost	Margin	y	x	cumulative x	$A_0 + A_c +$ cumulative y	Π
4	21	19	70	133×10^6	133×10^6	270	492,592
2	20	15	80	120×10^6	253×10^6	350	722,857
9	23	14	40	56×10^6	309×10^6	390	792,308
5	29	11	70	77×10^6	386×10^6	460	839,130
7	22	10	35	35×10^6	421×10^6	495	850,505
1	21	9	50	45×10^6	466×10^6	545	855,046
6	28	8	60	48×10^6	514×10^6	605	849,586
12	29	8	50	40×10^6	554×10^6	655	845,801
10	22	6	45	27×10^6	581×10^6	700	830,000
11	30	5	80	40×10^6	621×10^6	780	796,152
8	21	4	60	24×10^6	645×10^6	840	767,857
3	22	3	40	12×10^6	657×10^6	880	746,590

Table 7.14: Computations for $\hat{V} = (v_1, v_2, v_3)$

Using the sales of each product, the payments to each vendor by product can be computed, as shown in Table 7.17. Thus the total cost of purchase to the company will be \$1,223,121. Note that if we are interested in maximizing market share (instead of profit), the top five products would be 2, 11, 4, 5 and 6, with a combined market share of 64.28%. It is therefore clear that profit maximization and market share maximization may lead to different platform compositions.

Chakravarty and Balakrishnan (1998, 1999) have studied broader business models for platform design: Suppliers may be wholly owned subsidiaries or they may be independent, and a platform may support more than one product group—cars, sport-utility vehicles, and trucks. They report the following insights.

Vendor Group	Number of Vendors	Π	Vendor Relationship Cost	Platform Products	Net Profit
(v_1)	1	—	—	—	—
(v_2)	1	373,684	100,000	7, 8, 9, 10	273,684
(v_3)	1	—	—	—	—
(v_1, v_2)	2	818,947	200,000	1, 2, 4, 7, 9	618,947
(v_2, v_3)	2	396,226	200,000	4, 7, 8, 9, 10, 11	196,226
(v_1, v_3)	2	144,737	200,000	2, 6	−55,263
(v_1, v_2, v_3)	3	855,046	300,000	1, 2, 4, 5, 7, 9	555,046

Table 7.15: Costs and Profits by Vendor Groups

Product	Market Share	Sales Quantity	Profit Contribution
1	10.5%	10,526	94,737
2	16.9%	16,842	252,632
4	14.7%	14,736	280,000
7	7.5%	7,368	73,684
9	8.4%	8,421	117,895
Total	58.0%	57,893	818,948

Table 7.16: Market Share, Sales, and Profit

- Platform breadth (number of products supported), in an independent-supplier scenario, exceeds the corresponding value in the scenario with a wholly owned subsidiary.
- The total system profit (company and suppliers) in the independent-supplier scenario exceeds the corresponding value in the scenario with a wholly owned subsidiary.
- The company will be willing to pay more to the suppliers if it sells its products in a very competitive market. It will also buy relatively more from a general-purpose supplier.
- A component variant, useful in creating a new product, may not be used at current prices because of market dynamics. Suppliers may use this information to decide whether to discontinue the product or offer deep price discounts.
- A general-purpose supplier can leverage its versatility (breadth of its offering) to charge a premium for its product. In such a case the company would benefit by subsidizing a focused supplier to expand its offering.

Product	Payment to Vendors per Unit of Product		Total Payment to Vendors	
	Vendor 1	Vendor 2	Vendor 1	Vendor 2
1	15	6	157,890	63,156
2	15	5	252,630	84,210
4	3	18	44,208	265,248
7	12	10	88,415	73,680
9	—	23	—	193,693
		Total	543,144	679,977

Table 7.17: Payments to Vendors

7.6 TECHNOLOGY-DRIVEN PRODUCTS

Unlike consumer products, technology-based products normally fetch a high premium for product performance. Personal computers and telecommunications products are good examples. Instead of components, these products may be considered as bundles of technology chunks that keep evolving at a rapid pace. Consider a product such as the telephone and the changes in technology from a basic telephone to a cordless telephone (premium). The technology required for different products is shown in Table 7.18.

The relevant question, from a product design point of view, is whether or not to offer products corresponding to every level of improvement in technology. A trade-off exists among market share, the cost of new technology, the cost of switching technollgy, and product price. Krishnan, Singh, and Tirupati (1999) have pointed out that for technology-based products, consumers will be willing to switch to the next-generation product (within certain bounds) if the product they prefer is not available. In Table 7.18, for example, among the technologies shown, three technology levels of user interface board are included, out of at least ten such possible technologies. For a product with several technologies, with multiple technology levels for each, the product design problem becomes extremely complex. We discuss the solution for a simple product comprising a single

Technology	Corded Basic	Corded Premium	Cordless Basic	Cordless Premium
Tape—audio	Y	N	Y	N
Digital audio	N	Y	N	Y
Integrated digital audio	N	N	N	Y
Integrated caller ID (ICI)	N	N	N	Y
ICI with digital audio	N	N	Y	N
Motherboard 1	Y	N	N	N
Motherboard 2	N	Y	Y	N
Motherboard 3	N	N	N	Y
User interface board 1	Y	N	N	N
User interface board 2	N	Y	N	N
User interface board 3	N	N	Y	Y
Display board 1	N	Y	N	N
Display board 2	N	N	N	Y

Table 7.18: Technology Bundling

technology, such as the data acquisition product referred to in Krishnan, Singh, and Tirupati (1999).

Assume that for a single technology product there are five possible technology levels, denoted as τ_i, $i = 1$ to 5. Krishnan, Singh, and Tirupati have identified two types of costs for platform design: a fixed cost of component design, and a variable manufacturing cost per unit. The component design cost comprises two elements: the cost of designing unique components corresponding to a technology level τ_i, and the cost of adapting components from the existing technology level i to a new technology level j, a_{ij} ($j > i$).

The price at which a product can be sold, and the corresponding sales quantity, are assumed to depend on the technology level. The values of demand, unit price, and unit variable cost corresponding to the five technology levels are shown in Table 7.19. The one-time component design costs are shown in Table 7.20. Note that the cost of designing a unique component is assumed to be $15 million, irrespective of technology level. It is also assumed that if a customer's ideal point (see Chapter 4) is at technology level i, and if the product at technology level i is not offered, the customer would switch to the product at the next higher technology level. We represent this by the probability w_{ij} of switching from technology level i to technology level j. Thus if product i is not offered and if its demand is d_i, the effective demand at level $i + 1$ would be $d_{i+1} + w_{i,\,i+1}\, d_i$. The values of switching probabilities are shown in Table 7.21. Similarly, if products 1 to

i are not offered, demand for product $i+1$ would equal $d_{i+1} + \sum_{r=1}^{i} w_{r,\,i+1} \cdot d_r$.

The demand expression could be modified easily if customers switched in both directions.

Product at Technology Level i	Demand d_i	Unit Price p_i ($)	Unit Variable Cost c_i ($)
1	16,000	640	187.50
2	20,000	885	250.00
3	16,000	1,480	400.00
4	10,000	1,680	450.00
5	8,000	1,780	475.00

Table 7.19: Demand, Price, and Cost

From Technology Level i	To Technology Level j				
	1	2	3	4	5
0	$15M	$15M	$15M	$15M	$15M
1		$3M	$7M	$9M	$11M
2			$6M	$8M	$10.5M
3				$4M	$ 9.5M
4					$ 7.5M

Table 7.20: Cost of Component Design

The above platform design problem can be formulated as a longest-path problem. The five technology levels can be represented as a network, as shown in Figure 7.8. Node i in the network represents the product at technology level i ($i = 1$ to 5). Nodes S and E represent the start and end nodes, respectively. The following notation is used for the longest-path formulation.

$$g(i, j) = \text{additional profit from introducing product } j,$$
given that product i exists ($i < j$), and that no other products exist
in the technology performance interval τ_i to τ_j

Thus $g(i, j)$ can be thought of as the length of the arc that links nodes i and j directly.

$$f(S, j) = \text{maximum profit from all products offered in the technology}$$
performance interval τ_1 to τ_j (inclusive τ_1 and τ_j).

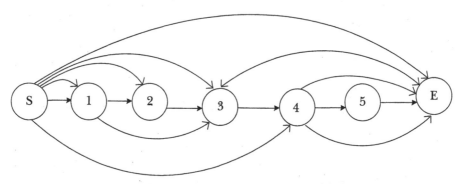

FIGURE 7.8: A SUBSET OF PATHS IN NETWORK REPRESENTATION

Our objective is to determine $f(S, E)$.

Since the profit $g(i, j)$ is additive, we can write the longest-path formulation as

$$f(S, j) = \underset{\tau_i < \tau_j}{Maximum} \{g(i,j) + f(S,i)\}$$

It is clear that $g(i, j)$ can be computed as

$$g(i, j) = (p_j - c_j)\,(\textit{effective demand, } D_j) - a_{ij}$$

where

$$p_j = \text{unit price of product } j$$

$$c_j = \text{unit cost of product } j$$

$a_{ij} = \text{cost of designing components of product } j,$
inclusive of cost of adaptation of components of product i

The effective demand D_j can be computed, as explained earlier, as

$$D_j = \sum_{r=i+1}^{j} w_{rj}\, d_j$$

where d_j is product j's demand (assuming no switching of consumers) and w_{rj} is the probability that a customer whose ideal point is at technology level τ_r will switch to product j if product r is not offered. Note that the probability that the customer does not buy any product when r is not available (i.e., balking) can be written as $1 - \sum_{j=r}^{n} w_{rj}$, where τ_n is the highest technology performance level being considered.

Thus to compute $g(1, 4)$, using Tables 7.19, 7.20, and 7.21, we have

$$D_4 = 20{,}000\,(0.2) + 16{,}000\,(0.4) + 10{,}000 = 20{,}400$$

$$a_{ij} = \$9 \textit{ million}$$

From Technology Level i	To Technology Level j				
	1	2	3	4	5
1	X	0.4	0.3	0.2	0
2		X	0.5	0.2	0.1
3			X	0.4	0.2
4				X	0.6

**Table 7.21: Probabilities of Customer Switching
to Higher-Level Technologies**

Therefore,

$$g(1, 4) = (1,680 - 450)(20,400) - 9,000,000 = 16,092,000$$

To do the longest-path computation, we first evaluate $f(S, 1)$ (= $-\$7,760,000$), and use this result to obtain $f(S, 2)$ = Maximum $\{g(S, 2), f(S, 1) + g(1, 2),\}$. We then evaluate $f(S, 3)$ using $f(S, 2), f(S, 1)$, and $g(S, 3)$, and so on. These computations are summarized in Table 7.22. It is clear from the table that $f(S, E)$ = \$29,504,000. The corresponding optimal product platform would be $\{3, 4, 5\}$.

The longest-path formulation can easily be modified for different scenarios. For example, the company may already be offering products at certain technology levels and may be interested in rationalizing its products by adding and/or discontinuing some products because of changes in demand and/or costs. For an existing product, the component design cost would be zero. A second scenario involves the presence of competing products. The customer now has an option to switch to a competing prod-

Destination Node j	$F(S, i) + g(i, j)$						$F(S, j)$
	$i = S$	$i = 1$	$i = 2$	$i = 3$	$i = 4$	$i = 5$	
1	-7,760,000	—	—	—	—	—	-7,760,000
2	1,764,000	1,940,000	—	—	—	—	1,940,000
3	18,274,000	13,320,000	13,220,000	—	—	—	18,264,000
4	14,028,000	8,332,000	14,112,000	26,564,000	—	—	26,564,000
5	10,056,000	6,296,000	13,886,000	27,034,000	29,504,000	—	29,504,000
E	0	-7,760,000	1,940,000	18,264,000	26,564,000	29,504,000	

Table 7.22: Computation of Longest Path

uct if the product corresponding to his/her ideal point is not offered. Thus if a competing product is at level τ_i and the company is considering introducing a product at level τ_{i+1}, it cannot expect any customers to switch from level τ_{i-1} to τ_{i+1} (assuming no product at level τ_{i-1}). The competing product at level τ_i will act as an absorbing barrier.

A related issue is whether a customer switching from τ_i to τ_{i+1} should be charged a unit price of p_i or p_{i+1}, assuming $p_i < p_{i+1}$. The argument for charging the lower price (p_i) is that it would increase the probability that the customer would make the switch, although the revenue per customer would decrease. Gupta and Krishnan (1998) and Fisher, Ramdas, and Ulrich (1998) have studied the decision-making issues in such scenarios.

7.7 MANUFACTURING COST OF VARIETY

The graph depicting the variation of profit with the number of products in the platform (section 7.5) is shown in Figure 7.9. Observe that profit is

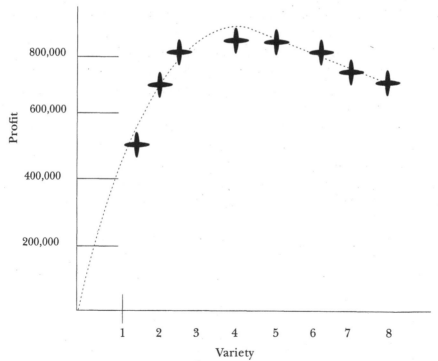

FIGURE 7.9: VARIETY VS. PROFIT

maximized for a variety measure of five (five products on the platform). That is, the cost of increasing variety beyond five outweighs the market revenue generated by it. The cost in this case was entirely due to payments made to suppliers, which assumes a business model based on outsourcing.

There is still a significant number of companies that market their own products and do their own manufacturing. For such companies, unlike in the outsourcing paradigm, the large investment that must be made in manufacturing processes may limit the product variety that can be offered. From a manufacturing point of view, there is a group of core components (common to all products) and a group of optional components. It is the optional components that create the sense of variety in the minds of customers. It is clear that economies of scale can be exploited in the manufacture of core components. For optional components, the investment in manufacturing processes may not be justifiable in all cases. What complicates this decision is that a component or a group of components may require more than one manufacturing process (such as turning, milling, forging, grinding, welding, etc.), each of which has unique needs in terms of manufacturing equipment, tools, and skilled manpower. Consider a piece of electronic test equipment with three attributes and two option levels for each attribute, as shown in Table 7.23. It is clear that eight products are possible, which will require processes as shown in Table 7.24.

Thus an investment in processes 1, 2, 3, 12, 22, 23, and 24 will enable the manufacture of products 1 and 2. To produce products 3 and 4 as well, there needs to be additional investment in processes 13 and 14, and so on. The relevant question would be whether or not there is sufficient demand for products 1, 2, 3, and 4 to justify the investment in processes.

Chakravarty and Baum (1992) have modeled this problem as a mixed-integer program, and we discuss a version of that formulation. Let y_j, p_j, and a_j be the production quantity, unit price, and unit variable cost of

Attribute	Attribute Level and Processes			
	Level 1	Processes	Level 2	Processes
Display	Line display	1, 2, 3	Graphic Display	2, 4, 5, 6
Data Input/Output Method	Connect to PC	1, 12	Connect to network	1, 13,14
Speed	Slow	12, 22, 23	Fast	12, 23, 24

Table 7.23: Attribute Level and Manufacturing Processes

Product (j)	Process Groups (K_i)		
1	(1, 2, 3,)	(1, 12),	(12, 22, 23)
2	(1, 2, 3),	(1, 12),	(12, 23, 24)
3	(1, 2, 3),	(1, 13, 14),	(12, 22, 23)
4	(1, 2, 3),	(1, 13, 14),	(12, 23, 24)
5	(2, 4, 5, 6),	(1, 12),	(12, 22, 23)
6	(2, 4, 5, 6),	(1, 12),	(12, 23, 24)
7	(2, 4, 5, 6),	(1, 13, 14),	(12, 22, 23)
8	(2, 4, 5, 6),	(1, 13, 14),	(12, 23, 24)

Table 7.24: Process Requirements

product j, respectively. Let F_k be the fixed cost (investment) of process k. Let $x_k = 1$ if the kth process is used, 0 otherwise. The optimization problem can be expressed as

$$Maximize \sum_{j \in J} (p_j - a_j)y_j - \sum_{k \in K} F_k \cdot x_k \qquad (7.10)$$

subject to

$$x_k \geq \sum_j y_j \cdot t_{jk}/L, \forall_k$$

$$a_j = \sum_k c_k t_{jk}, \forall_j$$

$$y_j \leq D_j, \forall_j$$

where

$$t_{jk} = \text{processing time of product } j, \text{ using the } k\text{th process}$$

$$c_k = \text{cost per unit time for using process } k$$

$$D_j = \text{demand of product } j$$

$$L = \text{a large number}$$

Since demand D_j will depend on the attractiveness of the options of individual attributes, we can use equation 7.8 to express D_j as

$$D_j = \frac{M A_j}{A_c + \sum_{j \in J} A_j} \tag{7.11}$$

where A_j is the attraction of the jth product, A_c the attractiveness of competing products, and M the market size. It can be easily established that in the optimal solution $y_j = D_j$ or $y_j = 0$. Hence using equation 7.11, the objective function 7.10 can be rewritten as

$$\text{Maximize } M \sum_{j \in J} \left(p_j - \sum_k c_k t_{jk} \right) \frac{A_j}{A_c + \sum_{j \in J} A_j} - \sum_{k \in K} F_k x_k \tag{7.12}$$

Consider a subset of S of processes. Let J_s be the subset of products that can be completely manufactured using only the processes in S. That is, if $j \in J_s$ and $k \in K - S$, then $t_{jk} = 0$. Hence for the subsets S, and J_s the problem becomes an *unconstrained* optimization, written as

$$\underset{\hat{J}_s}{\text{Maximize}} \quad \frac{M \sum_{j \in \hat{J}_s} \left(p_j - \sum_{k \in S} c_k t_{jk} \right) A_j}{A_c + \sum_{j \in \hat{J}_s} A_j} - \sum_{k \in S} F_k x_k \tag{7.13}$$

where $\hat{J}_s \subset J_s$.

Since for a given S, $\sum_{k \in S} f_k x_k$ is a constant, the objective function can be written as

$$\underset{\hat{J}_s}{\text{Maximize}} \quad \Pi(S, \hat{J}_s) = \frac{M \sum_{j \in \hat{J}_s} \left(p_j - \sum_{k \in S} c_k t_{jk} \right) A_j}{A_c + \sum_{j \in \hat{J}_s} A_j} = \Pi(S)$$

which is similar to equation 7.9 and can be maximized by ranking the products in set J_s by the ratio

$$r_j = p_j - \sum_{k \in S} c_k t_{jk}, j \in J_s$$

Hence the optimal value for $\Pi = \underset{S}{\text{Max}} \, \Pi(S)$ can be found by considering all possible subsets of K.

Chakravarty and Baum conducted extensive simulation with their mixed-integer optimization model to discern the sensitivity of profit and market share of products to the attractiveness of individual product attributes. They found (1) a much higher level of market share and profit can be obtained if the attractiveness of attributes that require the fewest number of processes is increased, and (2) required production capacity increases at a slower pace with market share if the attractiveness of the attribute with the fewest number of process requirements is increased.

A somewhat different manufacturing scenario has been studied by Raman and Chhajed (1995). In their formulation they incorporate features of both Chakravarty and Baum's (1992) model and Dobson and Kalish's (1988) model, which was discussed in Chapter 4. From Dobson and Kalish they include the following aspects: (1) price p_j, which, instead of being stipulated by marketing, becomes a decision variable, and (2) customers' reservation price (to model the behavior of customers who do not get to have their top-ranked products). Because of this they need to have decision variables for each individual customer. They incorporate the cost of investment in processes in a way similar to that in the Chakravarty and Baum model. They have developed efficient heuristics for solving small-size problems (up to ten customers, five attributes with three levels for each, and five processes). The Dobson and Kalish model differs from that of Raman and Chhajed in the way fixed costs are handled; the model includes a fixed cost for a product, irrespective of the processes used.

7.8 A STRATEGY FOR MANAGING VARIETY

By now it should be clear that in our discussion of platform concepts and models we have primarily concentrated on two drivers of product variety: product economics and customer economics. The third driver, system economics, is not treated in depth, as it is still in its infancy in terms of being a viable alternative in the computer and telecommunications industries. Hax and Wilde (1999) provide a framework linking the three strategies with an organization's capabilities. They categorize capabilities as operational effectiveness, customer targeting, and innovation. Hayes and Pisano (1996) describe operational effectiveness as the ability to improve performance in cost, quality, and cycle time through business practices such as agile manufacturing, total-quality-management (TQM) and virtual integration. Customer targeting is the ability to increase customer satisfaction through product offering, distribution, and forming alliances with other suppliers/manufacturers. Finally, innovation is the ability to visualize and realize products, processes, and systems of tomorrow using the knowledge base (and resources) of today. A 3 by 3 contingency matrix with nine cells, as in Figure 7.10 can be used to capture the interactions among strategies and capabilities.

Consider the strategy of product economics. It is clear that production cost, quality, differentiation, distribution cost, market share, rate of new product introduction, and time to market are all very important in improving product-related economy. But the capability that would be most synergistic with this strategy is, obviously, operational effectiveness. Indeed, this was the thrust of Japanese competition in the 1980s and early 1990s. Although operational effectiveness can contribute to customer economics and system economics, it requires different business models, such as collaborative partners and/or transparent IT interface, for it to happen. Overreliance on product economics as a strategy has perhaps been the root cause of the problems Japanese companies are facing today.

Capabilities	Strategies		
	Product Economics	Customer Economics	System Economics
Operational Effectiveness	Product cost, quality, and differentiation	Supplier alliance for total solution, customer's cost of use	Extended-enterprise effectiveness
Customer Targeting	Distribution channels, market share	Customer selection, total customer satisfaction	Selection of best partners
Innovation	Time to market, rate of product introduction	New services for customers to enhance satisfaction	New ways of splitting the value chain amongst partners

FIGURE 7.10: STRATEGIES AND CAPABILITIES

The strategy of customer economics is clearly most synergistic with the capability of customer targeting. The objective is to identify a few attractive customers and work with them to provide a total solution to their problems. Once an understanding with customers is reached, one can reengineer processes for operational effectiveness to suit the new business model, and innovate new services targeted to individual customers to lock in the customer. Such innovations may also involve reengineering products and processes to link seamlessly with the customer's value chain.

The main objective of the system economics strategy is to create proprietary standards in the system that everyone must follow (e.g., Microsoft's Windows software). Such standards are difficult to imitate if they are complex and/or if they involve inputs from several partners (called complementors) in the system. The challenge lies in developing an appropriate system architecture that determines the capabilities of partners and in setting up appropriate interfaces among partners, that define their roles, and protocols of interaction (Baldwin and Clark 1997). Competitive advantage accrues for the ability to quickly incorporate new complementors in the system to create a modified system standard as customers' needs or business conditions change.

REFERENCES

Anderson, E., C. Fornell, and R. Rust (1997), "Customer Satisfaction, Productivity, and Profitability: Differences Between Goods and Services," *Marketing Science,* vol. 16, no. 2, pp. 129–145.

Balakrishnan, N., A. Chakravarty, and S. Ghose (1997), "Role of Design Philosophies in Interfacing Manufacturing with Marketing," *European Journal of Operational Research,* vol. 103, pp. 453–469.

Baldwin, C., and K. Clark (1997), "Managing in an Age of Modularity," *Harvard Business Review,* September–October, pp. 84–93.

Baldwin, C., and K. Clark (1999), *Design Rules: The Power of Modularity,* MIT Press, Cambridge, Mass.

Chakravarty, A., and N. Balakrishnan (1998), "Achieving Product Variety Through Optimal Choice of Module Variations," working paper, Tulane University, New Orleans.

Chakravarty, A., and N. Balakrishnan (1999), "Product Design with Multiple Suppliers for Component Variants," working paper, Tulane University, New Orleans.

Chakravarty, A., and J. Baum (1992), "Coordinated Planning for Competitive Products and Their Manufacturing Operations," *International Journal of Production Research,* vol. 30, no. 10, pp. 2293–2311.

Chakravarty, A., and S. Ghose (1993), "Tracking Product-Process Interactions:

A Research Paradigm," *Production and Operations Management,* vol. 2, no. 2, pp. 72–93.

DeGroote, X. (1994), "Flexibility and Marketing/Manufacturing Coordination," *International Journal of Production Economics,* vol. 36, pp. 153–167.

Dobson, G., and S. Kalish (1988), "Positioning and Pricing a Product Line," *Marketing Science,* vol. 7, no. 2, pp. 107–125.

Farlow, D., G. Schmidt, and A. Tsay (1996), "Supplier Management at Sun Microsystems (A)," Stanford Case Study OIT-16A.

Fisher, M., K. Ramdas, and K. Ulrich (1998), "Component Sharing in the Management of Product Variety: A Study of Automotive Braking Systems," *Management Science,* vol. 45, no. 3, pp. 297–315.

Goldberg, J., and J. Zhu (1989), "Module Design with Substitute Parts and Multiple Vendors," *European Journal of Operational Research,* vol. 41, pp. 335–346.

Gupta, S., and V. Krishnan (1998), "Integrated Component and Supplier Selection for a Product Family," *Production and Operations Management,* vol. 8, no. 2, pp. 163–182.

Hax, A., and D. Wilde (1999), "The Delta Model: Adaptive Management for a Changing World," *Sloan Management Review,* winter, pp. 11–28.

Hayes, R., and G. Pisano (1996), "Manufacturing Strategy: At the Intersection of Two Paradigm Shifts," *Production and Operations Management,* vol. 5, no. 1, pp. 25–41.

Ho, T., and C. Tang (1998), *Managing Product Variety,* Kluwer Academic Publishers, Boston, Mass.

Kekre, S., and K. Srinivasan (1990), "Broader Product Line: A Necessity to Achieve Success?" *Management Science,* vol. 36, no. 10, pp. 240–251.

Krishnan, V., R. Singh, and D. Tirupati (1999), "A Model-Based Approach for Planning and Developing a Family of Technology-Based Products," *Manufacturing and Service Operations Management,* vol. 1, no. 2, pp. 132–156.

Moore, W., J. Louviere, and R. Verma (1999), "Using Conjoint Analysis to Help Design Product Platforms," *Journal of Product Innovation Management,* vol. 16, pp. 27–39.

Page, A., and H. Rosenbaum (1987), "Redesigning Product Lines with Conjoint Analysis: How Sunbeam Does It," *Journal of Product Innovation Management,* vol. 4, pp. 120–137.

Raman, N., and D. Chhajed (1995), "Simultaneous Determination of Product Attributes and Prices and Production Processes in Product-Line Design," *Journal of Operations Management,* vol. 12, pp. 107–125.

Robertson, D., and K. Ulrich (1998), "Planning for Product Platforms," *Sloan Management Review,* summer, pp. 19–31.

Suh, N. (1990), *The Principles of Design,* Oxford University Press, New York.

Ulrich, K., and S. Eppinger (1995), *Product Design and Development,* McGraw-Hill, New York.

8

Product Realization

8.1 PRODUCT REALIZATION FACTORS

A product platform, to be competitive, needs to be introduced quickly, without compromising product performance. This is because products that arrive to meet the needs of customers faster than competitors' will grow at a rapid pace, in terms of both market share and profitability. The Chrysler Corporation entered the 1990s with only two profitable product lines, minivans and jeeps, and needed a new product quickly. It made the crucial decision to cut product development time to 40 months from 60 months. It designed and developed the LH platform, with models such as Intrepid, Vision, New Yorker, and Concorde, in a record time (for an American company at that point) of 42 months, and saw its development cost drop by 30%. It soon went even further, developing the Neon in a mere 31 months. As far back as 1982 a Booz, Allen, and Hamilton survey of 700 companies had predicted that more than 30% of profits would come from new products (Kotler 1991).

Development of a new product involves several steps, including concept generation, product design, engineering analysis, process design, pilot production, and testing. These can be grouped into two broad categories: product-related (concept generation, product design, and engineering analysis) and production-related (process design, pilot production, and testing). In Figure 8.1, four major factors that influence product realization—product performance, development time, development cost, and changes in attribute preference—are shown. In Chapter 7 we discussed the interaction between customers' preference for product attributes and

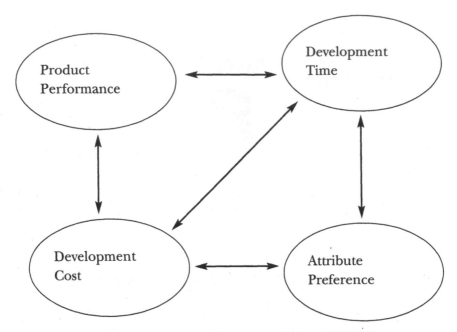

FIGURE 8.1: PRODUCT REALIZATION FACTORS

development and/or manufacturing costs, in the context of planning a platform. For any given platform, during the product realization phase, (which can run into years in certain industries) there are major concerns with not being able to correctly assess customers' attribute preferences and with innovations in technology that may affect product and process costs. It is clear that the longer the development cycle time, the greater will be the uncertainties associated with customers' preferences and new technologies. The question, when the product's design specification should be frozen, is therefore very important. While too early a design freeze may imply opportunity costs, a delay may lead to a corresponding delay in product launch and hence loss of market share. Note that a delay in design freeze would also increase development cost because of frequent redesigns and/or increased cost of production ramp-up.

Development time and cost are strongly influenced by how the development project is managed. Decisions related to project coordination, resource allocation, design reviews, prototyping, and product evaluation must be made. Product launch date is also a crucial decision variable, influenced by market conditions and the competitor's launch date.

8.2 PRODUCT DEFINITION

With the objective of an early product launch, most companies attempt to define product specifications as early as possible. With a fast pace of technology evolution and changes in customers' preferences, there is a high risk in defining the product too early. There are essentially two choices: to keep introducing products with incremental changes, and to periodically incorporate significant new technology in the product to create new markets. We consider these two options separately.

Product Attractiveness

Product prototypes have traditionally been used to understand whether and how products can be made to satisfy design specifications. Bhattacharya, Krishnan, and Mahajan (1998) suggest a novel way of using these prototypes to assess customers' true preferences. The development team presents customers with a prototype and asks for their feedback in terms of how the product attributes measure up to their expectations. The team incorporates customers' feedback in the next prototype and asks customers to evaluate it again. This process of modifying prototypes based on customers' feedback is continued until a satisfactory level of convergence between the product's attributes and customers' expectations is obtained. Freezing product definition, therefore, is equivalent to discontinuing the process of building prototypes and presenting them to customers. Bhattacharya, Krishnan, and Mahajan model this process to determine when to stop building additional prototypes. They postulate that the attractiveness of a product increases every time customers' feedback is incorporated in modifying the design specifications. Corresponding to the nth prototype, a product's attractiveness A_n is assumed to be the aggregate of a deterministic component μ_n and a random component τ_n. Bhattacharya, Krishnan, and Mahajan suggest that the demand D_n of the product at the nth prototype can be expressed as

$$D_n = M(\alpha A_n - \beta p) \qquad (8.1)$$

where

p is the unit price

M is the market size

α is the marginal increase in demand
(expressed as a ratio of the market size) with product attractiveness

β is the marginal decrease in demand
(expressed as a ratio of market size) with product's unit price

It is clear that there are two ways of coping with increasing values of n: delay the launch date of the product, and speed up pilot production and production ramp-up while holding the launch date fixed. Delaying the launch date will obviously decrease profit, depending on the competitor's launch date. The complexity of the decision increases with the presence of competitors; we will explore this in detail later. At this time we discuss the second option, that is, accelerating pilot production.

Using equation 8.1, profit $(\Pi_{n,\,p})$ can be written as

$$\Pi_{n,\,p} = M(\alpha\,A_n - \beta\,p)\,(p - c_n) \tag{8.2}$$

where c_n is the cost of pilot production. It is clear that c_n is an increasing convex function of n. If the manufacturer is risk-neutral, we can obtain a certainty equivalent of Π_n by replacing A_n with μ_n in equation 8.2. So

$$\Pi_{n,\,p} = M(\alpha\mu_n - \beta p)\,(p - c_n) \tag{8.3}$$

By equating $\partial\Pi_n/\partial p = 0$, we obtain the expression for optimal p, from equation 8.3, as

$$p^* = \frac{\alpha\mu_n + \beta c_n}{2\beta} \tag{8.4}$$

Since $p^* > c_n$, it is clear that $\alpha\mu_n > \beta c_n$.
Substituting p for p^* in equation 8.3, we rewrite Π_n as

$$\Pi_n = M(\alpha\mu_n - \beta c_n)^2 / 4\beta \tag{8.5}$$

Since n is an integer, it is clear that optimal value of n will satisfy

$$\Pi_{n-1} \le \Pi_n \ge \Pi_{n+1},$$

which, using equation 8.5, can be simplified to

$$\alpha \mu_{n-1} - \beta c_{n-1} \le \alpha \mu_n - \beta c_n \ge \alpha \mu_{n+1} - \beta c_{n+1} \tag{8.6}$$

and expressed as

$$\frac{\mu_n - \mu_{n-1}}{c_n - c_{n-1}} \ge \frac{\beta}{\alpha} \ge \frac{\mu_{n+1} - \mu_n}{c_{n+1} - c_n} \tag{8.7}$$

The specific value of n satisfying expression 8.7 can be determined easily.

From equation 8.5, a continuous-value estimate of optimal n can be obtained as

$$\frac{\partial \mu_n}{\partial n} \bigg/ \frac{\partial c_n}{\partial n} = \frac{\beta}{\alpha}$$

For the risk-averse scenario, Bhattacharya, Krishnan, and Mahajan have shown that the certainty equivalent of $\Pi_{n,p}$ in equation 8.2 can be written as

$$\Pi_{n,p} = M \left\{ \alpha \mu_n - \frac{M}{2} \sigma_n^2 r \alpha^2 (p - c_n) \right\} (p - c_n)$$

where r is the coefficient of risk aversion ($r > 0$). Note that it is possible to construct different models that maximize expected profit, subject to variance of profit not exceeding a certain threshold.

Analogous to equation 8.4, p^* can be written as

$$p^* = \frac{\alpha \mu_n + (M \sigma_n^2 r \alpha^2 + \beta) c_n}{M \sigma_n^2 r \alpha^2 + 2\beta} \tag{8.8}$$

Since $\alpha \mu_n > \beta c_n$, it follows that p^* in equation 8.8 will be smaller than the p^* in 8.4. That is, randomness in A_n is partially compensated for by a price reduction. Analogous to equation 8.5, we can write

$$\Pi_n = \frac{M (\alpha \mu_n - \beta c_n)^2}{2 (M \sigma_n^2 r \alpha^2 + 2\beta)} \tag{8.9}$$

Prototype	Mean Attractiveness	Ramp-up Cost	Standard Deviation
n	μ_n	c_n	σ_n
1	24.00	4.00	10.0
2	33.94	11.31	8.0
3	41.57	20.78	7.0
4	48.00	32.00	6.0
5	53.66	44.72	5.5
6	58.79	58.79	5.0
7	63.50	74.08	4.5
8	67.88	90.509	4.2
9	72	108.0	3.9

Table 8.1: Product Attractiveness and Cost

Finally, we express the optimality of n as

$$\frac{K_1\mu_n - \mu_{n-1}}{K_1 c_n - c_{n-1}} \geq \frac{\beta}{\alpha} \geq \frac{K_2\mu_{n+1} - \mu_n}{K_2 c_{n+1} - c_n} \tag{8.10}$$

where

$$K_1 = \left(\frac{M\sigma_{n-1}^2 r\alpha^2 + 2\beta}{M\sigma_n^2 r\alpha^2 + 2\beta}\right)^{1/2}$$

$$K_2 = \left(\frac{M\sigma_n^2 r\alpha^2 + 2\beta}{M\sigma_{n+1}^2 r\alpha^2 + 2\beta}\right)^{1/2}$$

As an example, let the values of μ_n and c_n be as in Table 8.1. Let $\alpha = 0.12$ and $\beta = 0.005$. The relevant values corresponding to expression 8.7, assuming $\sigma_n = 0$ are shown below.

N	$\mu_n - \mu_{n-1}$	$c_n - c_{n-1}$	$\dfrac{\mu_n - \mu_{n-1}}{c_n - c_{n-1}}$
1	—	—	—
2	9.94	7.31	1.3597
3	7.63	9.47	0.8057
4	6.43	11.22	0.5731
5	5.66	12.72	0.4450
6	5.13	14.07	0.3646
7	4.71	15.29	0.3080

Since $\beta/\alpha = 0.4166$, it is seen from the values of $(\mu_n - \mu_{n-1})/(c_n - c_{n-1})$ that $0.4450 > 0.4166 > 0.3646$. Hence the optimal value of n is 5. From equation 8.5, we have $\Pi_5 = 88.33M$.

We repeat the above computation steps for the case when σ_n decreases with n, as shown below.

n	K_1	$K_1\mu_n$	$K_1 c_n$	$\dfrac{K_1\mu_n - \mu_{n-1}}{K_1 c_n - c_{n-1}}$
1	—	—	—	—
2	1.25	42.425	14.138	1.817
3	1.143	47.508	23.749	1.090
4	1.166	55.968	37.312	0.871
5	1.091	58.543	48.789	0.628
6	1.10	64.669	64.669	0.551
7	1.111	70.555	82.303	$\boxed{0.500}$
8	1.071	72.72	96.935	0.403
9	1.07	77.538	116.307	0.374

Note that since $M\sigma_n^2 r\alpha^2$ is very large compared to β in this example, K_1 is approximated as σ_{n-1}/σ_n. With $\beta/\alpha = 0.4166$, we have $0.500 > 0.4166 > 0.403$. Hence the optimal value of n is 7. Observe that, product definition is delayed from $n = 5$ to $n = 7$. That is, the manufacturer's aversion to the risk of losing customers will result in a delay of product definition by two prototype-equivalent units of time. Letting $r = 0.5$, the value of Π from equation 8.9 can be evaluated to be $\Pi_7 = 76.67M/(1 + 1.458M)$. For a large value of M, Π_7 will approach a value of 52.59. That is, randomness in A_n places an upper bound on profit, when M is a large value.

Technology Innovation

It has been shown that performance of new technology improves with development time, though at different rates in different phases. The S curve, as in Figure 8.2, has been used to model increase in technology performance with time (Thompke 1998).

Three phases are commonly ascribed to technology evolution: emerging, pacing, and mature. As shown in Figure 8.2, the rate of performance improvement during the emerging phase is very high. The rate becomes moderate during the pacing phase and levels off during the mature phase.

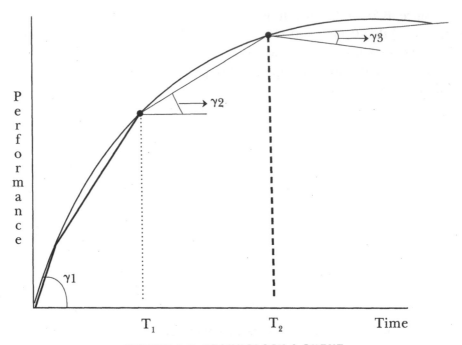

FIGURE 8.2: TECHNOLOGY S CURVE

For example, Internet technology is still in an emerging phase and is associated with a high growth rate but also a high level of uncertainty. Computer-chip technology can be considered to be in a pacing phase, as it has become the main driver of the computer industry's growth. Keyboard technology, on the other hand, has reached a mature phase, as no major changes are expected in the foreseeable figure. The S curve in Figure 8.2 can be approximated by three linear segments corresponding to time segments 0 to T_1, T_1 to T_2, and beyond T_2. The performance improvement rates are shown to be γ_1, γ_2, and γ_3, respectively.

The technology performance level R at time t can therefore be expressed as

$$R_t = \gamma_1 t, \, t \leq T, \tag{8.11}$$

$$R_t = \gamma_1 T_1 + \gamma_2 (t - T_1), \, T_1 < t \leq T_2 \tag{8.12}$$

$$R_t = \gamma_1 T_1 + \gamma_2 (T_2 - T_1) + \gamma_3 (t - T_2), \, T_2 < t \tag{8.13}$$

Unlike the previous case, where the product's attractiveness is enhanced internally, the technology acquisition option relies on buying technology from external sources and incorporating it into products.

Product technology helps to increase the attractiveness of the product. Therefore, corresponding to equation 8.1, the product's demand can be expressed as

$$D_t = M(\alpha R_t - \beta p) \tag{8.14}$$

The increase in product demand may, however, come with an increased cost of pilot production and/or delay in product launch. Technology-intensive companies such as BMW have preferred to delay product definition well into the pilot production phase (Pisano 1996). As a consequence, not all production and engineering problems can be resolved by the time the car goes for production ramp-up. This leads to frequent shutting down of the line during ramp-up for troubleshooting or for changing tools, and increases production ramp-up time by six months. BMW had to incur the cost of additional engineers to solve the problem of fine-tuning production processes to accord with the product design. This also increases the risk of imperfect cars reaching customers. Thus while technology innovation increased demand, as in equation 8.14, the cost of pilot production, production ramp-up, and repairs on sold units increases substantially.

Note that unlike D_n, D_t is a continuous function of t. The marginal profit of the product is equal to $p - c_t - a R_t$, where c_t is the cost of pilot production per unit and a is the marginal cost of increasing technology performance. In high-tech industries such as PC manufacturing, competition is based on market share and sales (as in equation 8.14) and not on marginal profit and price (p). Thus, corresponding to equation 8.2, we have

$$\Pi_{t,p} = M(\alpha R_t - \beta p)(p - c_t - a R_t) \tag{8.15}$$

and the optimal price p^* would be

$$p^* = \frac{(\alpha + a\beta) R_t + \beta c_t}{2\beta} \tag{8.16}$$

Π_t, corresponding to equation 8.5, would now be expressed as

$$\Pi_t = M\{(\alpha - a\beta) R_t - \beta c_t\}^2 / 4\beta \tag{8.17}$$

Let

$$y = 2\left(\frac{\beta\Pi_t}{M}\right)^{1/2} = (\alpha - a\beta)\,R_t - \beta c_t$$

Since R_t is concave and c_t is convex in t, it is clear that y is a concave function of t as well. Therefore the value of t corresponding to $dy\,/\,dt = 0$ will maximize the value of y.

Hence at the optimum

$$\frac{dR_t}{dt}\Big/\frac{dc_t}{dt} = \frac{\beta}{\alpha - a\beta}$$

From Figure 8.2, $\dfrac{dR_t}{dt} = \gamma$, $\gamma = \gamma_1, \gamma_2, \gamma_3$, and so on.

Hence

$$\frac{dc_t}{dt} = \frac{\gamma(\alpha - a\beta)}{\beta} \qquad\qquad (8.18)$$

Solving equation 8.18 determines the value of t, the optimal time to freeze technology definition.

　　As an example, let

$$c_t = 4t^{3/2}$$

$$T_1 = 25,\ T_2 = 100$$

$$\gamma_1 = 40,\ \gamma_2 = 30,\ \gamma_3 = 15$$

$$\alpha = 0.11,\ \beta = 0.05,\ a = 10$$

Therefore,

$$\frac{\gamma(\alpha - a\beta)}{\beta} = \gamma\,(0.11 - 0.05)\,/\,0.05 = 1.2\gamma$$

and

$$\frac{dc_t}{dt} = 6t^{1/2}$$

Hence, from equation 8.18,

$$6t^{1/2} = 1.2\gamma$$

That is,

$$t = 0.04\gamma^2 \tag{8.19}$$

First, assume $t \le T_1 (= 25)$. Then $\gamma = \gamma_1 = 40$. Substituting for γ in equation 8.19, we have

$$t = 0.04 \, (1,600) = 64 > T_1 (= 25)$$

It therefore follows that $t > 25$.

Next, assume $25 < t \le T_2 (= 100)$ and $\gamma = \gamma_2 = 30$. From equation 8.19, as before, $t = 0.04(900) = 36$. Since $25 < 36 \le 100$, the optimal solution is $t = 36$.

From equation 8.17,

$$\Pi_{36} = \{(0.11 - 0.05) \, R_{36} - .05c_{36}\}^2 \, / \, \frac{M}{4\beta}$$

From equation 8.12,

$$R_{36} = 40 \, (25) + 30 \, (36 - 25) = 1,330$$
$$c_{36} = 4(36)^{3/2} = 4 \times 36 \times 6 = 864$$

Hence

$$\Pi_{36} = \frac{(0.06 \times 1,330 - 0.05 \times 864)^2}{4(.05)} \cdot M$$

That is, $\Pi_{36} = 6,700M$. This is the maximum profit, obtained by freezing the technology definition of the product at $t = 36$.

Analogous to the random component of A_n, it follows that R_t can also have a random component. The certainty equivalent of the random technology performance can be found in a way similar to that of A_n, and expressions similar to those in equations 8.8 and 8.9 can be established.

8.3 PRODUCT LAUNCH DATE

In many industries it is not possible to fix the product launch date a priori. Unlike the high-tech industry with its novel products, the auto, aircraft, and appliance industries are required to fine-tune product launch because of the presence or launch of competitors' products. Product launch can be pushed back because of delays in product definition, pilot production (and production ramp-up), or both. Clark (1989) has estimated that the opportunity cost per day of delaying the launch of a passenger car is almost 100 times the unit cost. Because of the high probability of competing products being launched at any time, time to market has become an important "barometer" of competitiveness in such industries.

Optimal Launch Date

Product attractiveness or performance, which we discussed earlier, is the other critical factor affecting a product's profitability. If a product is launched too early, with an inadequate performance level, its sales will be low. There is therefore a clear trade-off between high sales for a short period and lower sales over a longer period. Another effect of reducing time to market is a reduction in the time to break even. Hewlett-Packard has used the so-called return map, as shown in Figure 8.3 to formalize the break-even concept for its pocket calculator.

We treat the issue of determining the optimal product launch date separately from how to *achieve* reduction in time to launch. Options such as eliminating some development activities (such as unnecessary reviews), overlapping the five development steps, doing each step faster, and investing in new technology (such as CAD/CAM and simulation), may all contribute to reduction in time to launch. We will discuss later the important decisions related to reducing the development cycle time.

We first present a simple adaptation of the technology model we discussed earlier. As shown in Figure 8.4, it is assumed that the performance of the product can be improved at a constant rate of k per unit time, up to the time T when it is launched. The development cost during this period is W per period. The product's market share is assumed to be determined by its performance R and the unit price at the time of launch. The product is sold at a constant rate until time T_a, when it is terminated (super-

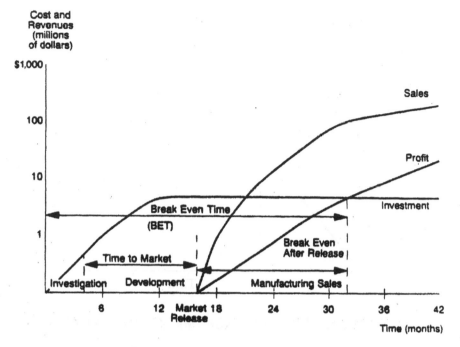

FIGURE 8.3: PROJECT RETURN MAP

(Source: Wheelwright and Clark 1992.
Reprinted by permission of Harvard Business Review, Harvard College.)

seded by other products). Unlike the previous model, however, we assume that the time allowed for pilot production is fixed, so that any delay in product definition delays the product launch time T. Also, unlike in the model in equation 8.15, the product has a finite life (from T to T_a). Hence we would now be interested in maximizing the total profit up to the time T_a.

The profit function corresponding to equation 8.15 can be written as

$$\Pi_{T,p} = M(\alpha R_T - \beta p)(p - c)(T_a - T) - WT$$

Since $R_T = kT$, we have

$$\Pi_{T,p} = M(\alpha kT - \beta p)(p - c)(T_a - T) - WT \qquad (8.20)$$

Note that $\Pi_{T,p}$ is concave in p as well as T. Hence the solution corresponding to

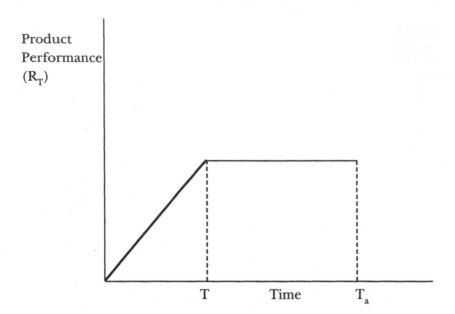

FIGURE 8.4: PRODUCT LAUNCH (MARKET MONOPOLY)

$$\frac{\partial \Pi}{\partial p} = 0 \text{ and } \frac{\partial \Pi}{\partial T} = 0$$

will ensure maximum profit.

Since

$$\frac{\partial \Pi}{\partial p} = 0 \Rightarrow \alpha k T - 2\beta p + \beta c = 0$$

and

$$\frac{\partial \Pi}{\partial T} = 0 \Rightarrow M\{\alpha k (T_a - 2T) + \beta p\} - \frac{W}{p - c} = 0$$

the optimal T can be found by solving the quadratic equation

$$3T^2 - \left[2T_a + \left(1 + \frac{3}{M}\right) \frac{\beta c T_a}{(\alpha k)^2} \right] T + \left[\frac{2\beta c T_a}{\alpha k} + \left(\frac{\beta c}{\alpha k}\right)^2 + \frac{4\beta W}{(\alpha k)^2 M} \right] = 0$$

$$(8.21)$$

and p can be determined as

$$p = \frac{\alpha k}{2\beta} T + \frac{c}{2} \tag{8.22}$$

Note that optimal price p increases linearly with product launch time (T), irrespective of the size of the window of opportunity $(T_a - T)$. Since p must exceed c, from equation 8.22, we have $T > \beta c/\alpha k$.

As an example, let $M = 10{,}000$, $\alpha = 0.01$, $k = 40$, $\beta = 0.15$, $c = \$10$, $W = 500$, and $T_a = 100$. From equation 8.22 we have

$$p = \frac{\alpha k}{2\beta} T + \frac{c}{2} = \frac{40}{3} T + 5$$

and from equation 8.21 we have

$$3T^2 - 201.5T + 75.1 = 0,$$

so that $T = 0.4$ or 66.8.

Since Π_T is minimumed at $T = 0.4$, (i.e. at optimal values of p),

$$T_{opt} = 66.8$$

$$p_{opt} = \$895$$

From equation 8.20,

$$\Pi_{T,p} = 10{,}000\,(4T - 0.15p)\,(p - 10)\,(100 - T) - 500\,T$$

$$= 10{,}000\left(267 - \frac{267}{2} - .75\right)(885)\,(33.25) - \frac{(500)\,(267)}{4}$$

$$= \$4{,}605{,}250$$

Note that if M is large, the value of T can be approximated from equation 8.20 as

$$T = \frac{2}{3}\left(T_a - \frac{\beta c}{\alpha k}\right) \tag{8.23}$$

$$p = \frac{1}{3}\left(\frac{\alpha k}{\beta} T_a + \frac{c}{2}\right) \tag{8.24}$$

Substituting the values for T_a, α, k, s, and c,

$$T_{opt} = \frac{2}{3}\left(100 = \frac{(.15)(10)}{(.1)(40)}\right) = 66.45$$

$$p_{opt} = \frac{1}{3}\left(\frac{(.1)(40)}{.15}(100) + \frac{10}{2}\right) = \$890.40$$

which compare favorably with T_{opt} and p_{opt} found earlier, without approximation.

Equations 8.23 and 8.24 can now be used to study the impact of k, the rate of improvement in product performance. Note that if k increases, both T and p increase. However, T converges to a value of $2\,T_a/3$, whereas p keeps increasing indefinitely. Substituting for T and p in equation 8.20, profit can be expressed as

$$\Pi_k = \frac{M}{108\beta}(2\alpha k T_a - 5\beta c)^2\left(T_a + \frac{2\beta c}{\alpha k}\right) - \frac{2W}{3}\left(T_a - \frac{\beta c}{\alpha k}\right)$$

from where the value of k maximizing Π_k can be determined if the nature of variation of W with respect to k is known.

General Model With Competition

Consider now the case where a competing company (company 2) introduces its product at time T_2 shown in Figure 8.5. Without loss of generality, we assume $T_2 > T_1$. The profits for the two companies can be expressed as

$$\Pi_1\,(p_1,\,T_1) = M(\alpha_1 k_1 T_1 - \beta_1 p_1)\,(p_1 - c_1)(T_2 - T_1) + $$
$$M(\alpha_1 k_1 T_1 - \alpha_2 k_2 T_2 - \beta_1 p_1 + \beta_2 p_2)\,(p_1 - c_1)\,(T_a - T_2) - W_1 T_1 \tag{8.25}$$

$$\Pi_2\,(p_2,\,T_2) = M(\alpha_2 k_2 T_2 - \alpha_1 k_1 T_1 - \beta_2 p_2 + \beta_1 p_1)$$
$$(p_2 - c_2)(T_a - T_2) - W_2 T_2 \tag{8.26}$$

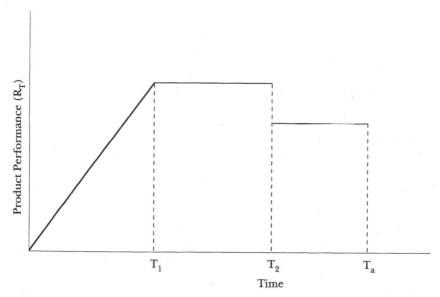

FIGURE 8.5: PRODUCT LAUNCH MARKET DUOPOLY

Assuming $T_2 = T_1 + t$, an equilibrium solution will be obtained where

$$\frac{\partial \Pi_1}{\partial p_1} = 0, \frac{\partial \Pi_1}{\partial T_1} = 0, \frac{\partial \Pi_2}{\partial p_2} = 0, \frac{\partial \Pi_2}{\partial t} = 0$$

We can conclude from equation 8.21 that the solution of Π_1 and Π_2, as in equations 8.25 and 8.26, would lead to algebraic equations of degree four. These equations can be solved iteratively to determine optimal values of T_1, T_2, p_1, and p_2. It is clear that the leader company (company 1) will optimize in its unit price and launch time. A follower company (company 2), on the other hand, will optimize in its unit price and the interval by which its launch follows the leader's product launch.

Special Cases

We consider two special cases in high-tech industries, such as personal computers and digital telephones. In these industries the unit price of a new version of a product does not change appreciably from the previous version, as the learning effect keeps the unit cost almost unchanged. In such a case, price is not a determinant of demand, and so it is not a decision variable. We can therefore modify equations 8.25 and 8.26 as

$$\Pi_1(T_1) = M\alpha_1 k_1 T_1 (p_1 - c_1)t +$$
$$M\{(\alpha_1 k_1 - \alpha_2 k_2)T_1 - \alpha_2 k_2 t\}(p_1 - c_1)(T_a - T_1 - t) - W_1 T_1 \quad (8.27)$$

$$\Pi_2(T_2) = M\{(\alpha_2 k_2 (T_1 + t) -$$
$$\alpha_1 k_1 T_1\}(p_2 - c_2)(T_a - T_1 - t) - W_2(T_1 + t) \quad (8.28)$$

Letting $\frac{\partial \Pi_1}{\partial T_1} = 0$, and after simplification, we have

$$T_1 = \frac{T_a}{2} + \frac{\{(p_1 - c_1)(\alpha_1 k_1 + \alpha_2 k_2) - (\alpha_1 k_1 - \alpha_2 k_2)\}}{2(\alpha_1 k_1 - \alpha_2 k_2)}$$
$$+ t - \frac{W_1}{2M(\alpha_1 k_1 - \alpha_2 k_2)} \quad (8.29)$$

Thus, if company 2 is a passive follower who imitates company 1's product and introduces it to market t periods after company 1's product launch, the optimal launch of product 1 will be expressed as in equation 8.29, with a known value of t.

If company 2, on the other hand, competes actively, it would set $\frac{\partial \Pi_2}{\partial t} = 0$ to maximize its profit. This would lead to an optimal value of t as

$$t = \frac{T_a}{2} - \frac{\alpha_1 k_1}{2\alpha_2 k_2} T_1 - \frac{W_2}{2M(p_2 - c_2)\alpha_2 k_2} \quad (8.30)$$

Equations 8.29 and 8.30 can now be solved simultaneously for T_1 and t. For a large value of M, equation 8.29 can be approximated as

$$T_1 = \frac{T_a}{2} + \frac{p_1 - c_1}{2} \cdot \frac{\alpha_1 k_1 + \alpha_2 k_2}{\alpha_1 k_1 - \alpha_2 k_2} \cdot t \quad (8.31)$$

$$t = \frac{T_a}{2} - \frac{\alpha_1 k_1}{2\alpha_2 k_2} T_1 \quad (8.32)$$

Observe that if $\alpha_1 k_1 > \alpha_2 k_2$ the optimal value of T_1 would increase if the profit margin $p_1 - c_1$ increases. In that case, company 1 would increase $p_1 - c_1$ to the extent that $T_1 + t$ exceeds T_a, ensuring that company 2's product never appears in the market.

If, on the other hand, $\alpha_1 k_1 < \alpha_2 k_2$, T_1 will decrease if t or $p_1 - c_1$ increases. This would imply that company 1 will launch its product early at a lower product performance rating, but it will sell its product for a longer period $(= T_a - T_1)$. It is clearly seen from equation 8.31 that now $T_1 < T_a / 2$.

An equilibrium solution will be obtained by solving equations 8.31 and 8.32 simultaneously. The optimal solutions can be written as

$$T_1 = \frac{1 + b_1}{1 + b_1 b_2} \cdot \frac{T_a}{2}, \; t = \frac{1 - b_2}{1 + b_1 b_2} \cdot \frac{T_a}{2} \tag{8.33}$$

where

$$b_1 = \frac{p_1 - c_1}{2} \cdot \frac{\alpha_1 k_1 + \alpha_2 k_2}{\alpha_1 k_1 - \alpha_2 k_2}, \; \text{and}$$

$$b_2 = \frac{\alpha_1 k_1}{2 \alpha_2 k_2}$$

It therefore follows, that

$$\frac{T_1}{t} = \frac{1 + b_1}{1 - b_2} \tag{8.34}$$

Since company 2 is a follower, it would like to make sure that $t < T_a - T_1$.

It can be verified from equation 8.33 that to ensure a nonnegative solution, one of the following conditions must hold: (a) $b_1 > 0$ and $b_2 < 1$, or (b) $b_1 < -1$ and $b_2 > 1$. Note, however, that if $b_2 > 1$, then $\frac{\alpha_1 k_1}{\alpha_2 k_2} > 1$, in which case $b_1 > 0$. Hence for equilibrium, $b_2 < 1$ and $b_1 > 0$. It can be verified that $b_1 > 0$ implies $\alpha_1 k_1 / \alpha_2 k_2 > 1$, and $b_2 < 1$ implies $\alpha_1 k_1 / \alpha_2 k_2 < 2$. Hence we must have $1 \leq \alpha_1 k_1 / \alpha_2 k_2 \leq 2$.

As an example, let $T_a = 100$, $p_1 - c_1 = 50$, $\alpha_1 = 0.1$, $\alpha_2 = 0.15$, $k_1 = 40$, and $k_2 = 20$. From the definitions,

$$b_1 = \frac{p_1 - c_1}{2} \cdot \frac{r + 1}{r - 1}, \; \text{where } r = \alpha_1 k_1 / \alpha_2 k_2 = 4 / 3$$

that is, $b_1 = 175$ and $b_2 = \frac{r}{3} = 2/3$. Hence

$$T_1 = \frac{1 + 175}{1 + \dfrac{350}{3}} T_a = \frac{(176)(3)}{353} \cdot \frac{100}{2} = 74.8$$

$$t = \frac{1 - \dfrac{2}{3}}{1 + \dfrac{350}{3}} (50) = \frac{50}{353} = 0.142$$

It is interesting to note that if company 2 is passive, it must ensure that $\alpha_2 k_2 > \alpha_1 k_1$, so as not to be driven out of the market. However, for an equilibrium solution to exist, when company 2 optimizes t, the condition $1 \le \alpha_1 k_1 / \alpha_2 k_2 \le 2$ must hold. That is, in the first case company 2's product development rate needs to be high. In the second case, where the follower company is not passive, it is not forced to maintain such a high rate of developing product performance.

Note, that if a competing product of performance rating R_2 *exists* at time zero, a third scenario must be considered. From equation 8.25, letting $T_2 = T_1 = T$ and $R_2 = \alpha_2 k_2$, we have

$$\Pi = M(\alpha kT - \alpha_2 R_2)(p - c)(T_a - T) - WT \tag{8.35}$$

$$\frac{\partial \Pi}{\partial T} = 0 \Rightarrow T = \frac{T_a}{2} + \frac{1}{2\alpha k}\left\{\alpha_2 R_2 - \frac{W}{M(p - c)}\right\} \tag{8.36}$$

Cohen, Eliashberg, and Ho (1996) consider a situation where the probability of purchase, instead of demand, is a function of the performance rating of the product (as in the logit model in Chapter 4). For such a situation, the market share of the product can be written as

Market share $= \dfrac{kT}{kT + R_2}$

assuming the product utility to be an exponential function of performance rating. Hence equation 8.35 can be rewritten as

$$\Pi(T) = \frac{MkT}{kT + R_2}(p - c)(T_a - T) - WT \tag{8.37}$$

$$\frac{\partial \Pi}{\partial T} = 0 \Rightarrow T = \frac{1}{k}\left[\left\{\frac{R_2\,(R_2 + kT_a)}{1 + \dfrac{W}{M\,(p - c)}}\right\}^{1/2} - R_2\right] \qquad (8.38)$$

8.4 TECHNICAL AND MARKET UNCERTAINTIES

Improvements in the performance of a new technology with time can be described by a S-curve, shown in Figure 8.2 (Christenson and Bower 1996). Product development is somewhat similar to technology innovation in that its performance can be improved over time. We have hitherto assumed a constant rate of improvement in product performance and have designated this constant rate as k. This is an approximation of the more likely case shown in Figure 8.2, with a variable rate of development.

What, then, determines the rate of performance improvement? Several factors, obviously, have an impact on this rate, whether constant or variable. Increasing the rate of investment in developing a product will quicken the pace of development. As shown in Figure 8.6, a higher rate of investment $(I_2 > I_1)$ will achieve a product performance R in a shorter

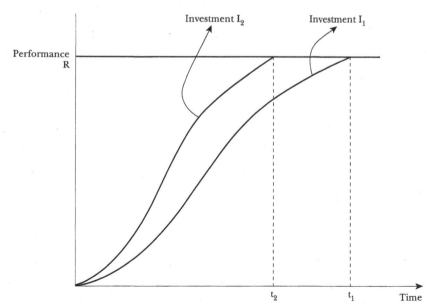

FIGURE 8.6: INVESTMENT AND DEVELOPMENT TIME

time $(t_2 < t_1)$. Obviously investment alone does not determine the development time, as two companies and/or two different products with the exact same investment rate may achieve totally different development times. Thus the performance level of the product and the efficiency of a company in utilizing its investment are also major factors. Finally, there is also a random factor that may prevent a development project from being completed on time even if the investment level, the product's performance level, and the efficiency of the company are all known. This randomness is caused by technological uncertainties related to R&D innovation. Therefore, in development projects with significant technical uncertainties, we cannot expect development times to be deterministic functions of the product's performance level and the rate of development (assumed hitherto). We can, however, use probabilities to describe the likelihood of completing a development project by a certain time. Kamien and Schwartz (1982), Reinganum (1981), and Loury (1979) have shown how exponential functions can be used for this purpose, in general. Ali, Kalwani, and Kovenock (1993) model the probability of completing a project by time τ using a negative exponential function.

It is clear that if the investment in the project is increased, the probability of its completion by τ will also increase. We would expect the same result if the company's efficiency in using the investment for product development increases. If the desired level of product performance increases, however, we would expect the probability to decrease. We use the above facts to modify the expression used by Ali, Kalwani, and Kovenock. We write the probability density function of product development time t as

$$f(t) = \frac{\epsilon_j x_j}{R} \, exp\left(-\frac{\epsilon_j x_j}{R} t\right) \tag{8.39}$$

so that

$$Prob\,(t \le \tau) = 1 - exp\left(-\frac{\epsilon_j x_j}{R}\tau\right)$$

where t = product development time, x_j = total investment in the development project by company j, ϵ_j = efficiency of company j in product development, and R = product performance level desired.

Note that while x_j and ϵ_j are company characteristics, R is a product characteristic. We may model an innovative product by assuming R to be high and a derivative product by assuming R to be low. We assume that sales revenue per period from the product is V. Obviously, V and R are related, but the nature of this relationship is of no direct consequence, as we will see soon. For the initial model, consider a scenario where two companies compete with the same product, one the leader (first to launch the product) and the other the follower. Let the product launch time for company j be T_j.

The technical uncertainty creates an uncertainty about being the leader or follower in the marketplace. In Figure 8.5, it is assumed that the product launch times, T_1 and T_2, of two competitors could be controlled with certainty by the respective companies. With technical uncertainties, however, even if the values of x_j and ϵ_j for company 1 exceed the corresponding values for company 2, T_1 may exceed T_2. All that can be said is that as ϵ_1 and x_1 increase, the probability of T_1 exceeding T_2 will be reduced.

We note from equation 8.25 that for a product with a given performance level and a given unit price, the revenue per period during the interval T_1 to T_2 is constant. Similarly, the revenue per period during the interval T_2 to T_a is also constant, but it is a fraction of the revenue per period during T_1 to T_2, as shown in Figure 8.7.

Denote these revenue rates as V and λV, where $\lambda \leq 1$. It is clear that there is a monopoly of company 1 from T_1 to T_2, and a duopoly exists from T_2 to T_a. It is also clear that if $T_1 > T_2$, company 1 will have a zero revenue during T_2 to T_1, and a revenue rate of $\beta_1 V (\beta_1 < 1)$ during T_1 to T_a.

Therefore, for the scenario corresponding to Figure 8.7a, the market revenue to company 1 can be written as

$$\text{Rev} = V(T_2 - T_1) + \lambda_1 V(T_a - T_2) \tag{8.40}$$

Similarly, corresponding to Figure 8.7b we have

$$\text{Rev} = \beta_1 V(T_a - T_1) \tag{8.41}$$

Since T_1 may or may not exceed T_2, we need to compute the expected revenue by combining equations 8.40 and 8.41. Note that the probability of company 1's product being completed during the interval T_1 to $T_1 + dT_1$ and company 2's product being completed during the interval T_2 to $T_2 + dT_2$ can

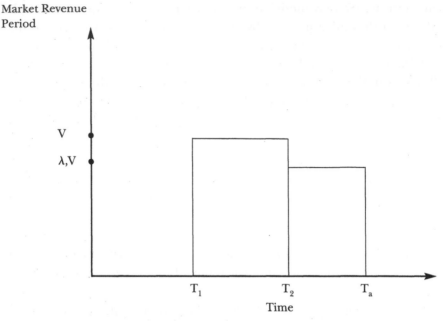

FIGURE 8.7a: COMPANY 1 AS THE LEADER

be written as in equation 8.42. Both companies develop the product to level R, since we have assumed that they compete with identical products.

$$Prob\left(T_1 \leq t_1 \leq T_1 + dT_1 \quad \text{and} \quad T_2 \leq t_2 \leq T_2 + dT_2\right) = \frac{\in_1 x_1 \in_2 x_2}{R^2}$$

$$exp\left(-\frac{\in_1 x_1}{R} T_1 - \frac{\in_2 x_2}{R} T_2\right) dT_1 \, dT_2 \quad (8.42)$$

Hence, the expected revenue to company 1 can be written as

$$E\left(Rev\right) = \frac{\in_1 \in_2 x_1 x_2}{R^2}\left[V \int_{T_1=0}^{T_a} \int_{T_2=T_1}^{T_a} \{(1 - \lambda_1)T_2 - T_1 + \lambda_1 T_a\} \, exp\right.$$

$$\left(-\frac{\in_1 x_1 T_1}{R} - \frac{\in_2 x_2 T_2}{R}\right)\Bigg\} dT_1 dT_2 + \beta_1 V \int_{T_2=0}^{T_a} \int_{T_1=T_2}^{T_a} (T_a - T_1) \, exp$$

$$\left(-\frac{\in_1 x_1 T_1}{R} - \frac{\in_2 x_2 T_2}{R}\right) dT_2 \, dT_1 \Bigg]$$

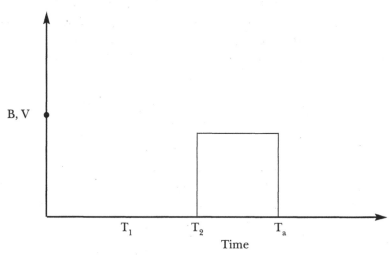

Market Revenue
Period

B, V

T_1 T_2 T_a

Time

FIGURE 8.7b: COMPANY 1 AS THE FOLLOWER

The solution after integration of right hand side a complex expression, shown in the Appendix to this chapter. For a large value of T_a or x_1 or x_2, this expression may be approximated as

$$E\left(Rev\right) = \frac{VR}{\left(\in_1 x_1 + \in_2 x_2\right)}$$

$$\left[\left\{T_a - \frac{R}{\in_1 x_1 + \in_2 x_2}\right\}\left(\frac{\in_1 \lambda_1 x_1}{R} + \frac{\in_2 \beta_1 x_2}{R}\right) - \right.$$

$$\left.\left(1 - \lambda_1\right)\left(\frac{\in_1 x_1}{\in_2 x_2}\right) - \beta_1\left(\frac{\in_2 x_2}{\in_1 x_1}\right)\right] \quad (8.43)$$

Therefore, the profit to company 1 is stated as

$$\Pi_1\left(x_1, x_2\right) = \frac{V_1 R}{\left(\in_1 x_1 + \in_2 x_2\right)}\left[\left\{T_a - \frac{R}{\in_1 x_1 + \in_1 x_2}\right\}\right.$$

$$\left.\left(\frac{\in_1 \lambda_1 x_1}{R} + \frac{\in_2 \beta_1 x_2}{R}\right) - \left(1 - \lambda_1\right)\left(\frac{\in_1 x_1}{\in_2 x_2}\right) - \beta_1\left(\frac{\in_2 x_2}{\in_1 x_1}\right)\right] - x_1 \quad (8.44)$$

Note that $\Pi_1(x_1, x_2)$ cannot be shown to be concave or convex in x_1 (for a given value of x_2), and an optimal solution for x_1 cannot be ascertained, in general. Observe in equation 8.44, however, that some terms in the equation are too small compared to others and can therefore be ignored. Dropping these terms, we may approximate $\Pi_1(x_1, x_2)$ as

$$\Pi_1(x_1, x_2) \approx VT_a \cdot \frac{\in_1 x_1 \lambda_1 + \in_2 x_2 \beta_1}{\in_1 x_1 + \in_2 x_2} - x_1 \tag{8.45}$$

Note that $\Pi_1(x_1, x_2)$, as in equation 8.45, is concave in x_1 and will have a maximum value when $\frac{\partial \Pi_1(x_1, x_2)}{\partial x_1} = 0$, which yields

$$x_1^* = a_1 x_2^{1/2} - b_1 x_2 \tag{8.46}$$

where

$$a_1 = \left\{ V_1 \, T_a (\lambda_1 - \beta_1) \cdot \frac{\in_2}{\in_1} \right\}^{1/2}, \, b_1 = \frac{\in_2}{\in_1}$$

Substituting the value of x_1 from equation 8.46 in equation 8.45 and simplifying, we have

$$\Pi_1(x_2) = \lambda_1 \, VT_a + \frac{\in_2}{\in_1} x_2 - 2 \left\{ VT_a(\lambda_1 - \beta_1) \frac{\in_2}{\in_1} \right\}^{1/2} \cdot x_2^{1/2}$$

Since $\Pi_1(x_2)$ is convex in x_2, it can be shown that $\Pi_1(x_2)$ will increase indefinitely with x_2 if

$$x_2 > VT_a(\lambda_1 - \beta_1) \in_1 / \in_2 \tag{8.47}$$

Company 2 will, however, optimize its profit $\Pi_2(x_1, x_2)$ independently. This yields an optimal value of x_2 as

$$x_2^* = \left\{ VT_a(\lambda_1 - \beta_1) \frac{\in_1}{\in_2} \right\}^{1/2} x_1^{1/2} - \frac{\in_1}{\in_2} x_1 \tag{8.48}$$

From equations 8.46 and 8.47, the Nash equilibrium solution can be deduced to be

$$x_1^* = x_2^* = V T_a (\lambda_1 - \beta_1) \cdot \frac{\epsilon_1 \epsilon_2}{(\epsilon_1 + \epsilon_2)^2} \tag{8.49}$$

$$\Pi_1^* = V T_a (\lambda_1 \epsilon_1^2 + \beta_1 \epsilon_2^2 + 2\beta_1 \epsilon_1 \epsilon_2) / (\epsilon_1 + \epsilon_2)$$

$$\Pi_2^* = V T_a (\lambda_1 \epsilon_2^2 + \beta_1 \epsilon_1^2 + 2\beta_1 \epsilon_1 \epsilon_2) / (\epsilon_1 + \epsilon_2)$$

Observe that if company 2 wishes to hurt company 1, it will set x_2 as in equation 8.47, which would exceed the Nash equilibrium value of x_2. Therefore $\Pi_1 (x_2) < \Pi_1^*$ when $x_2 = v_1 T_a (\lambda_1 - \beta_1) \epsilon_1 / \epsilon_2$. But at this value of $x_2 \neq x_2^*$, $\Pi_2 (x_2)$ will also be less than Π_2^*. Hence both companies would gain from cooperation to stabilize the Nash equilibrium. The optimal value Π_j^* increases with the efficiency ϵ_j of company j.

Choice of Products

Now consider the situation where both companies have the option of choosing one of two products. Denote companies by j and k and products by r and s. Expression 8.45 can now be generalized for the case when both companies develop the same product (r, r) or different products (r, s) as,

$$\Pi_j(r, r) = V_r T_a \cdot \frac{\epsilon_{rj} x_{rj} \lambda_{1r} + \epsilon_{rk} x_{rk} \beta_{1r}}{\epsilon_{rj} x_{rj} + \epsilon_{rk} x_{rk}}, r = 1, 2 \tag{8.50}$$

$$\Pi_j(r, s) = V_r T_a \cdot \frac{\epsilon_{rj} x_{rj} \lambda_{2r} R_s + \epsilon_{sk} x_{sk} \beta_{2r} R_r}{\epsilon_{rj} x_{rj} R_s + \epsilon_{sk} x_{sk} R_r}, r \neq s, r = 1, 2; s = 1, 2$$

$$\tag{8.51}$$

The four possible values of profit for company j ($j = 1, 2$) can be listed as $\Pi_j(1, 1)$, $\Pi_j(2, 2)$, $\Pi_j(1, 2)$, and $\Pi_j(2, 1)$, which can be expressed in the form of a payoff matrix, shown in Figure 8.8(a) and Figure 8.8(b).

It is clear from Figure 8.8 that if company j is risk-averse, it will choose to develop product r if

$$\text{Minimum } \{\Pi_j(r, r), \Pi_j(r, s)\} > \text{Minimum } \{\Pi_j(s, r), \Pi_j(s, s)\}$$

		Company 2	
		Product 1	Product 2
Company 1	Product 1	$\Pi_1(1, 1)$	$\Pi_1(1, 2)$
	Product 2	$\Pi_1(2, 1)$	$\Pi_1(2, 2)$

FIGURE 8.8a: PAYOFF TO COMPANY 1

and it will develop product s otherwise. It is also clear that product r will be the dominant solution for company j (risk-averse or not) if

$$\Pi_j(r, r) > \Pi_j(s, r) \text{ and } \Pi_j(r, s) > \Pi_j(s, s).$$

If both of the above conditions are violated, product s will be the dominant solution for company j. If one of the conditions holds (and the other is violated), there would be no dominant solutions, and company j's strategy will depend on what company k decides to do. In such a situation, company j can make an intelligent decision if it can assess the likelihood of company k's developing product r (or s).

Let

$$p_{kr} = \text{probability of company } k \text{ developing product } r$$

If company j develops product r, its expected profit can be written as

$$\text{Profit}(r) = p_{kr}\Pi_j(r, r) + (1 - p_{kr})\Pi_j(r, s)$$
$$= \Pi_j(r, s) + p_{kr}\{\Pi_j(r, r) - \Pi_j(r, s)\}$$

Similarly, if company j develops product s, we can write the profit as

		Company 1	
		Product 1	Product 2
Company 2	Product 1	$\Pi_2(1, 1)$	$\Pi_2(1, 2)$
	Product 2	$\Pi_2(2, 1)$	$\Pi_2(2, 2)$

FIGURE 8.8b: PAYOFF TO COMPANY 2

$$\text{Profit } (s) = p_{kr} \Pi_j(s, r) + (1 - p_{kr}) \Pi_j(s, s)$$
$$= \Pi_j(s, s) + p_{kr} \{\Pi_j(s, r) - \Pi_j(s, s)\}$$

Thus company j will adopt product r if

$$\Pi_j(r, s) + p_{kr} \{\Pi_j(r, r) - \Pi_j(r, s)\} > \Pi_j(s, s) + p_{kr} \{\Pi_j(s, r) - \Pi_j(s, s)\}$$

That is,

$$p_{kr} \{\Pi_j(r, r) + \Pi_j(s, s) - \Pi_j(r, s) - \Pi_j(s, r)\} > \Pi_j(s, s) - \Pi_j(r, s) \qquad (8.52)$$

which can be split into two sets of conditions:

(a)
$$\Pi_j(r, r) + \Pi_j(s, s) > \Pi_j(r, s) + \Pi_j(s, r)$$

and
$$p_{kr} > \frac{\Pi_j(s, s) - \Pi_j(r, s)}{\{\Pi_j(r, r) + \Pi_j(s, s)\} - \{\Pi_j(r, s) + \Pi_j(s, r)\}}$$

which implies that $\Pi_j(r, r) \geq \Pi_j(s, r)$, since $p_{kr} \leq 1$

(b)
$$\Pi_j(r, r) + \Pi_j(s, s) < \Pi_j(r, s) + \Pi_j(s, r)$$

and
$$p_{kr} < \frac{\Pi_j(s, s) - \Pi_j(r, s)}{\{\Pi_j(r, r) + \Pi_j(s, s)\} - \{\Pi_j(r, s) + \Pi_j(s, r)\}}$$

which implies that $\Pi_j(s, s) \leq \Pi_j(r, s)$.

Ali, Kalwani, and Kovenock (1993) have studied these and other strategies in more detail but only for the case when T_a is infinity (allowing for the time value of money). There detailed experimentation leads to interesting conclusions regarding the importance of the ratio V_r / V_s in determining product choice and the role of product development efficiency $\in_{jr}$ in product decisions.

8.5 OVERLAPPING DEVELOPMENT TASKS

Overlapping is the process of starting a downstream operation (e.g., manufacture of a product) before completing the upstream operation (e.g., designing of the product). There is an inherent risk in overlapping arising from factors such as incompatibilities among design phases, nonzero time to evolve design specifications through a series of prototypes called design iterations, and incorporation of engineering changes. Imai, Nonaka, and Takeuchi (1985) and Takeuchi and Nonaka (1986) were perhaps the first to report that faster development processes can be achieved by overlapping activities with one another. Both aeronautics and software industries have reported such improvements (Sabbagh 1996; Cusumano and Selby 1995; Hodemaker, Blackburn, and Wassenhove 1999). Clark and Fujimoto (1991) and Wheelwright and Clark (1992) have provided theoretical explanations of their empirical observations through concepts such as simultaneity ratio and intensive-communication. Finally, Eisenhardt and Tabrizi (1995), and Terwiesch and Loch (1999) used a sample survey tool and have reported in greater depth on the impact of overlapping in the computer industry, including the impact of fast uncertainty resolution in project development. Chakravarty (2000) studies three different modes of overlapping, and develops insights related to both cost and time of completion. A comprehensive bibliography can be found in Krishnan and Ulrich (1998).

Design incompatibilities may arise in design projects involving large systems (Eppinger et al., 1994). The design of large products such as cars, aircraft, and ships typically involves thousands of tasks; the tasks are divided into smaller segments, and the components in a segment are designed sequentially (segment by segment). None of these can be designed in isolation, because the parameter values chosen in one segment may impact the choices in other segments. The interrelationships in a set of design tasks are shown in Figure 8.9.

Observe in Figure 8.9, that the parameter values chosen for segment A (e.g., the design of the wheel) may need modification when a later segment B (e.g., the brake rotor) is designed. Therefore, if the manufacture of A is started before completing the design of B, some parts of A may have to be rebuilt. As in Loch and Terwiesch 1998, we conceptualize the design information communicated between segments A and B to be precise, so design iterations motivated by precision improvement will be unnecessary. The upstream design incompatibilities are resolved by multiple feedbacks between A and B. At the completion of design B, the set of modifications to the design of A is communicated downstream to manufacturing, where production of A might be in progress. The manufacture of A is preempted, and after all modifications are incorporated, the production of A is continued. For n segments (n is a decision variable), the design duration of the

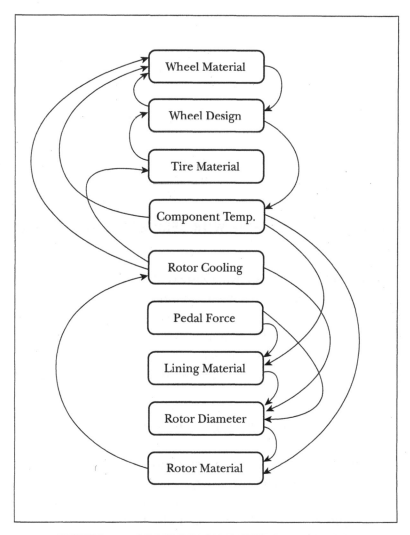

**FIGURE 8.9: DESIGN TASK INTERDEPENDENCE
FOR BRAKE SYSTEM DESIGN**

(Source: Eppinger et al. 1989. Adapted and printed by permission
of Springer-Verlag GmbH & Co., KG.)

project is divided into n intervals (which need not be of equal duration). Thus each interval except the first represents a duration over which the incompatibilities between two segments are "collected" in a single batch.

In Figure 8.10, three design segments (D_1, D_2, D_3), three build segments (B_1 (= B_{11} + B_{12}), B_2 (= B_{21} + B_{22}), B_3), and two reconciliation

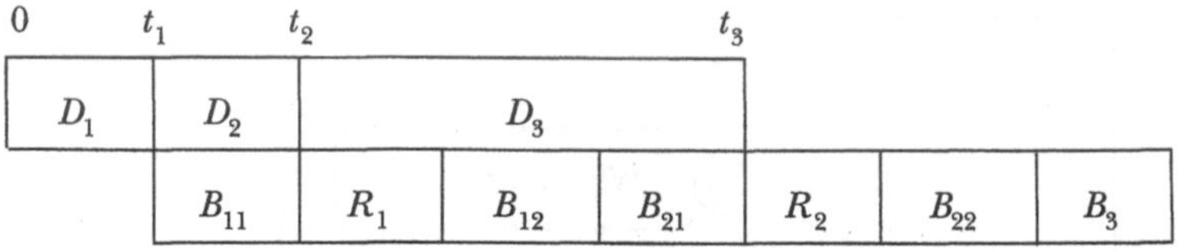

Legend

Design segments	D_1	D_2	D_3
Build segments	$B_{11} + B_{12}$	$B_{21} + B_{22}$	B_3
Batching intervals	0 to t_1	t_1 to t_2	t_2 to t_3
Reconciliation Work segments		R_1	R_2

FIGURE 8.10: OVERLAPPING WITH DESIGN INCOMPATIBILITIES

work segments (R_1, R_2) are shown. Observe that manufacture of B_1 is preempted at time t_2 to modify B_{11} based on modified D_1. The time it takes to modify B_{11} is shown as R_1. Work remaining on B_1 ($= B_{12}$) is then completed before starting on B_2, which may then be interrupted at t_3. The modification work on B_2 is shown as R_2.

The completion time of the project, with respect to Figure 8.10, can be written as

$$T = t_1 + B_1 + B_2 + B_3 + R_1 + R_2$$

Since design, manufacture, and modification (remanufacture) may be performed at different rates, we use γ_B (design rate/build rate) and γ_R (design rate/remanufacture rate) in accounting for the total time T. It is clear from Figure 8.10 that $B_i = \gamma_B \cdot D_i$. It also follows that the jth modification work R_j may have to be performed at time t_{j+1} on all manufactured work completed up to that time (whether or not it has been modified before). That is, R_j will be performed on $\sum_{i=1}^{j-1} B_i + B_j$. Thus we can write

$$R_1 = B_{11} \frac{\gamma_R}{\gamma_B} P_{12} = \left(\frac{\gamma_R}{\gamma_B}\right)(t_2 - t_1)\, P_{12}$$

where P_{12} is the probability of design incompatibility between segments 1 and 2. Similarly,

$$R_2 = (B_1 + B_{21}) \frac{\gamma_R}{\gamma_B} P_{23}$$

Since $B_{21} = t_3 - t_2 - R_1 - B_{12}$ and $B_{12} = \gamma_B t_1 - (t_2 - t_1)$, the expression for R_2 can be simplified to

$$R_2 = \left(\frac{\gamma_R}{\gamma_B}\right)(t_3 - t_1 - R_1)P_{23}$$

that is,

$$R_2 = \left(\frac{\gamma_R}{\gamma_B}\right)\left\{t_3 - t_1 - \left(\frac{\gamma_R}{\gamma_B}\right)(t_2 - t_1)P_{12}\right\}P_{23}$$

In general, we may write

$$R_k = \left(\frac{\gamma_R}{\gamma_B}\right)\left\{t_{k+1} - t_1 - \sum_{j=1}^{k-1} R_j\right\}P_{k,\,k+1} \tag{8.53}$$

where $R_0 = 0$, and $t_0 = 0$.

The expression for product development time with n design segments can thus be written as

$$T(n) = t_1 + \gamma_B t_n + \sum_{k=1}^{n-1} R_k \tag{8.54}$$

where R_k is defined as in equation 8.53.

It is clear that the probability of design incompatibility will depend upon the work content of the segment being designed and the work content of segments already designed. Thus in a two-segment design scenario the probability P_{12} will be proportional to the product of $(t_2 - t_1)$ and $(t_1 - 0)$. The probability can be written as

$$P_{12} = \frac{t_1(t_2 - t_1)}{\alpha t_2^2}, \quad \alpha > 0.25$$

where α is a measure of performance of the design team. A high value of α implies a low probability of design incompatibility. The above expression satisfies the observation by Eppinger et al. (1994) that in a design project most troublesome incompatibilities arise somewhere in the middle of the project. This is because the maximum value of P_{12} occurs when $t_1 = t_2/2$ and Max (P12) $= (t_2^2/4)/(\alpha t_2^2) = 1//4\alpha$. Since $P_{12} \leq 1$, it is sufficient to have $\alpha > 0.25$.

Optimality of Overlap

We first consider the case of $n = 2$. We have from equation 8.54,

$$T(2) = t_1 + \gamma_B t_2 + R_1$$

$$= t_1 + \gamma_B t_2 + \left(\frac{\gamma_R}{\gamma_B}\right)(t_2 - t_1)P_{12}$$

That is,

$$T(2) = t_1 + \gamma_B t_2 + \left(\frac{\gamma_R}{\gamma_B}\right)(t_2 - t_1) \cdot \frac{t_1(t_2 - t_1)}{\alpha t_2^2}$$

Collecting the terms with t_1, we rewrite $T(2)$ as

$$T(2, t_1) = t_1 + \frac{t_1(t_2 - t_1)^2}{\alpha t_2^2}\left(\frac{\gamma_R}{\gamma_B}\right) + \gamma_B t_2$$

It can be shown that

$$\frac{\partial T}{\partial t_1} = 1 + \left(\frac{\gamma_R}{\gamma_B}\right)\frac{(t_2 - t_1)(t_2 - 3t_1)}{\alpha t_2^2}$$

$$\frac{\partial^2 T}{\partial t_1^2} = 2 + \left(\frac{\gamma_R}{\gamma_B}\right)\frac{3t_1 - 2t_2}{\alpha t_2^2}$$

$T(2, t_1)$ is concave in t_1 for $t_1 < \frac{2}{3}t_2$, and it is convex when $t_1 > \frac{2}{3}t_2$. Also note that $\frac{\partial T}{\partial t_1} > 0$ when $t_1 = 0$. The gradient decreases gradually and turns negative before $t_1 = 2t_2/3$. Thus the minimum value of T, if it exists, will be obtained at $t_1 > 2t_2/3$. The two possible minimum points of T would be at $t_1 = 0$ and $t_1 = t_1^* > 2t_2/3$. At $t_1 = 0$, $T(2, t_2) = \gamma_B t_2$. Therefore for t_1^* to be the minimum point,

$$t_1^* + \frac{t_1^*(t_2 - t_1^*)^2}{\alpha t_2^2}\left(\frac{\gamma_R}{\gamma_B}\right) + \gamma_B t_2 < \gamma_B t_2$$

That is,

$$\left(\frac{\gamma_R}{\gamma_B}\right)\frac{(t_2 - t_1^*)^2}{\alpha t_2^2} + 1 < 0$$

which, obviously, is impossible.

Hence minimum $T(2)$ will be obtained at $t_1 = 0$, implying no overlapping. However, it may not be feasible to do the entire design as a single project. In reality, therefore, there would exist additional constraints such as $t_1 \geq a$ where a is given. The optimal value of t_1 will equal a if $T(2, a) < T(2, t_2 - t_a)$; otherwise $t_1 = t_2 - t_a$. For $n = 3$, we have a slightly different situation. It can be shown that $\frac{\partial T}{\partial t_1}$ can now be negative at $t_1 = 0$ in which case $T(3, t_1, t_2)$ will be convex for $t_1 < \frac{2}{3}t_2$ and concave for $t_1 > \frac{2}{3}t_2$. Let the stationary point of T $(t_1 < \frac{2}{3}t_2)$ be denoted by t_1^*. The minimum of T will be obtained at $t_1 = t_1^*$ or $t_1 = t_2$. Note that $\frac{\partial T(3, t_1, t_2)}{\partial t_1}$ can be simplified as

$$3t_1^2 - 4t_1 t_2 + t_2^2\left(1 + \frac{1}{t_3^2}\right) = 0$$

so that

$$t_1^* = \frac{t_2}{3}\left\{2 - \frac{\sqrt{t_3^2 - 3}}{t_3}\right\}$$

The optimal value of t_2 can similarly be obtained. To solve for optimal t_1 and t_2 simultaneously, however, will be algebraically tedious.

Engineering Change

Design incompatibilities can also be modeled as a stream of engineering changes (ECs) communicated to the downstream build operations (Terwiesch and Loch 1998). That is, information passed down at any time is precise, but ECs are generated by the design opertions at a rate μ after all preliminary information is communicated. The ECs can be addressed by manufacturing as and when they occur, or they can be batched and a predetermined number of ECs can be addressed at a time. The modification work (remanufacturing) associated with an EC occurring at a time τ, denoted as $R(\tau)$ in Figure 8.11, can be assumed to be proportional to the duration $(\tau - t)$, irrespective of whether ECs are batched. Note that $T - t$ is the overlap time.

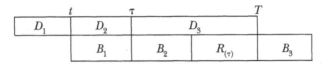

FIGURE 8.11: ENGINEERING CHANGE

Thus we may write $R(\tau)$ as

$$R(\tau) = k\mu(\tau - t)$$

where k is the time needed to resolve one EC. Hence the nominal modification work associated with ECs can be expressed as

$$R = \int_t^T k\mu(\tau - t)\, d\tau = \frac{1}{2}\mu k(T - t)^2 \qquad (8.55)$$

The extent of modifications needed for an EC is usually determined in a meeting of a cross-functional team. Since such meetings frequently cause delays, the total time of modifications, including meeting delays, can be written as

$$R' = R + M' \qquad (8.56)$$

where $M' = m/\mu$, if there is no batching of ECs; m being the delay caused by a meeting.

Assuming meeting time m to be constant irrespective of the number of ECs resolved in a meeting, it is apparent that batching the ECs may reduce the total project time. Also note that there will be another source of delay related to meetings. Since ECs would continue to arrive during the duration of the meeting ($= m$), an additional modification delay of $m\mu k$ must be included.

To determine an optimal batching policy for the ECs, assume that a meeting is called to address Q ECs at a time. It is clear that the total number of meetings required can be written as

$$n = \frac{(T-t)\mu}{Q} \tag{8.57}$$

Since during a meeting cycle the number of ECs will grow from zero to $Q-1$, the average delay caused by these ECs can be written as

$$\text{Delay per cycle} = k \cdot \frac{Q-1}{2} \tag{8.58}$$

Hence the total delay caused by meetings per unit time can be written, from equations 8.57 and 8.58, as

$$M = \frac{\mu m}{Q} + \frac{k(Q-1)}{2}$$

It is clear that $\frac{\partial M}{\partial Q} = 0$ will yield the optimal value of Q, which can be written as

$$Q = \sqrt{2\mu m/k}$$

and

$$M = \sqrt{2\,\mu km} - \frac{k}{2} \tag{8.59}$$

Since the total meeting delay M' would equal $M(T-t)$, we have from equations 8.56 and 8.59

$$\text{Total delay} = R + M(T-t)$$

That is,

$$\text{Total delay} = \frac{1}{2}\mu k(T-t)^2 + \left(\sqrt{2\mu km} - \frac{k}{2}\right)(T-t)$$

Therefore, the total project completion time will be written as

$$D = T + \gamma_B T + \frac{1}{2} \mu k (T - t)^2 + \left(\sqrt{2\mu km} - \frac{k}{2} \right)(T - t)$$

Since D is convex in t, the minimum of D is obtained at $\frac{\partial D}{\partial t} = 0$. After simplification, this yields

$$t = T + \left(\frac{m}{\mu k} \right)^{1/2} - \frac{1}{2\mu} - \frac{1}{\mu k} \tag{8.60}$$

so that overlap

$$T - t = \frac{k + 2}{2\mu k} - \left(\frac{m}{\mu k} \right)^{1/2} \tag{8.61}$$

It is clear that overlap decreases if m increases, implying that meetings are not run efficiently. It can also be verified that the overlap decreases with k, reaches a minimum at $k = 4 / \mu m$, and increases thereafter. That is, the overlap will be high if the time to fix an EC has an extreme value. Similarly, it can be shown that the overlap will be at a minimum with respect to μ when $\mu = (k + 2)^2 / mk$, and that it will increase for extreme values of μ on either side of $(k + 2)^2 / mk$. Finally observe that the duration between successive meetings, Q/μ, will be equal to $(2m/\mu k)^{1/2}$. Thus using equation 8.61, we may write

$$\text{Duration between meetings} = \frac{k + 2}{\sqrt{2} \, \mu k} - \sqrt{2} \, (\text{overlap})$$

Design Evolution

In designing compact products (instead of a system), where design specifications evolve gradually over time, the issues are somewhat different. Krishnan, Eppinger, and Whitney (1997) describe the problem of designing the handle of an automobile door panel. The specification of the door handle evolved gradually as additional information was obtained. In general, the specification can only be estimated to lie between two values (say, *a* and *b*) initially, where the degree of evolution ϕ can be said to be zero. When design evolution is complete (i.e., $\phi = 1.0$), the specification will have a unique value between *a* and *b*.

Consider a design task shown in parts a and b of Figure 8.12, where $\phi_0 = 0$ and $\phi_T = 1$. Let the manufacturing operation start at time *t* (Figure 8.12a).

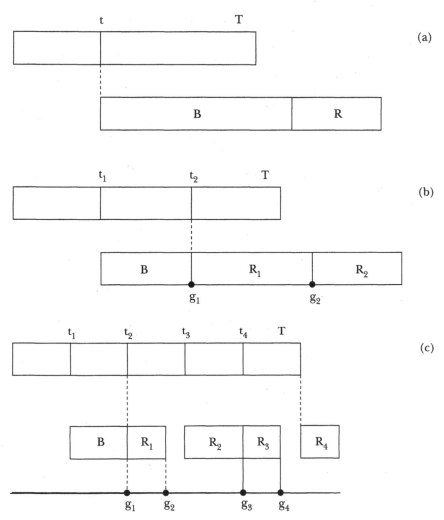

FIGURE 8.12: DESIGN EVOLUTION AND OVERLAP
(a)   SINGLE PROTOTYPE
(b)   TWO PROTOTYPES
(c)   FOUR PROTOTYPES

Let B be the duration of building the product, independent of the value of t. Thus if $t = T$, the total project duration would be $T + B$. However, if $t < T$, since design evolution is not complete, the build work done during B may have to be modified, and the extent of modification will be a function of the changes in specification due to evolution during the interval t to T. The project duration time would be equal to $T + R$ if the duration of $B < T - t$ and $t + B + R$ if the duration of $B > T - t$.

In the case of two build prototypes, it is clear in Figure 8.12 (b) that the project duration will be $t_1 + B + R_1 + R_2$, $t_2 + R_1 + R_2$, or $T + R_2$. Consider the single-prototype case. It is clear that the change in specification during t to T can be estimated as $x = (\phi_T - \phi_t)G$, where $G = \frac{b - a}{2}$. Let the modifications required in B be expressed as $R = f(x)$. Hence project duration D can be written as $D = t + B + f(x)$, if $t > T - B$. That is, $D = t + B + f\{(\phi_T - \phi_t)G\}$, and since $\phi_T = 1.0$, we have $D = t + B + f\{(1 - \phi_t)G\}$.

Krishnan, Eppinger, and Whitney suggest that ϕ_t will be a concave (increasing) function of t, and R will be a convex (increasing) function of x. Thus we may write $\phi_t = (t/T)^{1/2}$, and $f(x) = ax$. Hence $D = t + B + G^2(1 - \phi_t)^2$. That is

$$D = t + B + G^2 \left\{ 1 - \left(\frac{t}{T} \right)^{1/2} \right\}^2$$

It is clear that we can write $\frac{\partial D}{\partial t}$ and $\frac{\partial^2 D}{\partial t^2}$, after simplification, as

$$\frac{\partial D}{\partial t} = 1 + \frac{G^2}{T} - \frac{G^2}{(Tt)^{1/2}}$$

$$\frac{\partial^2 D}{\partial t^2} = \frac{G^2}{2(t^3 T)^{1/2}} > 0$$

Thus D is convex in t. Hence $\frac{\partial D}{\partial t} = 0$ will find the optimal t. That is,

$$T^{1/2} + \frac{G^2}{T^{1/2}} - \frac{G^2}{(t)^{1/2}} = 0$$

which leads to

$$t_{opt} = T \left(\frac{G^2}{T + G^2} \right)^2 \qquad (8.62)$$

$$D_{opt} = B + \frac{TG^2}{T + G^2}$$

If $D = T + R$, it can be shown, in a similar way, that $t_{opt} = T$ and $D_{opt} = T + B$. Note that if the initial uncertainty about design specification is high, G will be high, and the value of t_{opt} in equation 8.62 will approach T. The value of overlap, $T - t$, will approach zero. Thus initial uncertainty reduces overlapping.

Next consider the two-prototype case. The three possible objective functions $t_1 + B + R_1 + R_2$, $t_2 + R_1 + R_2$, and $T + R_2$ can all be shown to be convex in t_1 and t_2. The optimal values of t_1, and t_2 for the next objective function ($D = t_1 + B + R_1 + R_2$) can be shown to be

$$\left. \begin{array}{l} t_1 = T\left(\dfrac{G^2}{2T + G^2}\right)^2 \\[4mm] t_2 = T\left(\dfrac{T + G^2}{2T + G^2}\right)^2 \end{array} \right\} \tag{8.63}$$

Observe that the conditions for the above objective function to hold are $t_1 + B \geq t_2$ and $t_1 + B + R_1 \geq T$. If these conditions are violated, one of the other two objective functions would be considered, the range of its feasibility will be checked against the optimal t_1 and t_2, and so on. If no match is obtained between optimal t_1 and t_2 and the feasibility range of individual objective functions, the optimal values can be determined by solving $t_1 + B = t_2$ and $t_1 + B + R_1 = T$, for t_1 and t_2. The check for feasibility may become cumbersome if a large number of prototypes is being considered. Use of expressions like 8.63 would be viable if the number of prototypes considered is not large.

In situations where a large number of prototypes must be considered and/or when continuous functions $f(x)$ and ϕ_t cannot be established, a different approach for determining optimal overlaps must be considered. From Figure 8.12b it is clear that project duration D can be written as

$$D = \text{Maximum } (T, g_2) + f(\phi_T - \phi_2) \tag{8.64}$$

where g_2 is the completion time of the first prototype, started after time t_2 (R_2 in this case). The manufacturing time of the very last prototype is

$f(\phi_T - \phi_2)$. It is clear that $g_1 = t_1 + B$. Note that if there is exactly one prototype started before t_2, the expression for g_2 would be

$$g_2 = \text{Maximum}\ (t_2, t_1 + B) + f(\phi_2 - \phi_1).$$

However, since it is possible not to have any prototypes at all before t_2, by letting B start at t_2, an alternative value of g_2 would be

$$g_2 = t_2 + B$$

Hence, since we are interested in minimizing project completion time, we would have

$$g_2 = \text{Minimum}\ [(t_2 + B), \{\text{Maximum}\ (t_2, t_1 + B) + f(\phi_2 - \phi_1)\}]$$

which may be rewritten as

$$g_2 = \text{Minimum}\ [(t_2 + B), \{\text{Maximum}\ (t_2, g_1) + f(\phi_2 - \phi_1)\}]$$

We extend this with respect to Figure 8.12c, where it is clear that

$$g_3 = \text{Minimum}\ [(t_3 + B), \{\text{Maximum}\ (t_3, g_2) + f(\phi_3 - \phi_2)\}]$$

which can be generalized as

$$g_r = \text{Min}\left[(t_r + B) + \underset{1 \le m < r}{\text{Min}}\{\text{Max}\ (t_r, g_m) + f(\phi_r - \phi_m)\}\right] \qquad (8.65)$$

where $g_1 = t_1 + B$. The generalization of the objective function 8.64 would be written as

$$D = \underset{1 \le r < n}{\text{Min}}\{\text{Max}\ (t_n, g_r) + f(\phi_n - \phi_r)\};\ t_n = T \qquad (8.66)$$

The formulations in functions 8.66 and 8.65 can be solved as a two-step shortest path problem.

As an example, consider the design of a handle for an automobile door panel, as in Krishnan, Eppinger, and Whitney (1997). Expressing time in days, it is clear that $t_1 = 70$, $t_2 = 84$, $t_3 = 98$, $t_4 = 112$, and $t_5 = 126$. The authors specify the values of ϕ_t as $\phi_1 = 0$, $\phi_2 = 0.60$, $\phi_3 = 0.75$, $\phi_4 = 0.85$, and $\phi_5 = 1.0$. The extent of modification required in prototype (from the previous version) is expressed as ($a = 1$, $b = 21$), $x_{rm} = 10$ ($\phi_r - \phi_m$). The prototype modification (remanufacture) time (days) is defined as a step function:

$$f(x) = 1, \text{ if } x \le 2$$

$$f(x) = 21, \text{ if } x > 2$$

We first evaluate x_{rm} as shown in Table 8.2. The corresponding values of $f(x)$ are listed in Table 8.3. $B = 28$ days.

		r			
		2	3	4	5
	1	6.0	7.5	8.5	10.0
m	2	—	1.5	2.5	4.0
	3	—	—	1.0	2.5
	4	—	—	—	1.5

Table 8.2: Change in Design Specification

The values of g_1 to g_4 can now be computed as

$$g_1 = t_1 + B = 70 + 28 = 98$$

$$g_2 = \text{Min} \{84 + 28, \max(84, 98) + 21\} = 112, m^* = 0$$

		r			
		2	3	4	5
	1	21	21	21	21
m	2	—	1	21	21
	3	—	—	1	21
	4	—	—	—	1

Table 8.3: Prototype Modification Time (days)

$g_3 = \text{Min} \{98 + 28, \max (98, 98) + 21, \max (98, 112) + 21\} = 119, m^* = 1$

$g_4 = \text{Min} \{112 + 28, \max (112, 98) + 21,$
$\qquad\qquad \max (112, 112) + 21, \max (112, 119) + 1\} = 120, m^* = 3$

The values of D for values of r (1 to 4), using equation 8.62, are listed in Table 8.4.

It is clear in Table 8.4 that the minimum project duration is 127 days and that the prototyping will be done at t_1, t_3 and t_4. The prototypes and overlaps are shown in Figure 8.13.

		g_r	m^*	Max (T, g_r)	$f(\phi_s - \phi_r)$	D
	1	98	0	126	21	147
r	2	112	0	126	21	147
	3	119	1	126	21	147
	4	120	3	126	1	127

Table 8.4: Computation of D

8.6 DESIGN FACTORY

In today's competitive world most organizations tend to develop a large number of products concurrently. This places competing demands on

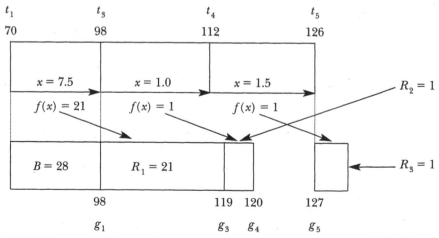

FIGURE 8.13: OPTIMAL OVERLAPPING FOR DOOR HANDLE DESIGN

shared resources. Design iterations caused by overlapping of tasks (discussed earlier) and design reviews implies additional burden on resources. The notion of a design factory implies managing product development in the same way the manufacture of products is managed. In contrasting the characteristics of manufacturing and product development, Adler and colleagues (1995) point out that manufacturing tasks are repeatable on multiple units of a few products and that manufacturing processes are largely standardized. The engineering tasks for product development, on the other hand, are mostly unique for each product. Therefore, the number of resource types required tends to be high, although the burden on a specific resource type may be low. This obviously creates unique problems of resource utilization. An increase in investment through shared resources may have multiple effects on product development: shorter launch time, increased performance level, and increased design conformance quality. We have already discussed the impact of investment on launch time and product performance earlier. Design conformance can also be increased by reviewing the project more frequently, which in turn calls for additional resources related to the task of design review and the design iterations such reviews entail.

Thus the three major issues in resource management are (1) aggregate capacity planning for all resource categories to minimize cost and maximize design conformance, (2) structuring of a design review strategy to obtain a certain level of design conformance at a minimum cost, and (3) matching and sequencing of processing times (development tasks) of products with available resources. We discuss the capacity planning issues first. Ahmadi and Wang (1999) have developed a model of capacity planning for design conformance, which they have applied to the development of the main engine of a space shuttle. Adler and co-workers develop a capacity planning model using spreadsheets for the plastics division of a chemical company. We briefly discuss these two approaches.

Consider a four-stage development project, shown in Figure 8.14.

Ahmadi and Wang define the design confidence level $\in_i$ at stage i as

FIGURE 8.14: RESOURCE ALLOCATION

$$\epsilon_i = 1 - e^{-\alpha x_1}$$

where x_i is the amount of resources (human, material, and technical) at stage i. ϵ_i can be interpreted as the percent of design specifications at stage i that are satisfied using the resources in stage i. Thus if there are n_i design specifications at stage i, we can express the confidence level at stage k as

$$E_k = \sum_{i=1}^{k} n_i \epsilon_i \Big/ \sum_{i=1}^{k} n_i$$

That is,

$$E_k = \sum_{i=1}^{k} \frac{n_i}{N} \epsilon_i$$

where

$$N = \sum_{i=1}^{k} n_i$$

It can be shown that E_k can be rewritten such that

$$E_k = \lambda_k \epsilon_k + \sum_{j=1}^{k-1} \lambda_{jk} E_j$$

where

$$\lambda_k + \sum_{j=1}^{k-1} \lambda_{jk} = 1, \lambda_{11} = 0, \lambda_1 = 1$$

Thus, with respect to Figure 8.14,

$$E_1 = \epsilon_1$$

$$E_2 = \lambda_2 \epsilon_2 + \lambda_{12} E_1$$

$$E_3 = \lambda_3 \epsilon_3 + \lambda_{23} E_2 + \lambda_{13} E_1$$

$$E_4 = \lambda_4 \epsilon_4 + \lambda_{34} E_3 + \lambda_{24} E_2 + \lambda_{14} E_1$$

If we let c_i be the cost per unit of resources at stage i (for the duration of its use), the total cost of resources can be written as $\sum_{i=1}^{k} c_i x_i$. Thus the capacity planning problem can be stated as

$$\text{Minimize } \sum_{i=1}^{k} c_i x_i$$

subject to

$$\sum_{i=1}^{k} \mu_i E_i \geq M \tag{8.67}$$

$$\sum_{i=1}^{k} \mu_i = 1$$

$$E_k = \lambda_k (1 - e^{-\alpha_i x_i}) + \sum_{j=1}^{k-1} \lambda_{jk} E_j \tag{8.68}$$

$$\lambda_k + \sum_{j=1}^{k-1} \lambda_{jk} = 1$$

$$x_i \geq 1, \text{ for all } i$$

Substituting the values of E_i from equation 8.68 into equation 8.67, we can rewrite 8.67 as

$$\sum_{i=1}^{k} (\phi_i)(1 - e^{\alpha_i x_i}) \geq M$$

It can be shown that if $k = 3$,

$$\phi_1 = \mu_1 + \lambda_{12}\mu_2 + (\lambda_{13} + \lambda_{12}\lambda_{23})\mu_3$$

$$\phi_2 = (\mu_2 + \lambda_{23}\mu_3)\lambda_2$$

$$\phi_3 = \lambda_3\mu_3$$

Thus the optimization problem can be rewritten as

$$\text{Minimize} \sum_{i=1}^{k} c_i x_i \tag{8.69}$$

subject to

$$\sum_{i=1}^{k} \phi_i (1 - e^{\beta_i x_i}) \geq M \tag{8.70}$$

$$\sum_{i=1}^{k} \mu_i = 1 \tag{8.71}$$

$$\lambda_i + \sum_{j=1}^{i-1} \lambda_{jk} = 1 \tag{8.72}$$

$$x_i \geq 1 \tag{8.73}$$

Ahmadi and Wang relax constraint 8.73 and solve equations 8.69 to 8.72 using Kuhn-Tucker conditions to obtain

$$x_i = \frac{\ell_n(\psi \alpha_i \phi_i) - \ell_n(c_i)}{\alpha_i}$$

where

$$\psi = \frac{\displaystyle\sum_{i=1}^{k} (c_i / \alpha_i)}{\displaystyle\sum_{i=1}^{k} \phi_i - M}$$

It can also be shown that if $x_i < 1$ for some i, then the solution will remain optimal if x_i is set equal to 1. A feasible solution with respect to constraint 8.73 will thus be obtained. That is, the optimal resource level at stage i is expressed as

$$x_i = \text{Maximum} \left\{ \frac{\ell_n(\psi \alpha_i \phi_i) - \ell_n(c_i)}{\alpha_i}, 1 \right\} \tag{8.74}$$

Note that the above procedure can be easily generalized to multiple development projects undertaken concurrently.

Adler and colleagues (1995) use a more direct approach to assess the capacity required for new product development. It is based on the average rates at which new product development requests arrive at the design factory, and the average processing rates at workstations for product engineers, process engineers, technicians, product managers, and application engineers. For a plastics division of a chemical company, they identified 17 major activities associated with new product development. Examples of such activities are identifying customer expectations, reviewing patents, developing manufacturing processes, determining market position, making slabs, testing slabs, making product, and testing product. For a given set of development projects, the average time required for each of the 17 activities at each of the workstations is estimated. The probabilities of design iterations between pairs of activities (if applicable) are estimated. Using the information on average processing time for each activity at each work station, and the probabilities of iteration, the authors use a spreadsheet to estimate the total amount of time that each resource group (workstation) spends on an average new product development project. These estimates can then be compared with the actual total hours available at the resource group.

The analysis using average values of arrival rates and processing times reveals possible bottlenecks at resource groups. However, it does not reveal why even with sufficient capacity (based on average values), it may create a work-in-process inventory of development projects at certain workstations. The fluctuations in arrival rates and processing rates may quickly lead to huge work-in-process inventories. A simple simulation model can be used to simulate the flow of products through the workstations as each of the seventeen activities. Design iterations are simulated using a Markovian model (explained later). Adler and co-workers used such a simulation model to conduct a detailed what-if analysis. Such analysis were related to (1) adding resources, (2) capping the number of projects in the system, (3) operating as a pull system, (4) resource pooling, and (5) centralized vs. decentralized coordination.

Design Reviews

A design review is analogous to an inspection station on a product line. The objective is to identify and rectify design flaws. Note that most of the value added through design up to the point when a design review takes place is lost if a flaw is detected. Hence frequent design reviews would minimize this loss. However, since each review can be expensive and cause delays, too-frequent reviews are not cost-effective. We are therefore interested in an optimal review strategy. We can state a review strategy in terms of a review period if reviews can be conducted at any time in the design process. If, on the other hand, reviews are conducted only at the end of a design stage, the interval between successive reviews may not be a constant. We examine the two cases separately.

Assume that design value added per period is v, as shown in Figure 8.15.

The design value lost is obviously $vT^2/2$ if a flaw is found in a design review at time T. It is clear that the probability of detecting a flaw will be an increasing function of the review interval T, approaching 1.0 when T approaches infinity. Hence we may express this probability as

$$\text{Probability of design flaw} = 1 - \frac{\alpha}{T + \alpha}$$

Hence

$$\text{Expected value lost} = \frac{vT^2}{2}\left(1 - \frac{\alpha}{T + \alpha}\right)$$

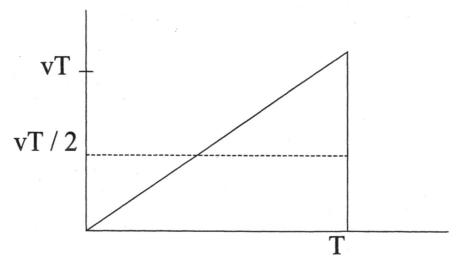

FIGURE 8.15: OPTIMAL DESIGN REVIEW INTERVAL

Let the cost of a design review = S. Therefore,

$$\text{Total cost per review} = S + \frac{vT^2}{2}\left(1 - \frac{\alpha}{T + \alpha}\right)$$

$$\text{Cost per period, } C = \frac{S}{T} + \frac{vT}{2}\left(1 - \frac{\alpha}{T + \alpha}\right)$$

Equating $\frac{dC}{dT} = 0$, we have the optimal review period from the solution of

$$\frac{S}{T^2} + \frac{\alpha^2 v}{2(T + \alpha)^2} - \frac{v}{2} = 0$$

For small values of α/T, we can approximate T as

$$T = \left(\frac{2S + \alpha^2 v}{v}\right)^{1/2} = \left(\frac{2S}{v} + \alpha^2\right)^{1/2}$$

and the upper bound of the total cost can be estimated as

$$c_u = \sqrt{v(2S + \alpha^2 v)}$$

It is clear that if the design is complex, involving a large number of engineers, v will be high and a more frequent review policy will be justified. This will also be the case if the probability of design flaws is high (i.e., α is small). If, on the other hand, the cost of a single review is high or if it causes significant delays, the frequency of reviews will be reduced. For more details of review policies of this type, refer to Ha and Porteus (1995).

When design reviews can be conducted only at the end of design stages, shown in Figure 8.14, the decision involves whether to do a review after completing the design at stage i ($i = 1$ to N). We can formulate this as a shortest-path problem, shown in Figure 8.16.

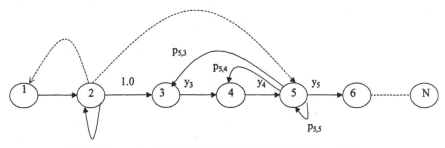

FIGURE 8.16: DESIGN REVIEWS (STAGES 2 AND 5)

The node i in Figure 8.16 represents the event of completion of design at stage i, and the arcs represent the transition of design effort from stage i to $i + 1$. If a design flaw is found during a design review at stage i, the flaw could have been introduced at any of the design stages r, where $r \leq i$. In Figure 8.16 it is assumed that design reviews are performed at the end of stages 2 and 5. The backward arrows from stage 5 indicate that the design flaws could be in any of the three stages 5, 4, and 3. The arc between stages 2 and 5 (broken line) indicate that there is no design review between these two stages. The shortest path problem can be stated as

$$f_n = \frac{\text{Minimum}}{n < m \leq N}\{c(n, m) + f_m\}$$

where

f_n = cost of all reviews between stages n and N (not including stage n)

$c(n, m)$ = cost of a review at stage m, given that no reviews take place in stages $n + 1$ to $m - 1$

To compute $c(n, m)$, let p_m be the probability of not finding any design flaws in the review at stage m. Let q_{mj} be the conditional probability of a design flaw in stage j (given that a flaw is detected at the review at stage m), so that

$$\sum_{j=n+1}^{m} q_{mj} = 1$$

Note that a design stage j (such as 3, 4, and 5) design may have to be repeated several times, as after each rectification of design flaws the design will be subjected to another review at stage 5. Let y_j denote the number of repetitions of stage j. From Figure 8.16 we get the following flow conservation equations:

$$y_3 = 1 + p_{5,3} y_5$$

$$y_4 = y_3 + p_{5,4} y_5$$

$$y_5 = y_4 + p_{5,5} y_5$$

where

$$p_{mj} = (1 - p_m) \cdot q_{mj}$$

The above set of equations can be solved simultaneously, and y_j ($j = 3, 4, 5$) can be expressed in terms of p_m and q_{mj} ($j = 3, 4, 5$).

The cost $c(n, m)$ can now be expressed as

$$c(n, m) = \sum_{j=n+1}^{m} y_j \theta_j + R_m$$

where θ_j is the cost of designing stage j and R_m is the cost of conducting a design review at stage m.

Similar analysis with respect to placing inspectors in a manufacturing line can be found in Chakravarty and Shtub 1987. The authors also include the cost of in-process inventory in the cost expression.

REFERENCES

Adler, P., A. Mandelbaum, V. Nguyen, and E. Schwerer (1995), "From Project to Process Management: An Empirically Based Framework for Analyzing Product Development Time," *Management Science,* vol. 41, no. 3, pp. 458–484.

Ahmadi, R., and R. Wang (1999), "Managing Development Risk in Product Design Processes," *Management Science,* vol. 47, no. 2, pp. 235–246.

Ali, A., M. Kalwani, and D. Kovenock (1993), "Selecting Product Development Projects: Pioneering vs. Incremental Innovation Strategies," *Management Science,* vol. 39, no. 3, pp. 255–274.

Bhattacharya, S., V. Krishnan, and V. Mahajan (1998), "Managing New Product Definition in Highly Dynamic Environment," *Management Science,* vol. 44, no. 11, pp. 550–564.

Chakravarty, A. (2000), "Overlapping Design and Build Cycles in Product Development," European Journal of Operational Research, in press.

Chakravarty, A., and A. Shtub (1987), "Strategic Allocation of Inspection Effort in a Serial Multi-Product Production System," *IIE Transactions,* vol. 19, no. 1, pp. 13–22.

Christensen, C., and J. Bower (1996), "Customer Power, Strategic Investment, and the Failure of Leading Firms," *Strategic Management Journal,* vol. 17, pp. 197–218.

Clark, K. (1989), "Project Scope and Project Performance: The Effect of Parts

Strategy and Supplier Involvement on Product Development," *Management Science,* vol. 35, no. 10, pp. 1247–1263.

Clark, K., and T. Fujimoto (1991), *Product Development Performance: Strategy, Organization and Management in the World Auto Industry,* Harvard Business School Press, Cambridge, Mass.

Cohen, M., J. Eliashberg, and T. Ho (1996), "New Product Development: The Performance and Time to Market Tradeoff," *Management Science,* vol. 42, no. 2, pp. 173–186.

Cusumano, M., and R. Selby (1995), *Microsoft Secrets: How the World's Most Powerful Software Company Creates Technologies, Shapes Markets, and Manages People,* The Free Press, New York.

Eisenhardt, K., and B. Tabrizi (1995), "Accelerating Adaptive Processes: Product Innovation in the Global Computer Industry," *Administrative Science Quarterly,* vol. 40, pp. 84–110.

Eppinger, S., D. Whitney, R. Smith, and D. Gebala (1989), "A Model Based Method for Organizing Tasks in Product Development," Research in Engineering Design, vol. 6, pp. 1–13.

Ha, A., and E. Porteus (1995), "Optimal Timing of Reviews in Concurrent Design for Manufacturability," *Management Science,* vol. 41, no. 9, pp. 1431–1447.

Hoedemaker, G., J. Blackburn, and L. Wassenhove (1999), "Limits to Concurrency," *Decision Sciences,* vol. 30, no. 1, pp. 1–18.

Imai, K., I. Nonaka, and H. Takeuchi (1985), "Manage the New Product Development Process: How the Japanese Companies Learn and Unlearn," in *The Uneasy Alliance,* K. Clark, R. Hays and C. Lorenz (editors), Harvard Business School Press, Cambridge, Mass.

Kamien, M., and N. Schwartz (1982), *Market Structure and Innovation,* Cambridge University Press, Cambridge, U.K..

Kotler, P. (1991), *Marketing Management: Analysis, Planning, Implementation and Control,* Prentice Hall, Englewood Cliffs, N.J.

Krishnan, V., S. Eppinger, and D. Whitney (1997), "Model-Based Framework to Overlap Product Development Activities," *Management Science,* vol. 43, no. 4, pp. 437–451.

Krishnan, V., and K. Ulrich (1998), "Product-Development Decisions: A Review of the Literature," working paper, Wharton School, University of Pennsylvania, Philadelphia.

Loch, C., and C. Terwiesch (1998), "Communication and Uncertainty in Concurrent Engineering," *Management Science,* vol. 44, no. 8, pp. 1032–1048.

Loury, G. (1979), "Market Structure and Innovation," *The Quarterly Journal of Economics,* vol. 93, pp. 395–410.

Pisano, G. (1996), "BMW: The 7 Series Project (A)," Harvard Business School Case 9-692-083. Cambridge, Mass.

Reinganum, J. (1981), "Dynamic Games of Innovation," *Journal of Economic Theory,* vol. 25, pp. 21–41.

Sabbagh, K. (1996), *Twenty-First-Century Jet*, Scribner, New York.

Takeuchi, H., and I. Nonaka (1986), "The New Product Development Game," *Harvard Business Review*, January/February, pp. 137–146.

Terwiesch, C., and C. Loch (1999), "Overlapping Development Activities," *Management Science*, vol. 45, no. 4, pp. 455–465.

Thompke, S. (1998), "Simulation, Learning and R&D Performance: Evidence from Automotive Development," *Research Policy*, vol. 27, pp. 55–74.

Wheelwright, S., and K. Clark (1992), *Revolutionizing Product Development*, The Free Press, New York.

The two integrals are evaluated separately as

$$I_1 = \int_{T_1=0}^{T_a} \int_{T_2=T_1}^{T_a} \{(1 - \lambda_1) T_2 - T_1 + \lambda_1 T_a\} e^{-r_1 T_1} e^{-r_2 T_2} \, dT_1 \, dT_2 \qquad (A1)$$

$$I_2 = \int_{T_2=0}^{T_a} \int_{T_1=T_2}^{T_a} (T_a - T_1) e^{-r_1 T_1} e^{-r_2 T_2} \, dT_1 \, dT_2 \qquad (A2)$$

where

$$r_1 = \frac{\epsilon_1 x_1}{R} \text{ and } r_2 = \frac{\epsilon_2 x_2}{R}$$

Integrate the RHS of equation A1 with respect to T_2 first, to obtain

$$I_1 = \frac{1}{r_2} \int_{T_1=0}^{T_a} \left[(-\lambda_1) T_1 e^{-(r_1+r_2)T_1} + \left(\frac{1-\lambda_1}{r_2} + \lambda_1 T_a \right) e^{-(r_1+r_2)T_1} + \right.$$
$$\left. \left\{ T_1 e^{-r_2 T_a} - (1 - \lambda_1) \left(T_a + \frac{1}{r_2} \right) e^{-r_2 T_a} \right\} e^{-r_2 T_1} \right] dT_1 \qquad (A3)$$

Integrating equation A3 with respect to T_1 and simplifying yields

$$I_1 = \frac{\lambda_1}{r_2 (r_1 + r_2)} \left(T_a - \frac{1}{r_2} - \frac{1}{r_1 + r_2} \right) + \frac{1}{r_2^2 (r_1 + r_2)} \qquad (A4)$$

Next, integrating the RHS of (A2) with respect to T_1, we have

$$I_2 = \int_{T_2=0}^{T_a} \left\{ \left(\frac{T_a}{r_1} - \frac{1}{r_1^2} \right) e^{-r_1 T_2} - \frac{T_2}{r_1} e^{-r_1 T_2} + \frac{1}{r_1^2} e^{-r_1 T_a} \right\} e^{-r_2 T_2} \, dT_2 \qquad (A5)$$

Integrating (A5) with respect to T_2, we have

$$I_2 = \frac{1}{r_1(r_1 + r_2)} \left\{ T_a - \frac{1}{r_1} - \frac{1}{(r_1 + r_2)} \right\} \tag{A6}$$

Thus

$$E(Rev) = r_1 r_2 V(I_1 + \beta_1 I_2)$$

Substituting the values of I_1, I_2 from equations A4 and A6, and for r_1 and r_2, we have

$$E(Rev) = \frac{VR}{\in_1 x_1 + \in_2 x_2} \left[\left\{ T_a - \frac{R}{\in_1 x_1 + \in_2 x_2} \right\} \right.$$
$$\left. \left(\frac{\in_1 \lambda_1 x_1}{R} + \frac{\in_2 \beta_1 x_2}{R} \right) - (1 - \lambda_1) \left(\frac{\in_1 x_1}{\in_2 x_2} \right) - \beta_1 \left(\frac{\in_2 x_2}{\in_1 x_1} \right) \right]$$

Supply Chains and Responsive Manufacturing

Appreciating an extended enterprise as a modular bundle of capabilities that evolves in time. Digital connectivity for process alignment is examined in terms of reengineering the value chain, new business models, and digital business community. Mathematical models of contracts, coordination, inventory sharing, flexibility, technology choice, and on-demand production are discussed.

CHAPTER

9

The Extended Enterprise: A Supply Chain Perspective

9.1 LEVERAGING THE VALUE CHAIN

As we have seen in Chapter 1, value is added to a product along a chain that spans raw material, components, and the final product delivered to the customer. The value-adding activities in this chain can generally be categorized as materials acquisition, component manufacturing, assembly, distribution, value-added resale, retailing, logistics service, financial service, warehousing, freight forwarding, shipping, and customer's activities. These may be owned by one or more entities, depending upon the nature and maturity stage of the industry. A value-adding entity in a chain may also serve several other value chains with single or multiple products. Hayes, Wheelwright, and Clark (1984) refer to such value chains, with their entities, as commercial chains. Lately such chains have come to be known as supply chains. They could also be called demand chains, supply webs, or simply networks.

In Chapters 1 through 3, we have seen how a given enterprise may be organized as domains or processes, and how such structures may be altered dynamically, in a pull mode. In this chapter and the following ones the emphasis will shift to managing relationships outside the enterprise by leveraging the total value chain. This also implies that the boundaries of the enterprise may alter as customers' needs change. Note also that a particular value chain may be realized by one of many possible supply chains linking the entities (with the enterprise as one of the entities). Consider a

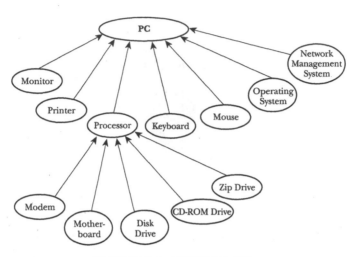

FIGURE 9.1: DESKTOP PC

desktop PC. The major components of the PC, as shown in Figure 9.1, are the monitor, printer, processor, keyboard, mouse, operating system, and network management system. Including the final assembly, there are eight major activities. An enterprise may select any subset of the eight activities to form its boundary provided it possesses the appropriate capabilities. Let us assume it decides to make only the processor. It is immediately obvious that a supply link between itself and the PC assembler must exist. There must also exist one supply chain(s) between the enterprise and its suppliers, since it needs to acquire several components to assemble the processor. Different choices of components for in-house manufacture will clearly lead to different supply chains.

The enterprise that (say) assembles only the processor, since it does not own the PC assembler (its customer) and/or the suppliers of components such as the motherboard, modem, disk drive, CD-ROM drive, and zip drive, must leverage all possible supply chains. It can do so in two different ways, which are not mutually exclusive: (1) better knowledge of PC demand (through the PC assembler) as well as the capabilities of suppliers, including decisions made by these entities with regard to their production, inventories, and contingencies, and (2) collaborative management, to minimize inefficiencies in handoffs with external partners. Making the supply chain transparent through information access and

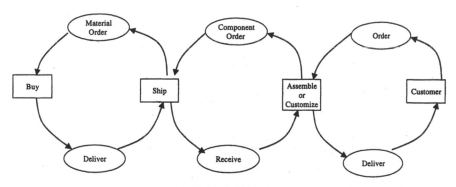

FIGURE 9.2: HANDOFF AND TRANSPARENCY

broadcasts and making the links seamless by designing appropriate interfaces are thus the two major enablers of an extended enterprise.

Consider the four-entity value chain shown in Figure 9.2. It is clear that there would be handoffs at two of the four entities: assemble and ship (Tyndall et al. 1998). A handoff implies an interaction between the internal functions of the entity (such as the rate of assembly) and the flows from external partners (such as customer orders and component delivery). Each such handoff represents potential mismatching, waste, information deterioration in the chain, customer-service degradation, and cost increase. These inefficiencies may be caused by communication breakdowns, breaks in authority, inventory and material buildups, and process inefficiencies. Process transparency is one of the ways to reduce handoff inefficiency. It implies complete knowledge of a partner's business processes and their updated status. For example, if a company uses contract manufacturers who have their own suppliers, it is not necessary for these suppliers to receive their orders only through the contract manufacturers. If the system is transparent, the OEM can transmit its order simultaneously to the contract manufacturers and their suppliers.

The structure of a supply chain need not be permanent. A manufacturer may bypass a distributor and supply customers directly. If a distributor is used, its role may be limited to generating customer orders, while the customers may be supplied directly by the manufacturer. Logistics services could be owned, or third-party logistics providers such as Ryder, Menlo, FedEx, UPS, and HUB could be used.

Thus there are several ways in which a customer order can be satisfied, each representing a chain in a network. Dynamic partnerships may be created to leverage the value chain, and to extend it in some cases. The question is, which of the possible chains represents the best value for customers and for each of the participating companies? The answer to the question lies both in the competitiveness of individual companies and in the choice of the chain that may link them. There are many issues that must be resolved. These include product and supply chain interactions, structure of the chain, and infrastructure; flows of cash, information, and material; ownership in the chain; volatility of demand from downstream to upstream; supplier capabilities; flexibility; and e-business. We next discuss these issues from a strategic perspective.

9.2 PRODUCT AND SUPPLY CHAIN ARCHITECTURES

It is clear from Figure 9.1, that what an enterprise chooses not to manufacture directly determines the extensiveness of the supply chain it must manage. Thus there is a strong interaction between the product portfolio of an enterprise and the associated supply chains. The nature of this interaction, however, depends upon the extent to which component outsourcing is feasible.

If a product's design is highly integrated, so the components of the product are tightly coupled, it will not be easy to outsource the manufacturing of components to several suppliers. An integral product architecture would be desirable if components must be tightly synchronized or if they are in close spatial proximity. In contrast, components in a modular structure can be individually upgraded, and they may not share functions. They would also have standardized interfaces among them. The carpenter's hammer, where the solid head drives nails and the claw head removes them, would be a good example of integral product design. A home stereo system where customers mix and match receivers, speakers, compact disc players, and other components from different manufacturers would be a good example of modular architecture (Fine 1998).

Supply chains can similarly be conceptualized as integral (vertical) or modular (horizontal). Fine (1998) defines an integral supply chain as one with close geographical, organizational, cultural, and electronic proximity.

Geographical proximity of suppliers is well understood and has been exploited in the context of just-in-time (JIT) manufacturing. Organizational proximity is somewhat abstract and can be measured by the extent of fragmentation of ownership and managerial control. Cultural proximity captures business practices and company values. Electronic proximity could be a substitute for geographical proximity. Electronic data interchange (EDI) e-mail, Internet/intranet, and video conferencing are some of the enablers of electronic proximity.

In an integral supply chain, business procedures are built up by leveraging the tacit knowledge (see Chapter 1) of individual partners. Therefore it is not easy to substitute one supplier for another, as the whole system of coordination might have to be reengineered. There is also a high degree of dependency among partners, and the processes of individual partners must be aligned with the total system. While there is little doubt of the value of electronic proximity, opinions may differ on organizational proximity. Total ownership of the chain and/or cultural homogeneity may not always be desirable.

A modular supply chain would allow fragmented ownership and cultural diversity. In fact, companies such as Dell Computer and Cisco Systems thrive on reducing ownership in the chain. By not requiring geographical proximity, a modular supply chain can broaden its scope and increase flexibility in its operations through quick restructuring. Multiple interchangeable suppliers for key components are permitted.

While most Japanese companies, such as Toyota in the auto industry, have relied upon integrated supply chains, the supply chains in today's PC industry are highly modular. Computer components including disk drives, motherboards, semiconductors, and modems are sourced throughout the world. In the auto industry, DaimlerChrysler has a single ownership but its suppliers are widely distributed geographically. Such supply chains could never succeed without electronic proximity. They are, however, not dependent on cultural homogeneity. Perhaps the telecommunications supply chain has become the most modular of all. The major products are audio, video, and data. Competition is based on speed of transfer, reliability of product integrity (especially for on-line audio such as telephone service), and price. The major technology enablers are wireline (telephone), wireless (cellular phone), cable, and the Internet backbone. There is a large number of suppliers such as hardware providers (routers, switches, terminals), service providers (access providers, brokered peering, Internet exchanges), backbone providers (AT&T, British Telecom, NTT, Deutsche Telekom, France Telecom), and software service providers (network management, Internet settlement regime, applications providers) There is a massive change in supply-chain modularity taking place at breathtaking pace. Some of the major moves have been AT&T's acquisition of TCI to get into the cable business, Lucent Technologies' acquisition of Ascend to launch a new product called "voice over Internet," AOL's move to become more integral by acquiring the Netscape browser, and Quest's merger with US West to get closer to its customer base. The collection of media headlines shown in Figure 9.3 explains it all.

9.3 PRODUCT AND SUPPLY CHAIN BUSINESS MODELS

The integral and modular architectures of product and supply chain generate four major business models. These are shown as four boxes, labeled A, B, C, and D, in Figure 9.4.

Three attributes—cost of ownership, manufacturing synergy, and management control—can be used to differentiate among the four business models. The cost of ownership is almost entirely determined by the architecture of the supply chain; it is high for an integral architecture and low for a modular structure. Similarly, manufacturing synergy will be high in an integral supply chain whether the product is integral or modular. High synergy implies high handoff efficiency. Note, however, that manufacturing synergy in box B (integral product and modular supply chain) will be much worse than in any of the other boxes, as it would be hard to obtain integrality of components if they are manufactured by different vendors. Finally, management control for coordination of production will generally increase if the product is modified from integral to modular and/or the supply chain is altered from integral to modular. Thus in business model A the ownership cost would be high, manufacturing synergy (yield) would be high, and required management control would be low. In business-model C, ownership cost is high, manufacturing synergy is at a medium level, and management control would be at a medium level. In business model B, ownership cost is low, manufacturing synergy would be poor, and management control would be at a medium level. Finally, in business model D ownership cost is low, manufacturing synergy is low, and

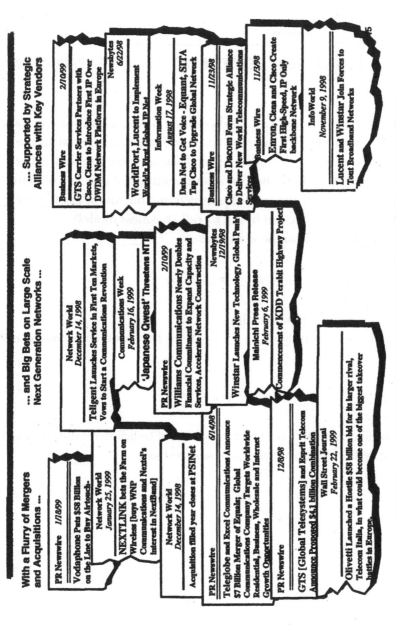

FIGURE 9.3: TELECOMMUNICATIONS SUPPLY CHAIN MODULARITY

		Supply Chain	
		Integral	Modular
Product	Integral	A	B
	Modular	C	D

FIGURE 9.4: INTERACTIONS BETWEEN PRODUCT AND SUPPLY CHAIN

management control is low. Let i and j denote product and supply chain structures, respectively. The value 1 (i or j) indicates an integral architecture and the value 2 (i or j) indicates modular architecture. Let w_{ij}, α_{ij}, and β_{ij}, respectively, denote the cost of ownership, cost of a *lack* of manufacturing synergy (yield) per unit of production, and cost of management control per unit of production in the i, j cell (business model).

Then cost in the i, j cell can be written as

$$C_{ij} = w_{ij} + \alpha_{ij} V + \beta_{ij} V$$

where V is the volume of production.

For the integral supply chain ($j = 1$), let the ownership cost be denoted as

$$w_{11} = w_{21} = W$$

For the modular supply chain ($j = 2$), the cost of building relationship will be a surrogate for ownership and can be stated as

$$w_{12} = w_{22} = \gamma W, \gamma \le 1$$

Hence we rewrite C_{ij} as

$$C_{11} \text{ (business model } A) = W + (\alpha_{11} + \beta_{11}) V$$

$$C_{12} \text{ (business model } B) = \gamma W + (\alpha_{12} + \beta_{12}) V$$

$$C_{21} \text{ (business model } C) = W + (\alpha_{21} + \beta_{21}) V$$

$$C_{22} \text{ (business model } D) = \gamma W + (\alpha_{22} + \beta_{22}) V$$

We can now analyze the four business models. For an integrated product architecture, we would prefer an integral supply chain (business model A) to a modular supply chain (business model B) if $C_{11} < C_{12}$, which can be simplified to

$$\frac{W}{V} < \frac{(\alpha_{12} - \alpha_{11}) + (\beta_{12} - \beta_{11})}{1 - \gamma} \qquad (9.1)$$

From our foregoing discussion, we know that $\alpha_{11} < \alpha_{12}$ and $\beta_{11} < \beta_{12}$. Since $\gamma \leq 1$, business model A can be justified for a moderate to high value of V, and business model B can be justified for a very low value of V (in relation to W). Similarly, the choice of business model C over D will depend upon whether

$$\frac{W}{V} < \frac{(\alpha_{22} - \alpha_{21}) + (\beta_{22} - \beta_{21})}{1 - \gamma} \qquad (9.2)$$

Note that $\alpha_{21} < \alpha_{22}$, but $\beta_{21} > \beta_{22}$. Hence the numerator *can be negative*. In scenarios where the cost of management control is much higher than the benefits of manufacturing synergy, an integral supply chain for a modular product (business model C) may not be justifiable for any value of V.

For the converse case, when the supply chain architecture is given, the choice of product architecture is dependent on only the values of α_{ij} and β_{ij} (W, V, and γ cancel out of the equations). Thus for an integral supply chain, an integral product (model A) is preferable to a modular product (model C) if

$$\alpha_{11} + \beta_{11} < \alpha_{21} + \beta_{21}$$

and for a modular supply chain, an integral product is preferable to a modular product if

$$\alpha_{12} + \beta_{12} < \alpha_{22} + \beta_{22}$$

Since $\alpha_{11} < \alpha_{21}$, and $\beta_{11} < \beta_{21}$ business model A will dominate over business model C. Similarly, $\alpha_{12} > \alpha_{22}$ and $\beta_{12} < \beta_{22}$, so there is no clear dominance between business models B and D. This implies that if most of the supply chain is owned, it becomes a sound strategy to encourage innovation and design of integral products. For a modular supply chain (as in an extended enterprise), a modular product can be justified only if the cost of management control can be controlled. With electronic proximity this is becoming more of a reality.

It follows that of the four business models, models C and D (i.e., with a modular product) can be very unstable, as model C may not be justifiable for any value of V in certain scenarios. IBM in the late 1970s and early 1980s employed model C with an integral supply chain for its PC business. For a modular product, such as a PC, it is hard to create entry barriers against nimble manufacturers who specialize in fewer modules. As a result, IBM quickly lost its manufacturing control advantage, and the industry mutated from model C to model D. Microsoft of today appears to be leading the PC industry back to model C from model D. It is doing so by leveraging its add-on modules, such as its Office suite and its Internet browser on its Windows platform. It is attempting to create an integral product from modules that have been developed seemingly independently (often adapted from other competitors). Thus, as Fine (1998) notes, there is an ongoing mutation back and forth between business models C and D.

Electronic Proximity

Two other supply chain architectures have become reality lately because of the advent of the Internet and extranets: transaction web architecture and virtual enterprise architecture. In the transaction web architecture, shown in Figure 9.5, the supply chain is modular. However, there is total connectivity among the partners, creating greatly enhanced electronic proximity. A traditional supply chain is usually demand-driven or supply-driven, where the flow of information and material is linear (i.e., sequential), as shown in Figure 9.6. In transaction web architecture there is a one-to-one link between each pair of partners in the supply chain. Since the web has total connectivity, note that a large number of paths are possible for designing, producing, and delivering a product in a collaborative environment. As Greis and Kasarda (1997) point out, the connectivity enables both spatial and temporal coordination of material flows from multiple sites. That is, components produced by several suppliers at several locations can be delivered to a customer at its site, in keeping with the customer's assembly schedule. Dell Computer has used this idea to have UPS pick up PCs from Dell's factory in Texas and monitors from Sony's facility in Mexico, mix and match them, and deliver them according to specific

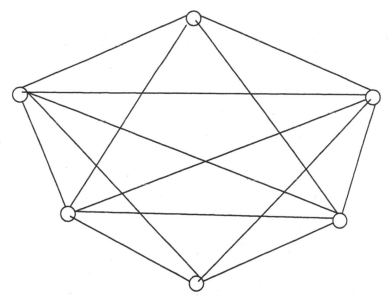

FIGURE 9.5: TRANSACTION WEB SUPPLY CHAIN ARCHITECTURE

customers' demand schedules (Magretta 1998). Multilateral transactions and negotiations can take place among groups of service providers simultaneously. Greis and Kasada call it the economy of conjunction. The connectivity and electronic proximity, in effect, transform a modular supply chain into a more integral architecture without requiring ownership of the chain.

Virtual enterprise architecture is a special case of the transaction web where the membership of the web can be established dynamically, based on the needs of customers and the capabilities of the suppliers. The architecture provides for addition and termination of partners. To make such architecture successful, however, the processes of partners need to be aligned so as to remain seamless at all times. Other issues such as procedures for dynamic work distribution, supplier capability updating, and progress monitoring need to be resolved. We defer the discussion of these topics to the next chapter.

Hybrid Architecture

Consider a three-component and four-supplier system, as shown in Figure 9.7. Assume that components C_1 and C_2 have a more integral architecture

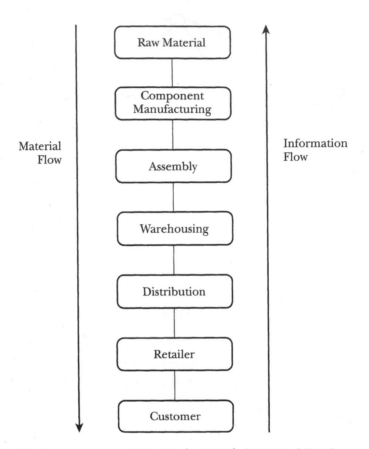

FIGURE 9.6: TRADITIONAL (LINEAR) SUPPLY CHAIN

with respect to the product (i.e., they may be part of an assembly) and that C_3 is more modular. Since supplier S_1 can supply both C_1 and C_2, it may be prudent for the OEM to own supplier S_1 and have C_3 procured from either S_3 or S_4. Thus S_1 with C_1 and C_2 would activate business model A, whereas S_3 (or S_4) would operate in business model D, creating a hybrid of two supply chain models for the same product.

If we assume instead that C_1 and C_3 together have a more integral architecture and C_2 is modular, it may make sense to own S_3 (to supply C_1 and C_3) and to form a modular architecture with S_2 for C_2. Thus, depending upon the degree of integrality with groups of components, a large number of hybrid architectures of the supply chain can be visualized. The optimal choice of hybrid will, of course, be based on the values of costs of

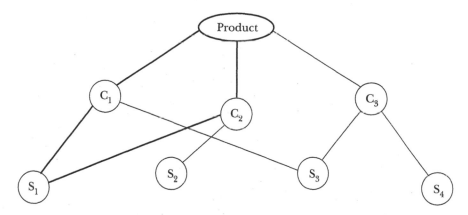

FIGURE 9.7: HYBRID ARCHITECTURE

ownership, manufacturing synergy (yield), and management control (co-ordination), discussed earlier.

9.4 SUPPLY CHAIN DYNAMICS

There are usually several tiers in a supply chain (shown in Figure 9.6), with customers at the downstream end and manufacturers at the upstream end. The dynamics of connectivity and lead times tend to amplify certain effects in the chain as one moves upstream. The first effect, which has come to be known as the bullwhip effect, is essentially the amplification of demand volatility. The second effect, known as clock speed, is the amplification of the interval of transition to new products or new business models.

Forrester (1961) was perhaps the first to note that the variance of demand at a manufacturing facility far exceeded the variance of customer demand. Forrester used a special simulation of the production and supply system to observe the bullwhip effect, and called the phenomenon "industrial dynamics." The "beer distribution game" (Sterman 1989), which has become popular in business education, comprises a brewery, a distributor, a wholesaler, and a retailer. The game shows how a small change in consumer demand can produce large swings in orders and inventories at the

FIGURE 9.8: AMPLIFICATION OF DEMAND VOLATILITY

distributor and the brewery. A typical case of spikes in demand from consumers to manufacturers is shown in Figure 9.8. Lee, Padmanabhan, and Whang (1997a) and Metters (1997) report such behavior from industries as diverse as food, computers, household products, and pharmaceuticals.

For the specific case of a two-tier linear supply chain (comprising consumer, retailers, and a manufacturer), Lee, Padmanabhan, and Whang (1997b) show that the bullwhip effect can occur in any of the following four scenarios. First, customer demand, as perceived by the retailer, may be nonstationary. That is, if there is a surge in demand in the current period, the retailer adjusts its order for next period by incorporating a fraction of observed demand volatility. Note that in a stationary demand scenario, the retailer would ignore demand surges and place an order equal to the current period's observed demand. This, of course, assumes that there is no fixed setup cost associated with order placement and that the manufacturer is not capacity-constrained. A nonzero fixed setup cost for ordering leads to the second scenario in which a bullwhip effect may occur. This happens because the retailer, to minimize setup cost, orders less frequently by batching the orders. In the third bullwhip scenario the manufacturer has a limited capacity and so may use a rationing policy based on retailers' actual orders. This motivates the retailers to inflate their orders, leading to a bullwhip effect. Finally, the retailer may face an uncertain fluctuation in the manufacturer's prices and may hedge against it by placing higher-than-required orders when a low price is anticipated and lower-than-required orders when a high price is anticipated.

Nonstationary Demand

It is clear that while in the stationary demand case the order-up-to level (where product availability is increased to a set level) will not vary with time, in the nonstationary demand scenario it would. Letting S_t be the order-up-to level determined in period t and D_t be the demand in period t, it is apparent that the retailer's order quantity Q_2 in period 2 can be written as

$$Q_2 = D_1 + S_2 - S_1$$

Lee, Padmanabhan, and Whang estimate the value of S_2 (using Heyman and Sobel 1984) as

$$S_2 = \frac{\alpha(1 - \alpha^{L+1})}{1 - \alpha} D_1 + d \sum_{k=1}^{L+1} \frac{1 - \alpha^k}{1 - \alpha} + z\sigma \sqrt{\sum_{k=1}^{L+1} \sum_{i=1}^{k} \alpha^{2(k-i)}} \quad (9.3)$$

where L is the interval between placing an order and its arrival at the retailer, the value of z is read off a unit-normal probability distribution chart for a certain probability value, and α and d describe the serially correlated demand at the retailer. Based on Kahn 1987, we can express

$$D_t = d + \alpha D_{t-1} + \epsilon_t, 0 < \alpha < 1 \quad (9.4)$$

where ϵ_t is independent and identically normally distributed (i.i.d.) with a mean value of zero and variance of σ^2. Lee, Padmanabhan, and Whang further establish that

$$Var(Q_2) = Var(D_1) + \frac{2\alpha(1 - \alpha^{L+1})(1 - \alpha^{L+2})}{(1 + \alpha)(1 - \alpha)^2} \cdot \sigma^2 \quad (9.5)$$

Thus clearly $Var\,(Q_9) > Var\,(D_1)$, and the bullwhip effect (BW) is

$$BW = \frac{2\alpha(1 - \alpha^{L+1})(1 - \alpha^{L+2})}{(1 + \alpha)(1 - \alpha^2)} \tag{9.6}$$

It can be verified that BW increases with L (since $\alpha < 1$), with a minimum value of $2\alpha\sigma^2$ when $L = 0$ and an upper bound of $\frac{2\alpha}{(1 + \alpha)(1 - \alpha)^2}\sigma^2$. It is clear that the bullwhip effect is caused by two factors: order arrival lead time L, and the presence of tiers in the supply chain. It also follows that while BW increases with L, it cannot exist without tiers. That is, if the manufacturer supplies directly to the customer, BW will equal zero, irrespective of the value of the lead time. Companies such as Dell have recognized this fact in their direct-selling model, in which they do away with all middle-level retailers and resellers. Finally, note that increased nonstationarity, indicated by a high value of α (< 1), will increase the value of BW. The above result can be easily generalized to a supply chain with n tiers by noting that the bullwhip effect of each tier is additive, with tier-specific values of α, σ, and L.

Thus the bullwhip effect can be controlled by controlling the value of L, but its onset cannot be avoided in a supply chain with tiers if the individual tiers neglect to use the information that comes from being a part of a chain. In a sequential (linear) chain, the blindsiding that leads to the bullwhip effect can be mitigated by collaborative planning, forecasting, and replenishment (CPFR) among tiers. Inventories need to be treated not as a single-tier effect but as a multi-echelon phenomenon. The notion of echelon inventories with transparency among tiers is the way to tackle this problem. This implies that demand information will be made available to all tiers that need it, and inventory positions at all downstream tiers will be used to make ordering decision at any individual tier. This, in effect, transforms a sequential supply chain into a transaction web, as shown in Figure 9.5. Cisco Systems and others have operationalized these concepts as so-called dynamic replenishment, which is somewhat like a transaction web. We shall discuss the working of such systems in detail in the next chapter.

Order Batching

Lee, Padmanabhan, and Whang have discussed three cases of order batching: (1) retailers order every R periods, but their orders are not coordinated with other retailers in any way, (2) all retailers order at the same time, and (3) retailer orders are evenly distributed in time (period R). For the first case, it can be shown that the variance of total orders from retailers in any period can be expressed as

$$V = N\sigma^2 + m^2 N(R - 1) \tag{9.7}$$

where N is the number of retailers, m is the retailer's mean order, σ^2 is the variance, and R is the interval between successive orders of any retailer. It is clear that $V > N\sigma^2$, which is the total variance in an unbatched scenario. Note that V increases with R (batching period) and is equal to $N\sigma^2$ when $R = 1$ (no batching).

For case 2, when all retailers order together, the variance of retailers' total orders can be expressed as

$$V = N\sigma^2 + m^2 N^2 (R - 1) \tag{9.8}$$

which again exceeds $N\sigma^2$.

Finally, in case 3, the variance is established as

$$V = N\sigma^2 + m^2 k (R - k) \tag{9.9}$$

where $N = I. R + k$, I being the largest interger value of the quotient of N/R.

Optimal order batching is usually determined with the objective of minimizing inventory and fixed ordering costs (Chakravarty and Martin 1991; Banerjee 1986; Dada and Srikanth 1987; Lee and Rosenblatt 1986; Lal and Staelin 1984). The cost of the bullwhip effect (additional inventory) should be added to the objective function along with the inventory and setup costs. This would, obviously, reduce the value of the optimal batching inverval.

Lee, Padmanabhan, and Whang have also established bullwhip results for the other two scenarios. However, since these do not lead to meaningful expressions for demand variance, as in the cases discussed, we do not include them here. The summary of their result, however, for the capacity-constrained case is that the order size exceeds the corresponding order size in the classic "news vendor" solution. For the price fluctuation case, they essentially establish that the order-up-to level will not be fixed (as we saw in the case of nonstationary demand).

Clock Speed

The term *clock speed,* as used by Fine (1998), is a measure of the rate at which products, processes, and supply chains evolve. Fine suggests two metrics: the rate of new product introduction is the product clock speed, and the rate of capital equipment obsolescence is the process clock speed. We may add to it the rate of change in connectivity (through addition or deletion of entities or links) as a metric of supply chain clock speed. The clock speed may vary from industry to industry, and it may be different at the upstream and downstream ends of the same supply chain. In the telecommunications industry, cable manufacturers are at the upstream end of the supply chain and customers such as stock market investors or music lovers are at the downstream end. A financial services provider (such as Intuit or E*Trade) would be at an intermediate level. It is clear that the rate of change in the technology of cable (manufactured by Lucent Technologies) is much slower than the rate of new applications designed by financial services providers such as E*Trade and logistics services providers such as FedEx and UPS. The clock speed at the upstream industries may therefore act as a bottleneck for the rate of innovation at downstream industries. One way of alleviating this problem would be for a downstream industry to share the cost of increasing the clock speed of an upstream industry. A mechanism needs to be structured by individual companies or a consortium of companies to make this happen. There is a long way to go, as not even the framework of such a mechanism has yet been set in place.

9.5 CAPABILITIES IN THE CHAIN

A supply chain in an extended enterprise can be looked upon as an assembly of capabilities. The questions that need to be answered for individual companies and for the extended enterprise as a whole are (1) how it should align its capabilities with the capabilities of its suppliers, and (2) how it should realign its capabilities with the needs of its customers. The emphasis in the first case is on what products it should produce in-house and what it should outsource. In the second case the emphasis is on what new capabilities it should develop—focused capability dedicated to a few customers (customer-specific), or versatile capability (general-purpose) that can be used to supply a large number of customers.

The two major dimensions of capability can be described as the knowledge (know-how) and the capacity (time and money) that are required to make a product. The mix of these capabilities would most likely vary from product to product. Thus a company may find itself in one of the following four scenarios: it possesses the knowledge and capacity, it possesses only the knowledge, it possesses only the capacity, or it possesses neither the knowledge nor the capacity. The current capabilities of a company in the context of three subsystems of a car—engine, transmission, and electronics—are shown in Figure 9.9. It is clear from Figure 9.9 that since the company possesses adequate knowledge and capacity for engine production, it will make engines in-house. It will outsource the majority of transmission systems but will keep transmission design in-house. It will continue to manufacture some transmission systems in-house, but it can bring this subsystem fully in-house by adding capacity anytime it so desires. It will outsource all electronics, as it does not possess the know-how to make them. The company may be setting a trap for itself if the electronics system ever becomes a critical component of the product, however. Note that the product in this case, in the context of the three distinct components, has a modular architecture. For integral products, the company will be less likely to outsource.

The classic case of IBM and Microsoft is very relevant in this context. IBM decided to outsource the disk operating system (DOS) for its PC to

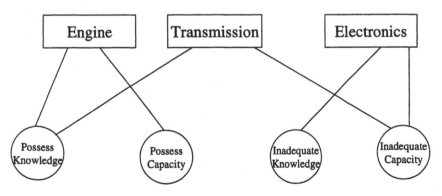

FIGURE 9.9: CAPABILITIES AND COMPONENT PRODUCTION

Microsoft in the early 1980s, as IBM had a capacity problem. Microsoft developed this expertise and soon found an independent market for its operating system (Windows), and so it became the main driver of the PC industry (with Intel). IBM spent about $1 billion in its attempt to develop OS/2 to kill Windows, but its effort failed.

Fine and Whitney (1996) describe Toyota's sourcing policy, which uses a strategy similar to what is shown in Figure 9.9. That is, Toyota ensures that it has adequate knowledge and capacity for in-house production of critical components such as car engines and, to a lesser degree, transmission systems. While the resource allocation issue is clear for critical components such as the engine, it is not so clear for the components that are less critical.

The decision problem faced is shown in Figure 9.10.

The eight components of a product, shown as circles, are arranged in decreasing order of criticality from left to right. Let the components be divided into three groups, so that those to the left of boundary B_1 are produced totally in-house, those to the right of boundary B_2 are totally outsourced, and those between B_1 and B_2 are partially outsourced for capacity. We are interested in the optimal positioning of the boundaries B_1 and B_2. In the optimal position, the marginal gain associated with a move of B_1 or B_2 to the left or right would be negative. In their current position in Figure 9.10, components 7 and 8 are analogous to the engine in Figure 9.9, components 4, 5, and 6 are analogous to the transmission, and com-

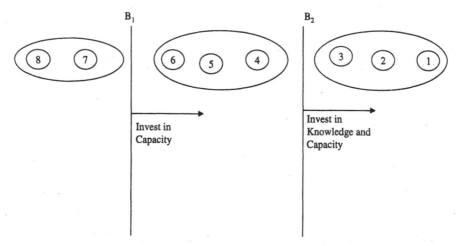

FIGURE 9.10: OUTSOURCING VS IN-HOUSE PRODUCTION

ponents 1, 2, and 3 are analogous to the electronics. Therefore, moving boundary B_1 to the right of component 6 would imply that component 6, which was being outsourced (fully or partially) for capacity reasons, will be brought fully in-house. This would obviously require additional investment in capacity. Conversely, moving B_1 to the left of component 7 would imply that component 7 would be outsourced. Moving boundary B_2 to the right, on the other hand, would imply that the knowledge work on component 3 (such as product design, engineering specification, and process planning) would be done in-house, requiring additional investment in knowledge building. We analyze these moves of boundaries B_1 and B_2 as follows.

Let

w_j = content of knowledge needed for component j

v_j = capacity needed for component j

x_j = knowledge content planned for component j ($x_j \leq w_j$)

y_j = current in-house capacity for component j ($y_j \leq v_j$)

$\rho_j = (y_j / v_j)$ = proportion of component j done in-house

p_j = probability of component j's criticality increasing significantly

I_j = cost of unit capacity increase for component j

E_j = cost of unit capacity increase, in an emergency, for component j

c_j' = cost of outsourcing per unit capacity

c_j = cost of in-house production per unit capacity

K_j = cost of increasing knowledge content of component j, available in-house, by one unit

F_j = cost of increasing knowledge content of component j by one unit in emergency mode

For boundary B_1, consider a move to the right (just beyond component 6). The costs are:

$$\text{Investment cost} = I_6 (v_6 - y_6)$$

$$\text{Production cost} = v_6 \, c_6$$

$$\text{Total cost, } T = I_6 \, (v_6 - y_6) + v_6 c_6 \qquad (9.10)$$

Now consider the cost of not moving B_1. If the criticality of component 6 increases,

$$\text{Investment cost} = E_6 \, (v_6 - y_6)$$

$$\text{Production cost} = v_6 \, c_6$$

If the criticality of component 6 does not increase,

$$\text{Investment cost} = 0$$

$$\text{Production cost} = \rho_6 \, v_6 \, c_6 + (1 - \rho_6) \, v_6 c_6'$$

Therefore, the expected total cost is

$$T_1 = p_6 \, \{E_6 \, (v_6 - y_6) + v_6 \, c_6\} + (1 - p_6)\{\rho_6 v_6 c_6 + (1 - \rho_6) \, v_6 \, c_6'\} \qquad (9.11)$$

The boundary will be moved if $T < T_1$. That is, $\qquad\qquad\qquad (9.12)$

$$I_6 \, (v_6 - y_6) + v_6 c_6 < p_6 \, \{E_6 \, (v_6 - y_6) + v_6 \, c_6\}$$
$$+ (1 - p_6)\{\rho_6 \, v_6 \, c_6 + (1 - \rho_6) v_6 \, c_6'\} \qquad (9.13)$$

which can be simplified, by substituting $y_6 \, / \, v_6$ for ρ_6, to

$$p_6 > \frac{I_6 + c_6 - c_6'}{E_6 + c_6 - c_6'} \qquad (9.14)$$

Thus, in general, the boundary B_1 will be kept moving to the right as long as

$$p_j > \frac{I_j + c_j - c_j'}{E_j + c_j - c_j'} = 1 - \frac{E_j - I_j}{E_j + c_j - c_j'} \qquad (9.15)$$

Now consider moving boundary B_2 to the right. It is clear that there will be an additional investment of $K_j \, (w_j - x_j)$ corresponding to $I_j \, (v_j - y_j)$ and $E_j \, (v_j - y_j)$. Also note that now $\rho_j = 0$. Hence, corresponding to condition 9.13 for component 3, we can write

$$I_3 (v_3 - y_3) + K_3 (w_3 - x_3) + v_3 c_3 < p_3$$
$$\{E_3 (v_3 - y_3) + F_3 (v_3 - x_3)\} + (1 - p_3)v_3 c_3' \quad (9.16)$$

which can be simplified to

$$p_3 > \frac{(v_3 - y_3) I_3 + (w_3 - x_3) K_3 + v_3 (c_3 - c_3')}{(v_3 - y_3) E_3 + (w_3 - x_3) F_3 + v_3 (c_3 - c_3')} \quad (9.17)$$

which can be generalized to

$$p_j > \frac{(v_j - y_j) I_j + (w_j - x_j) K_j + v_j (c_j - c_j')}{(v_j - y_j) E_j + (w_j - x_j) F_j + v_j (c_j - c_j')} \quad (9.18)$$

Observe that since $F_j > K_j$ the RHS of expression 9.18 will be much smaller than the RHS of expression 9.15. Therefore, it will be harder to satisfy 9.15 compared to 9.18. This would imply that it is less risky for the firm to outsource when it has inadequate capacity (but not inadequate knowledge). We can also express a measure of risk, in this context, as

$$\text{Risk} = p_j - \text{RHS of expression 9.15 or 9.18}$$

This is also the conclusion arrived at by Fine and Whitney (1996). Note, however, that this risk would be much higher in an industry with a fast clock speed. With a rapid rate of change in technology, the knowledge possessed by the company soon becomes obsolete. This knowledge is not usually updated if the company is not in the business of making the product. Thus in an industry with a slow clock speed, such as automobiles, Toyota can feel safe in outsourcing transmission systems because of its knowledge of the component. In the computer industry, which has a faster clock speed, IBM quickly lost whatever knowledge advantage it had over Microsoft. In such industries, advantage needs to be created through other means. Dell Computer is a classic example, as it chose supply chain knowledge (not product knowledge) as the basis for its outsourcing strategy. Hewlett-Packard's decision to outsource disk drives, which is a critical component in terms of technology, may appear strange. But the company reasoned that its knowledge of markets (and customers) for disk drives was critical knowledge, so it kept that in-house and outsourced the design and manufacture of disk drives.

	Capacity- Dependent	Capacity- Independent
Product/Process Knowledge	Toyota	Toyota, Merck
Market Knowledge	Hewlett-Packard	Joint Venture
Supply Chain Knowledge	Dell, FedEx, UPS	Joint Venture

FIGURE 9.11: KNOWLEDGE-BASED OUTSOURCING

In Figure 9.11, the interactions of the three types of knowledge with capacity is shown. The companies that decided to outsource under different scenarios are identified. Note that a lack of market knowledge or supply chain knowledge usually leads to joint ventures, and there are many examples of these, especially in international business.

Developing New Capabilities

In contrast to the above discussion, where the customer's (firm's) perspective was emphasized, we now discuss the supplier's perspective of the capability issue. We begin from the basic framework of Parker (1998), where relevant, and generalize and/or extend it in the context of the extended enterprise.

Consider the framework shown in Figure 9.12. The supplier can de-

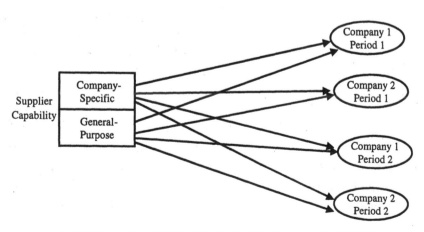

FIGURE 9.12: SUPPLIER CAPABILITY DEVELOPMENT

velop capabilities that can be specific to either company 1 or company 2, or they can be general-purpose capabilities that may be used by either company. The supplier capability is, therefore, shown divided into two compartments in Figure 9.12. Companies 1 and 2 are competitors, but they use different technologies, denoted as α_i and β_i for the ith company $(i = 1, 2)$, where α_i and β_i interact with the supplier's firm-specific capability and general-purpose capability, respectively, to yield productive outputs. Period 1 represents the present, and period 2 represents the future. It is assumed that technology does not change significantly from period 1 to 2 (but this assumption can easily be relaxed).

Let

x_{it} = supplier's capability specific to company i, developed in period t,

y_t = supplier's general-purpose capability, developed in period t,

z_{it} = company i's yield (output) in period t

We assume that the supplier has fixed resources in any period and that these can be allocated for developing company-specific or general-purpose capabilities independently of its development efforts in previous periods. It is also assumed that there is no cost of switching resources from a company-specific capability development project to a general-purpose capability development project in any period. Finally, the interaction between the company's technology and the supplier's capability (which determines yield) is assumed to be multiplicative.

Without loss of generality, we assume that company 2 is not active in period 1. Company 1's output in period 1 can be expressed as

$$z_{11} = \alpha_1 x_{11} + \beta_1 y_1$$

In period 2, the supplier has the following choices: stay with company 1 or switch to company 2. If the supplier stays with company 1, we have

$$z_{12} = \alpha_1 (x_{11} + x_{12}) + \beta_1 (y_1 + y_2)$$

If the supplier switches to company 2,

$$z_{22} = \alpha_2 x_{22} + \beta_2 (y_1 + y_2)$$

Note that since the supplier has fixed resources, y_t will be a decreasing function of x_{it}. We denote this relationship as

$$y_t = f(x_{it})$$

The function of $f(x)$ defines an exchange curve between the two types of capabilities and is shown in Figure 9.13. The curve with dotted lines in Figure 9.13 is a continuous-curve approximation of the relationship.

It is clear that the supplier's switch to company 2 in period 2 would depend on how much it expects to earn from such a switch, which in turn would depend on how much and what kind of capability the supplier develops in each period. Parker (1998) analyzes three contractual agreements: (1) a long-term contract between company 1 and the supplier, (2) an incentive contract whereby company 1 agrees to pay a portion of the surplus to the supplier so that the supplier does not switch to company 2, and (3) no contract, whereby company 1 pays a certain portion of total yield (not surplus) to the supplier, and the supplier is free to switch to company 2 in period 2.

Long Term Contract

With a long-term contract, the supplier's (and the firm's) objective can be expressed as maximization of

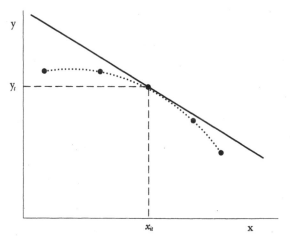

FIGURE 9.13: EXCHANGE CURVE OF CAPACITY TYPES

$$\Pi_a = (z_{11} + z_{12})$$

with respect to x_{11}, y_1, x_{12}, and y_2. The above expression can be simplified to

$$\Pi_a = 2\{\alpha_1 x_{11} + \beta_1 f(x_{11})\} + \alpha_1 x_{12} + \beta_1 f(x_{12}) \qquad (9.19)$$

Since $f(x)$ is a concave function, Π_a is concave in x_{11} and x_{12}. Therefore $\frac{\partial \Pi_a}{\partial x_{11}}$ and $\frac{\partial \Pi_a}{\partial x_{12}}$ will yield the optimal solution

$$\left.\begin{array}{l} \dfrac{\partial f(x_{11})}{\partial x_{11}} = -(\alpha_1/\beta_1) \\[4mm] \dfrac{\partial f(x_{12})}{\partial x_{12}} = -(\alpha_1/\beta_1) \end{array}\right\} \qquad (9.20)$$

Since $\frac{\partial f(x)}{\partial x} < 0$, it is clear that the point in Figure 9.13 where the slope of the exchange curve (absolute value) equals the value α_1/β_1 will be the optimal combination of x_{it} and y_t.

For a specific shape $f(x) = (1 - x^n)^{1/n}$ (i.e., $x^n + y^n = 1$), we can write the optimal value of x_{11} as

$$x_{11} = x_{12} = \left(\frac{A}{A+B}\right)^{\frac{1}{n}}, \ A = \alpha_1^{\left(\frac{n}{n-1}\right)}, \ B = \beta_1^{\left(\frac{n}{n-1}\right)}$$

And

$$y_1 = y_2 = \left(\frac{B}{A+B}\right)^{\frac{1}{n}}$$

Incentive Contract

With an incentive contract, let company 1 agree to pay the supplier $\frac{1}{k}$th of the surplus of output in each period, over and above the supplier's reservation price R in period 1 and z_{22} in period 2.

$$\Pi_b = R + \frac{(z_{11} - R)}{k} + z_{22} + \frac{(z_{12} - z_{22})}{k}$$

which expands to

$$\Pi_b = \left(1 - \frac{1}{k}\right) R + \frac{1}{k} \{2\alpha_1 x_{11} + \alpha_1 x_{12} + 2\beta_1 f(x_{11}) + \beta_1 f(x_{12})\} +$$

$$\left(1 - \frac{1}{k}\right) \{\alpha_2 x_{22} + \beta_2 f(x_{11}) + \beta_2 f(x_{22})\}$$

and the optimal value of x_{11} can be expressed as

$$\left. \begin{aligned} \frac{\partial f(x_{11})}{\partial x_{11}} &= -\frac{2\alpha_1}{2\beta_1 + (k-1)\beta_2} \\ \frac{\partial f(x_{12})}{\partial x_{12}} &= -\frac{\alpha_1}{\beta_1} \end{aligned} \right\} \tag{9.21}$$

If $x^n + y^n = 1$,

$$\left. \begin{aligned} x_{11} &= \left(\frac{A_1}{A_1 + B_1}\right)^{\frac{1}{n}}, A_1 = (2\alpha_1)\left(\frac{n}{n-1}\right), B_1 = \{2\beta_1 + (k-1)\beta_2\}\left(\frac{n}{n-1}\right) \\ x_{12} &= \left(\frac{A_2}{A_2 + B_2}\right)^{\frac{1}{n}}, A_2 = \{\alpha_1\}\left(\frac{n}{n-1}\right), B_2 = \{\beta_1\}\left(\frac{n}{n-1}\right) \end{aligned} \right\}$$

$$\tag{9.22}$$

No Contract Scenario

For the case of no contract, let company 1 pay a portion s_t of output to the supplier in period t. It is clear that for the supplier not to switch to company 2, company 1 must ensure that $s_1 z_{11} > R$ and $s_2 z_{12} > z_{22}$. Since period 2 is the last period, it is clear that supplier cannot do any worse than z_{22}. Hence the supplier's total income would be

$$\Pi_c = s_1 z_{11} + z_{22}$$

that is,

$$\Pi_c = \alpha_1 s_1 x_{11} + (\beta_1 + \beta_2) f(x_{11}) + \alpha_2 x_{22} + \beta_2 f(x_{22})$$

and the optimal values can be computed from

$$\frac{\partial f(x_{11})}{\partial x_{11}} = -\frac{\alpha_1 s_1}{\beta_1 + \beta_2}, \frac{\partial f(x_{12})}{\partial x_{12}} = -\frac{\alpha_2}{\beta_2} \tag{9.23}$$

If $x^n + y^n = 1$, we have

$$\left.\begin{aligned}
x_{11} &= \left(\frac{A_1}{A_1 + B_1}\right)^{\frac{1}{n}}, A_1 = \{\alpha_1 s_1\}\left(\frac{n}{n-1}\right), B_1 = (\beta_1 + \beta_2)\left(\frac{n}{n-1}\right), \\
x_{12} &= \left(\frac{A_2}{A_2 + B_2}\right)^{\frac{1}{n}}, A_2 = \alpha_2\left(\frac{n}{n-1}\right), B_2 = \{\beta_2\}\left(\frac{n}{n-1}\right)
\end{aligned}\right\} \tag{9.24}$$

For the general shape of the exchange curve (Figure 9.13), it follows that

$$\left|f'(x)\right| > \left|f'(u)\right| \Rightarrow x > u$$

To compare the values of x_{it} under different contractual arrangements, we denote x_{it} as x_{it}^L, x_{it}^I, and x_{it}^N corresponding to the long-term contract, incentive contract, and no-contract scenarios. Comparing expressions 9.20 and 9.21, it is clear that

$$x_{it}^L > x_{it}^I \text{ if } \frac{\alpha_1}{\beta_1} > \frac{2\alpha_1}{2\beta_1 + (k-1)\beta_2}$$

which implies, since $\beta > 0$, that $x_{11}^L > x_{11}^I$. In a similar way, we can establish that $x_{12}^L = x_{12}^I$. Comparing expressions 9.21 and 9.23, we have

$$x_{11}^I > x_{11}^N, \text{ if } \frac{2\alpha_1}{2\beta_1 + (k-1)\beta_2} > \frac{\alpha_1 s_1}{\beta_1 + \beta_2}$$

which simplifies to

$$\frac{\beta_1}{\beta_2} > \frac{(k-1)s_1}{2(1-s_1)} - \frac{1}{(1-s_1)}, s_1 \neq 1$$

Since $s_1 < 1$, a sufficient condition for $x_{11}^I > x_{11}^N$, it can be shown if $k < 3$, $x_{11}^I > x_{11}^N$ for any value of s_1. Similarly, $x_{12}^I > x_{12}^N$ if $\frac{\alpha_1}{\beta_1} > \frac{\alpha_2}{\beta_2}$.

We can summarize the above as

(a) $x_{11}^L > x_{11}^I$, and $x_{11}^L > x_{11}^N$

(b) $x_{11}^I > x_{11}^N$, if $\frac{\beta_1}{\beta_2} > \frac{(k-1)s_1 - 2}{2(1-s_1)}$, $s_1 \neq 1; k < 3$

is a sufficient condition.

$$(c) \quad x_{12}^L = x_{12}^I$$

$$(d) \quad \text{if } \frac{\beta_1}{\beta_2} < \frac{\alpha_1}{\alpha_2}, \text{ then } x_{12}^I > x_{12}^N$$

It is clear that in period 1 the supplier will be more committed to company 1 under a long-term contract, and an incentive contract may not dominate over no contract in terms of commitment. In period 2, the supplier's commitment to company 1 cannot be assumed under any contractual arrangement.

The case when company 1 optimizes the value of s_1, is somewhat more complex to analyze. It can be formulated as a two-person game, where both company 1 and the supplier optimize different objective functions, but with some binding constraint. Thus the supplier's problem, as before, would be

Maximize Π_c (supplier) $= s_1 z_{11} + z_{22}$ with respect to x_{11} and x_{12}

Company 1's objective is

Minimize Π_c (company 1) $= s_1 z_{11} + s_2 z_{12}$ with respect to s_1 and s_2

However, it must ensure that

$$s_1 z_{11} + s_2 z_{12} \geq R + z_{22} \tag{9.25}$$

so that the supplier stays with company 1.

For any value of s_1, we can compute x_{11} and x_{12} using expression 9.24. Hence we can evaluate $s_1 z_{11}$ and z_{22}. A feasible solution corresponding to this s_1 would exist if

$$s_1 z_{11} \geq R$$

and

$$s_2 = \frac{R + z_{22} - s_1 z_{11}}{z_{12}} \geq 0$$

To search for optimal s_1, compute s_2 for two values s_1 and s_1', and denote z_{it} as $z_{it}(s_1)$ or $z_{it}(s_1')$. If

$$s_1' z_{11}(s_1') + s_2' z_{12}(s_1') < s_1 z_{11}(s_1) + s_2 z_{12}(s_1')$$

update s_1 to s_1', and so on.

There are investment risks associated with uncertainties from the rapid pace of technology change (clock speed). Thus a supplier in the telecommunications industry may feel less secure making a large investment in capability compared to its counterpart in the automobile industry. This risk is amplified further if the supplier faces a large number of competitors. Finally, the supplier needs to be aware of the dynamics between the price it charges and demand for its capability. We have not included these issues in our analysis above. New theories need to be developed to enable such analysis.

9.6 OWNERSHIP AND COORDINATION

Manufacture of a product, from the procurement of raw material to delivery to the customer, involves several steps that include component manufacture, building subassemblies, and the final assembly of the product. A company may decide to do all component manufacturing, assembly, and testing by itself at a single location. This would be an example of the integral supply chain architecture we discussed before. The company may want to take advantage of location economics by locating component manufacturing at sites with inexpensive labor and material and locate assembly operations closer to customers. While there would be savings in labor, material, and the cost of transporting finished goods, new cost categories—transportation of components, in-transit inventory of components—would be introduced. In addition, an elaborate coordination of material flows would be required to ensure that the right components in the requisite quantities arrive at the right assembly locations at the right time.

If the company does not possess expertise in the transportation business, it may subcontract all logistics to a third-party logistics provider, who would deliver raw material, components, subassemblies, and the final assembly from their respective sources to their destinations. Depending upon the contract, the logistics providers may assume responsibility for all inventories in the pipeline as well.

In scenarios where a large number of diverse components are involved, the company may outsource a large chunk of its component manufacturing to subcontractors, to concentrate on assembly operations. This would also help the company divest its holdings in component plants, although its total procurement cost may increase. Though the total profit and return on sale may decrease, return on investment (or return on assets) may show a significant increase. The company may procure components from an independent supplier, or it may work out contractual arrangements with contract manufacturers and/or an OEM. A contract manufacturer procures its own raw material and manufactures the component according to the company's (customer) specifications. The customer usually pays for the contract manufacturer's inventory. An OEM, on the other hand, manufactures the component under a technology license and is responsible for its own procurements and inventories. Note that use of contract manufacturers and OEMs may not be limited to just component manufacturing; subassemblies and final assembly can also be outsourced.

Finally, the company may interpose distributors in the supply chain (inbound or outbound). The major motivations for an outbound distributor would be to increase the product's market penetration and to transfer most of its outbound logistics and inventory costs to the distributor. The distributor's margin may increase the product's unit price, which can be offset to some degree by access to a wider market and the distributor's economy of scale in transaction volume. The company may also reduce its corporate taxes by transferring the inventories to distributors.

In Figure 9.14, the material flows in a modular supply chain include the company's two factories, an OEM, a contract manufacturer, a distributor, and a third-party logistics provider. It is clear from Figure 9.14 that outsourcing increases the number of entities and hence the number of handoffs (the number of links in the system). A handoff implies an additional coordination of material, information, and/or cash flows, and hence it is a potential source of inefficiency. Using contract manufacturers, OEMs, and/or distributors typically reduces the number of suppliers the company needs to coordinate, leading to a reduction in handoffs. The risk, however, is that the company may be severing its links with suppliers and/or customers, and as a result, a future alliance of contract manufacturers and distributors can supplant the company. The Wintel alliance (Windows and Intel) is a good example of how IBM lost control of its sup-

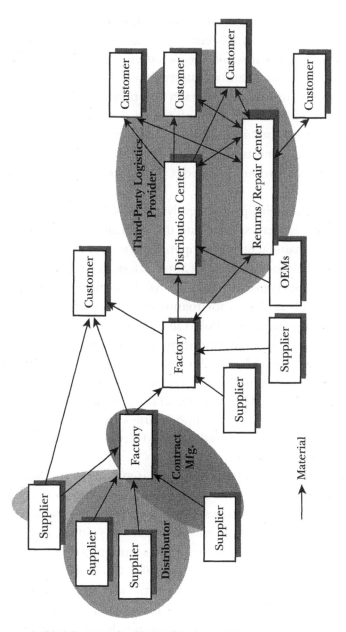

FIGURE 9.14: SUPPLY CHAIN MATERIAL FLOW

ply chain by not including such consideration in their strategy. Thus the number of entities in a supply chain is a function of how the entities are defined in terms of their scope (i.e., what exactly their role is). The number of feasible supply chain solutions may be large, and the economic and/or strategic implications of each solution may be different. Since a handoff may be a source of inefficiency (resulting from misaligned processes, inventory buildup, shortages, delivery delays, and nonconforming specifications), the trade-off between the inefficiencies of handoffs and the return on investment decides the optimal number of entities and their characteristics.

How much ownership in a supply chain is appropriate obviously depends upon many factors, including the criticality of specific capabilities, business focus, manufacturing competitiveness, and new business opportunities. Polaroid Corporation chose Sony as a strategic partner to automate the assembly of its cameras. However, as the criticality of manufacturing technology (such as design for manufacturability, application engineering, and tooling) increased, Polaroid decided to own these capabilities. They restructured Sony's partnership so that Sony essentially became a machine tool supplier. AT&T reduced its ownership of manufacturing and technology by spinning off Lucent Technologies so as to increase its focus on customer business. More recently AT&T has increased its ownership by acquiring TCI to pursue new opportunities in cable television. Lucent Technologies, similarly, acquired Ascend Corporation to pursue opportunities in Internet telephony.

In the telecommunications industry Cisco Systems and Nortel are moving aggressively to reduce ownership for manufacturing effectiveness. The cited reasons (Kaiser Associates 1998) are reducing production cost, accelerating time to market, accelerating time to volume, reducing capital investment, reducing fixed overhead costs, and improving inventory management. It is estimated that by the year 2002 Cisco will outsource 100% of an important component called circuit pack (CP), and Nortel will outsource 90% of CP. Cisco Systems has partnerships with Flextronics, Celestica, and Jabil for its manufacturing needs, and estimates that this arrangement slashes its manufacturing cost by about 30%.

In an extended enterprise with little ownership, various contractual terms among partners need to be worked out carefully. The objectives in such contracts are to minimize waste and encourage creativity by reducing inventories, reducing delays, and improving quality. Since entities that are not owned may not be in close geographical proximity, collaboration may turn out to be expensive and/or time-consuming. For example, involving suppliers and customers in product design may have to be facilitated over electronic or video networks. The other advantages from nonownership accrue through creative procurement such as vendor-managed inventory, real-time monitoring of partner performance, and evaluation of partners against certain targets.

These advantages are realized through economies of scale, the potential for better control (though not if the system becomes bureaucratic), and the availability of low-cost options with owned entities. For example, commonalities in components of products, if managed well, can provide manufacturing competitiveness. The option to be able to locate component and assembly plants at appropriate sites may be lost if the plants are not owned. Merck Pharmaceuticals, for example, has been able to leverage its ownership to increase its competitiveness in a big way.

Coordination

The major issues in the coordination of an extended enterprise are transparency of operations in the entities not owned, alignment of processes, common standards (specification data), information flow, transfer prices, and the interunit payment system. A traditional view of information flow is shown in Figure 9.15.

The typical information categories shown are forecasting, order execution, order engineering, order processing, and new product introduction. These are all related to transaction processing, and they would be of interest to all entities in the chain. A transaction web (Greis and Kasarda 1997) would thus be ideal in an extended enterprise (ownership minimized). In addition to the common information categories, each entity would have its unique information categories that are specific to its own needs. Thus the entity-specific information systems may not be synergistic with one another and with the needs of the extended enterprise. Conflicts may arise in terms of what is reported, how often it is reported, the units of measurement used, and when and how often updating takes place. The information required for coordinating the global enterprise in terms of due dates, inventories, production processes, scheduling, and costs may not be directly available in local systems of entities. They may not be deducible,

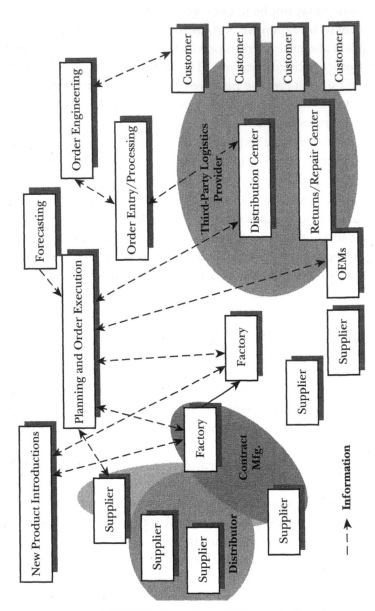

FIGURE 9.15: SUPPLY CHAIN INFORMATION FLOW

and this is one of the major problems. Second, the independent entities may not be willing to reveal their proprietary information to other partners. Separation of proprietary information from other information such as demand, which is required to minimize the bullwhip effect, can become expensive. Third, the entities may be using different application software, such as ERP and invoicing software. If this is the case, creation of a seamless interface among the entities becomes difficult. Fourth, the entities may be at a different levels of technological savviness—some may be at the level of using electronic data interchange (EDI), whereas others may be using intranets routinely as decision support tools. Fifth, coordination implies setting transfer prices that the entities can then use to plan their own operations. Transfer prices that would be fair to all may not be perceived as such by all. In any case, determination of such prices is by no means trivial. Sixth, coordination implies compensatory actions if targets are not met. There needs to be a clear understanding as to how to go about implementing decisions that may mean expediting, reducing costs, refocusing operations, procuring new technology, reducing inventory, and increasing agility. Finally, there needs to be an understanding about sharing best practices and other knowledge among the partners.

The flow of funds among partners is another infrastructural dimension. Because of multi-party transactions, the funds must also flow among multiple parties. Delays in cash transfers downstream may produce bullwhip-like effects in cash transactions upstream. This may become very complex if the float in payment terms (included in invoices) varies significantly from transaction to transaction. If funds are transferred electronically, appropriate security measures must be provided. The extended enterprise must also resolve the issue of accountability in the chain. For example, uncertainties in due date slippage upstream may produce another bullwhip effect downstream. This, coupled with value added downstream, may cause havoc with penalty payments. The issue is how to apportion the impact of such delays among all entities. The problem is especially thorny, as some of the downstream entities might have been involuntarily drawn into amplifying delays.

Vendor-managed inventory (VMI) is one way of sharing responsibility. VMI requires vendors to maintain a consignment of inventory at the customer site. The vendor monitors how the customer draws material out of the consignment and, based on this usage pattern, structures a replenishment strategy for the consignment. The size of the consignment is determined so as to provide the customer with a certain degree of protection from stock-outs.

9.7 SUPPLY CHAIN STRATEGIES

Corresponding to the product design strategies in Chapter 7, four possible strategies can be conceptualized in the context of a supply chain. These would be (1) supply chains for individual products, (2) supply chains for individual market segments, (3) supply chains for total customer solutions, and (4) supply chains for customer lock-in.

For individual product supply chains, the focus is on the economy of scale of the product and on the option to quickly disband the supply chain when the product is no longer viable. Saturn cars are a good example of such a supply chain. General Motors wanted to create a clear differentiation between Saturn and other models in terms of performance, use of technology, quality, and delivery. Saturn's car design was very different from that of other models, and the manufacturing practices (use of work teams, use of flexible manufacturing technology, and special supplier relationships) were all uniquely designed. The supply chain itself was very simple, as it involved a single plant, a number of well-coordinated component suppliers, and a number of dealerships with special incentives. The major decisions were where should the plant be located and which component suppliers should be chosen.

Supply chains for individual markets may be designed for a group of products. In the telecommunications industry, for example, companies specifically design supply chains for systems such as data-networks-systems (DNS). Such a supply chain is shown in Figure 9.16. Note that there could be several variations of the DNS product for customers in different countries. The circuit pack assembly and the final assembly could be located in any of several countries. The suppliers for a plant may be in the same country where the plant is located or in different countries. Thus there can be a number of variations to the supply chain shown in Figure 9.16, depending on the choice of plant locations, the choice of suppliers and their locations, and the choice of plants to supply customers in different countries.

Supply chains for a total customer solution is aimed not just at selling

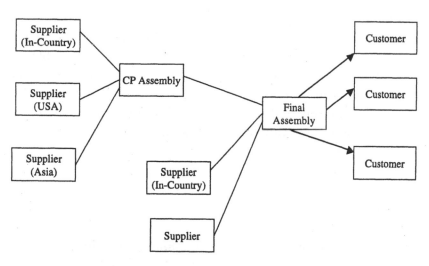

FIGURE 9.16: SUPPLY CHAIN FOR A MARKET

the product but at ensuring that customer maximizes profits from use of the product. This has at least two implications. First, it ensures that the cost of switching over from the customer's current product or system to the new product or system is minimized. Second, in scenarios where the company's product forms only a subset of components of the system used by the customer, the company would create a competitive advantage for itself by making sure that the other components (the ones the company does not make) do not hold up the customer's use of the system. To minimize switchover cost, the company's product may have to be modified and/or product installation would have to be smooth, without major downtime for the customers. Companies such as Lucent Technologies have created special functions called design for installation (DFI), similar to design for manufacturing (DFM). Installation may require many other special components, and these components (though inexpensive) may hold up installation by several days if the supply chain does not deliver them to the site at the right time and in the right quantities.

To ensure that the customer receives other components of the system that the company does not make, the company can create a consortium or partnerships with other component manufacturers or service providers. Cisco Systems, a manufacturer of Internet routers, recently created a partnership with KPMG, a management consulting company, to provide services in network management and other areas to customers who need those services to use Cisco's routers (*New York Times* 1999a). Lucent Technologies, on the other hand, has acquired International Network Services, a fast-growing consulting company, in a bid to move beyond merely making complex communications equipment and expand its ability to help its corporate clients install and use it (*New York Times* 1999b). The convergence of voice and data with Cisco's advantage in data communications and Lucent's advantage in voice communication is driving tremendous demand for services and support. The Cisco/KPMG deal is a typical example of an extended enterprise, for which issues such as seamless interfacing and transparency (discussed earlier) become critical in designing an appropriate supply chain.

Possible supply chains in the telecommunications industry are shown in Figure 9.17.

For example, if the customer requires data services, he/she can go through the local phone company, through a cellular phone company (not common yet), through the Internet, or directly via private lines. The data will be routed through packet switching and can use either the ATM (autonomous transfer mode) or the Internet protocols for transfer. Finally, the backbone support can be provided by a telephone network, wireless network, optical network, cable network, or private network. There are multiple companies with expertise in each type of service, such as access providers, switching/routers, protocols, and backbones. These companies

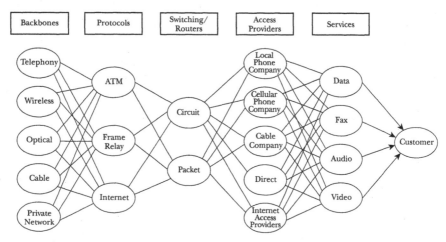

FIGURE 9.17: TELECOMMUNICATIONS INDUSTRY: SUPPLY CHAIN FOR A TOTAL CUSTOMER SOLUTION

tend to be of three types: hardware equipment (Lucent, Compaq, Cisco, Nortel, etc.), software providers (International Network Services and others), and network management and portals (AOL, Yahoo, MSN, etc.). It is clear that no single company can dominate all the services required in the total supply chain in Figure 9.17. Individual companies are therefore forming alliances, and then breaking them to form new alliances as new technologies emerge. If, for example, new technology emerges (as seems likely) that enables transmission of live conversation using packet switching (as is done with data transmission), the supply chains shown in Figure 9.17 will be modified rapidly, and new possibilities may emerge. Individual companies need to formulate clear strategies as to which part(s) of the supply chain are synergistic with their current and future capabilities. Such a road map would be based upon the dynamics among service providers and technology in terms of alliances, contractual provisions, acquisitions, and capabilities.

Supply chains that lock in customers and lock out competitors are based on products that establish a dominant design or on systems tailored for a group of customers that the customers find very hard to switch out of (because it would be expensive or entail a loss of learning). Microsoft's Windows and Intel's Pentium family of processors have together created such a dominant design. If a customer wants to access a majority of the applications currently available, he/she needs to buy the Windows operating system and run them on one of the Pentium processors.

To make it hard for customers to switch out, companies may increase their ownership of distribution channels and technical support capabilities. A supply chain that supports customized services can help to lock out competitors. Other ways of locking out competitors would be the establishment of a continuous stream of new products (which can create enormous barriers to entry) and the creation of product patents.

REFERENCES

Banerjee, A. (1986), "A Joint Economic Lot Size Model for Purchaser and Vendor," *Decision Sciences*, vol. 17, pp. 292–311.

Chakravarty, A., and G. Martin (1991), "Operational Economies of a Process Positioning Determinant," *Computers and O.R.*, vol. 18, no. 6, pp. 515–530.

Dada, M., and K. Srikanth (1987), "Pricing Policies for Quantity Discounts," *Management Science*, vol. 33, pp. 1247–1252.

Fine, C. (1998), *Clockspeed: Winning Industry Control in the Age of Temporary Advantage*, Perseus Books, Reading, Mass.

Fine, C., and D. Whitney (1996), "Is the Make-Buy Decision Process a Core Competence?" working paper, Center for Technology, Policy, and Industrial Development, Massachusetts Institute of Technology, Cambridge, Mass.

Fisher, M. (1997), "What Is the Right Supply Chain For Your Product?" *Harvard Business Review,* March–April, pp. 105–116.

Forrester, J. (1961), *Industrial Dynamics,* John Wiley and Sons, New York.

Greis, N., and J. Kasarda (1997), "Enterprise Logistics in the Information Era," *California Management Review,* spring, pp. 55–77.

Hayes, R., S. Wheelwright, and K. Clark, (1984), *Restoring Our Competitive Edge: Competing Through Manufacturing,* The Free Press, New York.

Heyman, D., and M. Sobel (1984), *Stochastic Models in Operations Research,* vol. 1, McGraw-Hill, New York.

Kahn, J. (1987), "Inventories and the Volatility of Production," *American Economic Review,* vol. 77, no. 4, pp. 667–679.

Kaiser Associates (1998), "Telecommunication Competitiveness Report," Kaiser Associates, Vienna, Va.

Lal, R., and R. Staelin (1984), "An Approach for Developing an Optimal Discount Pricing Policy," *Management Science,* vol. 30, pp. 1524–1539.

Lee, H., P. Padmanabhan, and S. Whang (1997a), "The Bullwhip Effect in Supply Chains," *Sloan Management Review,* spring, pp. 93–102.

Lee, H., P. Padmanabhan, and S. Whang (1997b), "Information Distortion in a Supply Chain: The Bullwhip Effect," *Management Science,* vol. 43, no. 4, pp. 546–558.

Lee, H., and M. Rosenblatt (1986), "A Generalized Quantity Discount Pricing Model to Increase Supplier's Profits," *Management Science,* vol. 32, pp. 1177–1185.

Magretta, J. (1998), "The Power of Virtual Integration: An Interview with Dell Computer's Michael Dell," *Harvard Business Review,* March–April, pp. 73–84.

Metters, R. (1997), "Quantifying the Bullwhip Effect in Supply Chains," *Journal of Operations Management,* vol. 15, pp. 89–100.

New York Times (1999a), "Cisco Systems to Invest $1 Billion in the Accounting Giant KPMG," August 8.

New York Times (1999b), "Lucent to Buy Consulting Company," August 11.

Parker, G. (1998), "Building Supplier Capability Through Project Choice," working paper, Sloan School of Management, Massachusetts Institute of Technology, Cambridge, Mass.

Sterman, J. (1989), "Modeling Managerial Behavior: Misperceptions of Feedback in a Dynamic Decision Making Experiment," *Management Science,* vol. 35, no. 3, pp. 321–339.

Tyndall, G., et al. (1998), *Supercharging Supply Chains,* John Wiley and Sons, New York.

10

Electronic Chains of Suppliers and Customers

10.1 DIMENSIONS OF DIFFERENTIATION

We have seen in Chapter 9 that integral and modular supply chain architectures have different strengths. While an integral architecture has strengths in terms of manufacturing synergy and process alignment, a modular architecture would be desirable if investment in ownership and supply chain restructuring are major issues. The ideal architecture would therefore be a virtual integration of a modular supply chain. Though the concept is not totally new, it has not been possible to realize it mainly because of the high cost of creating seamless interfaces between partners. Internet technology—with its offshoots, intranets and extranets, and supported by other technologies such as EDI and groupware—has now become the engine of such virtual integration. In Chapter 9 we briefly described ownership, process alignment, and electronic proximity in the context of an extended supply chain. To this we may add as a fourth dimension adaptability, which possesses interesting interactions with other three dimensions. Our main objective in this chapter is to understand how digital connectivity or electronic proximity may alter the nature of interactions within and between enterprises.

It is becoming clear that digital connectivity, far from being a passive technology that merely enables fast and reliable data transfer, has major strategic implications for the entire value chain. These come from being able to redefine values in a value chain by adding or deleting entities, and from the law of *increasing* returns obtained from connectivity (economic externalities). Ownership provides competitive advantage in terms of control and economies of scale, and process alignment reduces inefficiencies due to handoffs. Adaptability, on the other hand, provides the capability for restructuring the enterprise in an altered competitive environment.

There are obviously interactions among the dimensions of ownership, process alignment, and adaptability in a digitally connected supply chain. There are also many unanswered questions: What are the dominant enterprise structures in this context? Are virtual enterprises viable options? How does digital connectivity add value? What is this new value, and what are the cost-benefit implications? What are the new business models, and how do they enhance competitiveness? How have companies reengineered their value chains? How does one engineer collaboration in a digital business community? In the rest of this chapter we address these issues with examples drawn from industries, government, and researchers.

10.2 DIGITAL CONNECTIVITY

To understand the interaction among the four dimensions of ownership, process alignment, digital connectivity, and adaptability, we would require a four-dimensional representation. Since that is not possible, we discuss interactions between two dimensions at a time. We consider three dominant scenarios: traditional, high digital connectivity, and aligned processes (with high digital connectivity).

The interaction between ownership and process alignment in a traditional scenario is shown in Figure 10.1.

It is clear that if the business processes are not aligned, the entities in a value chain can be autonomous, whether owned or not. A low level of

Ownership		Low	High
	Low	Shopping mall	?
	Medium	Outsourcing (contractual)	Japanese *keiretsu*
	High	Decentralized business units	Centralized (hierarchical) control

Process Alignment

FIGURE 10.1: TRADITIONAL ENTERPRISE ARCHITECTURE

process alignment also indicates a lack of collaboration among the value chain entities. As in Figure 10.1, a shopping mall that brings different companies (even competitors) under a common roof represents low ownership by individual companies and almost negligible process alignment among them.

At the other extreme, most of the value chain can be owned by a single company. This does not imply, however, that decision making would be centralized. Lucent Technologies has been a conglomerate (until the recent spinoffs) supporting several autonomous business units. The business units have been targeted to different market segments of the telecommunications industry: switching and access products, optical networking products, microelectronics, business communication products (telephones), and data communication products. Although the business units are fully owned by Lucent, they have operated together in the past almost like stores in a mall. General Motors, on the other hand, has been a more hierarchically integrated company, where decision making has been the responsibility of regional managers in North America, Europe, and Asia. Several aspects of creating efficient interfaces in a decentralized firm have been discussed in Chapter 2.

Outsourcing is a way of reducing ownership in a value chain. In Chapter 9, we saw how outsourcing has implications for knowledge development and production capacity. In a firm that outsources, process alignment will be low if production economics is the driving force for outsourcing. In a Japanese *keiretsu* such as Toyota's, on the other hand, the company takes pains to develop long-term relationships with its supply partners.

The early form of digital connectivity, using EDI, has been around for some time. Senders and receivers (of data) are required to use specific equipment connected to phone lines, which transmit data in a special format (i.e., data need to be converted to an agreed-upon standard format before transmission and converted back to the company-desired format at the other end). Equipment specificity and special data transfer protocols have been limiting factors in the wider use of EDI. Once the connections are made, however, data can be transmitted at a very high speed, and transmission is very secure.

The Internet, which enables instant connectivity between businesses, governments, and individuals, made its debut in the mid-1990s. It is like another industrial revolution, as it has enabled changes in the business models that companies use to organize themselves, to engage in relationships, and to conduct their most basic transactions with customers and suppliers. A recent survey by Booz, Allen, and Hamilton (1999) has discovered several trends in the way digital connectivity is reshaping the economic landscape. The major findings of the survey were that (1) it is possible to re-create the value chain in real time, often by cutting out the intermediaries; (2) the balance of power has shifted toward customers; (3) the pace of business is becoming increasingly faster; (4) functional boundaries, inside and outside the enterprise, are dissipating rapidly; and (5) knowledge is becoming a key competitive asset.

Callahan and Pasternack (1999) report that important changes in business models related to suppliers, customers, and intermediaries have already taken place. Electronic payments and invoices are replacing the mix of telephone and fax, and the extended enterprise is replacing stand-alone entities; real-time information exchange is replacing periodic (manual or computer) information exchange; customers are directly accessing manufacturers; and customers have electronic access to product information and vendors' service capabilities.

The major changes in supply chain architecture in a digitally connected environment are shown in Figure 10.2. Observe that the shopping mall model (Figure 10.1) will now be more like an electronic marketplace. Anyone can be a buyer or seller, but not many transactions are based on mutual trust. The supply network linking producers and customers is shown in Figure 10.3. Note that customers 2 and 3 can buy everything they need from producer 1, including those items made by producer 2. But they can also buy them directly from producer 2 if they wish. Thus the choices of how to buy and sell in such an open electronic market are abun-

Ownership		Low	High
	Low	Electronic marketplace (stock market)	Alliance
	Medium	Extended enterprise, modular (horizontal)	Extended enterprise, integral (vertical)
	High	Decentralized business units	Centralized enterprise

Process Alignment

FIGURE 10.2: DIGITALLY CONNECTED ENTERPRISE ARCHITECTURE

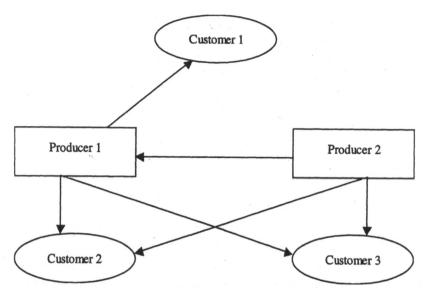

FIGURE 10.3: ELECTRONIC MARKETPLACE

dant. The company called eBay (formed in 1995) is an example of an Internet-based electronic marketplace that facilitates online auctions and trade. Electrical utilities have set up an electronic market called OASIS for trading transmission capacity. Some companies (e.g., IBM) have set up Web sites to auction not just inventory but also future production capacity.

With high ownership in a value chain, the processes of a business unit can be made transparent to other units whether or not the units are managed centrally (Malone, Yates, and Benjamin 1987). In a decentralized structure, as we have seen in Chapter 1, each business unit can use its own tacit (local) information and access the social (global) information of other units to optimize its own processes. Individual business units must be interfaced seamlessly with one another for this to happen. In a centralized system, ERP built in conjunction with EDI or an intranet can be used for resource optimization. Ownership of resources can be reduced by permitting some business units to be independent. This results in an extended enterprise architecture, shown in Figure 10.2 and discussed at length in Chapter 9. With a low process alignment the architecture of the extended enterprise is modular (or horizontal), and with a high process alignment it is integral (or vertical).

10.3 NEW BUSINESS MODELS

It is clear that with digital connectivity (i.e., electronic proximity), a modular architecture can become very competitive. Usually the company that has substantially more ownership than others acts as the leader of the extended enterprise. Ticoll, Lowy, and Kalakota (1998) term such a leader an aggregator, as shown in Figure 10.4. Note that unlike the open electronic marketplace (Figure 10.3), customers in a modular business model may place orders only through the aggregator.

The E*Trade extended enterprise has been created to channel other companies' products (mutual funds, insurance, securities) to end users. The extended enterprise also aggregates a variety of knowledge-based services including share prices, news feeds, and research reports. E*Trade's partners in the extended enterprise are mutual fund families, stock exchanges, investment advisors, brokers, and banks. They contribute to the extended enterprise in terms of widening the range of products offered by E*Trade, increasing revenues through listing partners on their own Web sites, strengthening the marketing channel, and increasing revenues from product offerings. E*Trade acts as the leader and aggregator. It has farmed out marketing and value-added financial services to independent partners. It acts as a back office offering one-stop shopping, real-time customer access, online trading, and access to external research. Wal-Mart is another company that positions itself between producers and customers and acts as an aggregator.

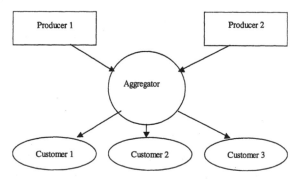

FIGURE 10.4: MODULAR BUSINESS ARCHITECTURE MODEL

High business process alignment in an extended enterprise is motivated by operational efficiency. A network of suppliers is shown in Figure 10.5. Major examples of companies that have successfully leveraged process alignment are Dell Computer, Cisco's single-enterprise business model, and the reengineered Chrysler Corporation (now Daimler-Chrysler). The Chrysler division has substantially increased process alignment in its value chain by integrating its suppliers' processes into its product design, production, and logistic processes. The suppliers of car interiors conduct their own market research and are responsible for integrated designs. Thus the Chrysler division has started to move away from a traditional outsourcing structure and toward an extended enterprise with high process alignment.

Dell Computer has developed and implemented an extended enterprise with high process alignment. It has stitched together a business with

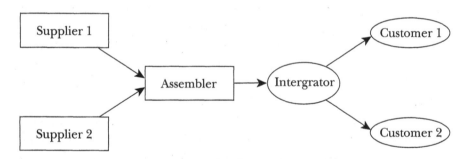

FIGURE 10.5: EXTENDED ENTERPRISE–HIGH PROCESS ALIGNMENT

partners who are treated as if they are inside the company (Magretta 1998a). Information, including database access, is shared in real time. Suppliers assign their engineers to Dell's design teams. If a corporate customer has a problem, Dell stops shipping the product while design flaws are fixed in real time. Third-party shipping is coordinated with customer orders. Logistics companies such as UPS pick up appropriate numbers of computers from Dell's facility in Texas and monitors from Sony's factory in Mexico, match the computers with the monitors, and deliver them to customers.

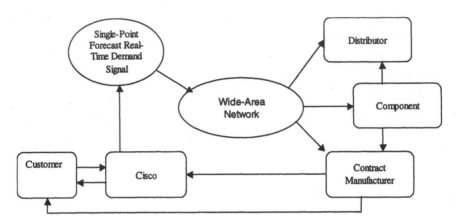

FIGURE 10.6: CISCO'S DYNAMIC REPLENISHMENT MODEL

Cisco Systems' dynamic replenishment model is perhaps the best example of an extended enterprise with high process alignment. As shown in Figure 10.6, customers place orders with Cisco, and the orders are then transmitted to Cisco's partners in the extended enterprise through a wide area network (WAN). The customer can be supplied by Cisco and/or by one or more of Cisco's partners directly. In building such an enterprise, Cisco had to open up its internal systems and information to customers, prospective partners, suppliers, and employees. Three basic assumptions at the core are (1) the relationship with key constituents is as much of a competitive differentiator as core products and services, (2) a critical element in a relationship is the way a company shares its information and systems, and (3) connectivity by itself is not adequate; there must exist supporting business processes in the network.

To obtain a high level of process alignment with minimum ownership, groups of companies in different industries have begun to form alliances. The manufacturing assembly pilot (MAP) in the auto industry and the Java alliance in the software industry are two examples. Three major competitors together with their suppliers (a total of 16 companies) formed the MAP alliance (Hoy 1999, Kalogeridis 2000). The objective is to improve the quality of information flowing through the supply chain and to rapidly move material through the tiers, as shown in Figure 10.7. Order lead time from the lowest tier up to the top tier was reduced from 28 days to just 11 days. EDI is the primary method of communication in this alliance. In contrast, the Java alliance was put together by four key companies: Sun, Microsystems Netscape (now owned by AOL), Oracle, and IBM; its primary

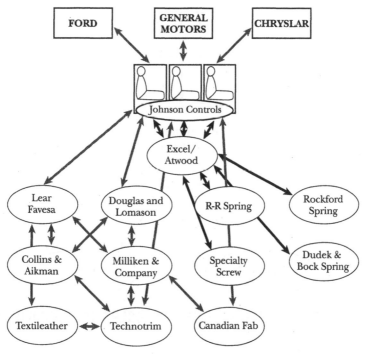

FIGURE 10.7: MAP ALLIANCE

focus is to promote Java software and its open computing architecture. The other participants in this business community include Intel, Kleiner-Perkins, Symantec, Microsoft, Corel, Andromedia, banks, and financial firms. Sun contributes leadership to the alliance, and other participants contribute in terms of hardware, venture capital, application development tools, market segments, and Internet browser tools. Collaboration among the participants is required to make Java usable. There is a growing market in trading Java applets and objects.

Ticol, Lowy, and Kalakota (1998) point out that for such alliances to succeed, issues such as governance, rules of collaboration, and rule compliance must be addressed. Alliance governance must find ways to motivate and retain alliance partners, monitor brand image and market profile, and improve the effectiveness of business processes, standardization, and knowledge creation. Rule making must include stakeholder interests, degree of control in the alliance, and rule-making processes. Rule compliance involves penalties/incentives for noncompliance/compliance, identification procedures for noncompliance, compliance enforcement protocols, competitive implication of rule enforcement, legal issues

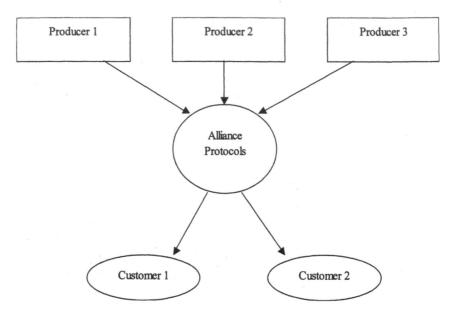

FIGURE 10.8: ALLIANCE OF PRODUCERS AND SUPPLIERS

related to enforcement, and the question of whether compliance would be voluntary. A possible supply network for an alliance is shown in Figure 10.8.

Agility in a Digital Enterprise

For agility, the norms in an extended enterprise and the rules in an alliance must evolve in time with technology and new business models. The interactions between the dimensions of ownership and agility are shown in Figure 10.9. Agility can be incorporated by permitting an alliance member

	Low	High
Low	Alliance	Multiple Overlapping Alliances
Medium	Extended Enterprise	Virtual Enterprise
High	Centralized	?

Ownership (vertical axis label), Agility (horizontal axis label, Low to High)

FIGURE 10.9: AGILITY IN A DIGITAL ENTERPRISE

to have membership in multiple alliances, with well-defined rules for each alliance. This would enable a company to increase or decrease its involvement in any alliance as needed. Alliances need not be permanent and may restructure themselves as needed. Such an architecture, called virtual enterprise (VE) or relationship enterprise, is a temporary network of independent companies put together rapidly to exploit fast-evolving business opportunities. Individual companies often find that they (or their partners in a current extended enterprise) do not have all the skills and competencies necessary to deal with the entire range of customers' emerging needs. An extended enterprise may add a nonalliance company that has the needed expertise, and/or terminate relationships with a current alliance company. By such restructuring, it becomes possible to satisfy new demands. A virtual enterprise can be set up for a specific combination of products and customers, and it can be dissolved smoothly when the market for that product or service declines. A VE is project-based, and so it works best in niche markets; it has a short life cycle. The information exchange in a VE is obviously far more complex than that in a traditional extended enterprise.

10.4 DIGITAL VALUE

It is generally recognized that business value is rapidly being created by digital connectivity (Margherio 1998). These values, however, are not all of the same kind, and have different strategic significance. Business values can be created in terms of reducing cost of purchase, selling more to a larger set of customers, increasing customer satisfaction by delivering new business services, reducing inventories in supply chains, and decreasing time to market. Companies need to be aware how digital connectivity should be managed in order to produce a certain kind of value: on-time delivery, low cost, or some other metric. Companies such as Dell, Cisco, and GE are so committed to creating value in this way that they have put a large portion of their ordering and customer service operation on the Internet.

Buying

The buying process can be complex, involving several subprocesses. First, the customer must identify a set of suppliers who can make the item. He/she must then select one of them based on price, quality, and other performance factors. The third step involves transmission of engineering specifications and detail drawings to the vendor, who makes product samples for customer approval. The customer then transmits a purchase order to the vendor for a specific quantity. When the supplier ships the product, he/she includes an invoice for payment. The customer verifies the invoice against the purchase order before making payment. The entire sequence of activities may take several weeks in certain industries.

At GE the requisitioning process itself (up to the point of issuing purchase orders) took more than seven working days before they converted it to a digital system (Fabias 1997). The Trading Process Network (TPN 1999), now employed by GE, requires its factories to transmit requisitions electronically. TPN automatically pulls the correct drawings and attaches them to electronic requisition forms. The bid package is sent to suppliers around the world using the Internet. Suppliers are also notified of incoming requisition packets by e-mail or EDI. GE can evaluate a supplier's bid in a single day. The procurement process, as shown in Figure 10.10, in-

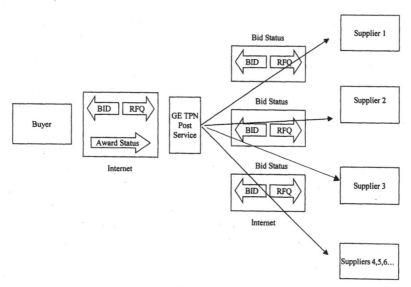

FIGURE 10.10: TRADING PROCESS NETWORK

volves the following steps. The buyer prepares an RFQ (requisition for quotation) on a PC, selects a list of potential suppliers, and posts it on TPN Post Service. TPN Post Service invites selected suppliers to bid. Suppliers retrieve relevant information from TPN Post Service, prepare their bids, and post the bids (not visible to other bidders). The buyer then downloads and analyzes the bids and selects the most desirable suppliers.

GE reports that it has reduced procurement-related labor cost by 30%, and 60% of procurement personnel have been redeployed. Material costs have declined by 20% as suppliers have become more competitive. GE used TPN for more than $1 billion worth of procurement in 1997, and it expected its TPN procurement to exceed $6 billion by the year 2002. It predicts that it will save about $700 million per year through e-procurement.

Implementation of a TPN-type procurement system takes about two to three months and requires all supply partners to possess EDI capability. It costs approximately $1 million in the first year, plus a transaction cost based on business volume. The whole system is based on the Oracle database architecture. Successful e-procurement applications have resulted in the following performance improvements:

Material cost/sales revenue ratio	decreases
Global effective tax rate	decreases
Inventory-to-sales-revenue ratio	decreases
Payables-to-sales-revenue ratio	increases
Gross margin	increases
Return on sales	increases
Cash-to-cash cycle time	decreases
Cash-flow-to-sales-revenue ratio	increases
Provisioning lead time (sourcing, making, delivery)	decreases
Upside flexibility	increases
Material availability to customer	increases

Selling

Digital connectivity has generated tremendous value for the online seller. According to one estimate, the online business-to-consumer market will account for $26 billion per year in revenue by 2002. Most other estimates put this number between $7 billion and $100 billion, depending on the categories of goods included. Virtual stores report lower operating costs than their physical counterparts. Such stores are service-oriented (providing advertising, site maintenance, accounting, liaison with suppliers and customer relationship management) and may not even take possession of the goods. The Cendant Corporation reports that it made a profit of $10 per customer in its first year using the Internet as compared to a loss of $9 per customer using telesales. In the second year it made a profit of $40 using the Internet against a profit of $30 using telesales. In 1997 Cendant sold more than $1.2 billion worth of goods and services over the Internet.

The most common way of online selling is to set up a virtual store (Web page) on the Internet. Electronic catalogs replace store shelves. The catalog usually includes a photograph of the product with a product description, pricing, and product size information. Third-party reviews may be added to assist buyers in their choice decision. A shopping cart icon is added for customers to add their selection, and credit card information is obtained to complete a sale.

Electronic stores can also keep track of what an individual customer purchases. The purchasing pattern is analyzed, using an intelligence agent, to reveal what the customer is likely to buy in the future. Amazon.com informs a customer by e-mail when a particular new book matches the customer's purchase pattern, and sends him/her reviews of that book.

Rocks (1999) details how value is created for Office Depot and its customers through selling on the Internet. Office Depot's cost of processing an order is cut from $2 to $1. It wins new customers and retains them longer. A typical customer cuts the cost of issuing a purchase order from $175 to $75. Phone calls to and from purchasing departments are reduced by 60%. Customers can also get a peek into Office Depot's inventory and reduce waiting time for goods. The customer's inventory is cut, as stocks can be replenished quickly using the Internet.

Ford Motors and Microsoft announced an online auto ordering system that allows customers to "build" their own cars online (Lohr 1999). This system has the capability of tracking orders all the way to dealers and Ford's own factories, so that the supply chain can be finely coordinated with customer orders.

The technical capabilities required for setting up an electronic store are:

- Graphical Web interface creation
- Scalable database to accommodate growing business
- Catalog creation

- Catalog indexing and search
- Facilities for quickly changing product descriptions, business processes, and prices
- Transaction capabilities, including credit card authorization
- Facilities for cross selling and banner ads
- Links to corporate data sources
- Links to EDI, ERP, and other middleware

Inventories

As we have seen in Chapter 1, if problems are communicated as they arise and the appropriate actions are taken, inventories in the system can be reduced significantly. If demand suddenly rises or if one factory cannot meet its production schedule, the company must become aware of it in time to increase production in another factory. For example, IBM has implemented its Advanced Planning System (APS). It saw its inventory turns increase by 40 percent and sales volume rise by 30 percent in the first year of its implementation. This has resulted in lower investment and inventory costs and has produced a saving of $500 million per year for IBM. Manufacturers, wholesalers, and retailers are working together to create standards and guidelines for better forecasting and ordering. The venture is known as collaborative planning, forecasting, and replenishment (CPFR). These standards would determine how to forecast future demand for products and how to share information on product availability.

Retailers and suppliers can post forecasts for their latest sets of products using CPFR (VICS 1998). Software compares the forecasts generated by collaborating companies and flags differences in those that exceed a normal safety margin, which are then reconciled between the parties concerned. According to one estimate by Ernst and Young, CPFR can reduce inventories by up to $350 billion in the U.S. economy.

Depending upon the business model used, digital connectivity has created opportunities for the manufacturer's inventory cost to be absorbed by the distributor or the retailer. A distributor, because of the nature of its business, may be better able to control the cost of carrying inventory. Thus a distributor's ownership of a manufacturer's inventory may benefit the whole system by lowering carrying costs. This scenario may also create the right incentives for the distributor to keep its inventories low. On the supplier's side, vendor-managed inventory (VMI) gives ownership of the inbound inventory to the supplier. We shall discuss this in detail later in this chapter.

Time to Market

Digital connectivity with suppliers and customers has enabled companies to transmit and receive purchase orders, invoices, and shipping notifications with short lead times. Product blueprints and specifications are also being shared through extranets or virtual private networks (VPN). This cuts product design and development time drastically. Design teams of manufacturers and suppliers can now collaborate using the Internet or extranets.

The auto industry has taken a leading role in the use of electronic connectivity for product design and development. As we saw in Chapter 8, determination of customer-desired attributes is an interactive process, and the Internet could be an ideal vehicle to obtain customer feedback on product prototypes. Both Ford Motors and Chrysler Corporation (before its merger with Daimler-Benz) have made great strides in linking groups of designers, engineers, suppliers, and manufacturers. Digital connectivity has reduced task times from weeks and months to just few days. Electronic sharing of information allows the team members to work on development projects simultaneously. Transfer of CAD and CAM files eliminates the need to physically build prototypes of each product group. Changes to the components can be made without building sample tooling and parts.

Auto manufacturers and their major suppliers can communicate production scheduling information using EDI, groupware, or extranets. The assembly plants download their weekly build plans to the suppliers. The breakdown of the number of components required at different plants is also communicated electronically. When the components are ready and loaded for transportation, the supplier notifies the plants of the components' delivery dates. The plants may then adjust their production schedules accordingly. Timely information from suppliers has helped auto manufacturers to increase their inventory turns from 10 to 130 times per year.

The Automotive Network Exchange (ANX) (Hoy 1999; Kalogeridis 2000) is a managed virtual private network that runs on the Internet and links manufacturers and suppliers worldwide. ANX has replaced all direct connections, fax, and mail with a single network. This has significantly lowered transmission costs for the parties in the network. ANX is expected to be the link between the big three automakers and the new auto exchange called Covisint. ANX has the potential to substantially reduce product development time and manufacturing cycle time.

Customer Service

Uncertainty about the arrival of a purchase can be very frustrating to a customer. Telephone inquiries to a supplier may result in a series of transfers from one department to another. This consumes time and money for the customer as well as the seller. Logistics providers are helping their business partners solve such problems using the Internet. A customer can access a company's Web page, enter the order number, and find out whether the product can be expected to arrive by a certain date. This whole exercise may take only a few minutes of the customer's time, and none of the seller's.

Customer service can also be improved by putting order tracking and technical support online. Cisco Systems estimates that it has improved the productivity of its *customers* by as much as 300 percent, resulting in a saving exceeding $125 million in its own customer service cost.

10.5 REENGINEERING OWNERSHIP IN THE VALUE CHAIN

In Chapter 9 we briefly discussed the notion of consolidation in a value chain with respect to the PC industry. That is, an enterprise must determine the set of the activities it is most competent in and would like to undertake in-house. It was also explained what roles manufacturing capacity and business knowledge play in establishing the boundaries of an enterprise. While component criticality and intellectual property are long-term strategic issues, short-term issues such as provisioning also require a company to consider a dynamic reengineering of its value chain.

Cisco Systems has leveraged the dynamics in its value chain by reengineering it in three different ways: supplier assembly, channel assembly, and in-house assembly (Cisco Systems 1998). In the supplier assembly model, shown in Figure 10.11, all subassemblies and the final assembly are completed by suppliers, some of whom are contract manufacturers. After final assembly is complete, the product is shipped either to a distributor or directly to a customer. Cisco never takes possession of the product and so does not incur any costs related to inventory, manufacture, or distribution. Thus Cisco's primary contribution to the value chain is the design of product. Cisco is also responsible for the selection of contract manufacturers, some raw material suppliers, and distributors. Cisco employs this business model for products that are simple in terms of the engineering knowledge and manufacturing know-how involved.

For more complex products, such as interface cards and memory devices for network routers, Cisco uses the channel assembly model (Figure 10.12). While the contract manufacturer carries out all subassembly work, final assembly becomes the responsibility of the distributor, as the product now must be configured according to actual customer demand. Since the distributor is in close contact with the customer, it is expected to possess better knowledge of the customer's preferences and delivery lead times. This model is similar to the idea of delayed product differentiation. As in the previous model, Cisco does not take possession of the product. For extremely complex products, Cisco uses the in-house assembly model (Figure 10.13). These products include ATM switches, complex routers, and new products in the development stage. Contract manufacturers finish all subassemblies, Cisco completes the final assembly, and distributors deliver the products to the customer. Here Cisco must take possession of products and incur final assembly and inventory costs on them.

In Cisco's single-enterprise system, all suppliers, contract manufacturers, distributors, value-added resellers (VARs), and customers, share information via a common information backbone designed for the extended enterprise. Demand forecasts generated at any point in the extended enterprise are made visible to all entities in the value chain. Suppliers such as Jabil receive 12-month demand forecasts, which are updated every week. A wide area network (WAN) that uses software from Oracle, I2, and Manugistics forms the information backbone. An information backbone typically costs $100 million to build and requires an operating cost of $10 million per year. It has saved Cisco $500 million in two years, and has reduced the order interval (order-to-delivery cycle) by 2.5 days on average.

In the direct-business model of Dell Computer, the emphasis is on re-

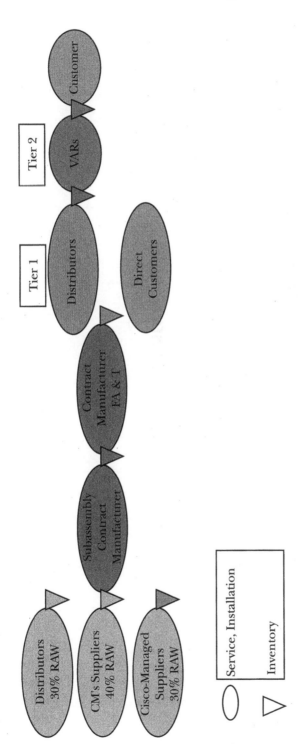

FIGURE 10.11: SUPPLIER ASSEMBLY

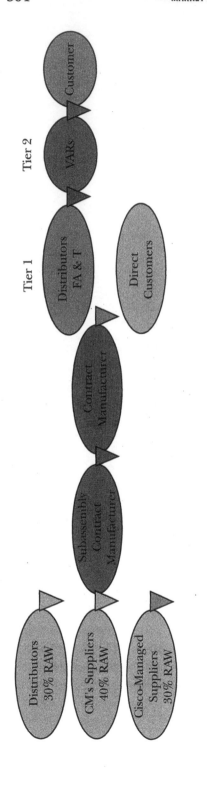

FIGURE 10.12: CHANNEL ASSEMBLY MODEL

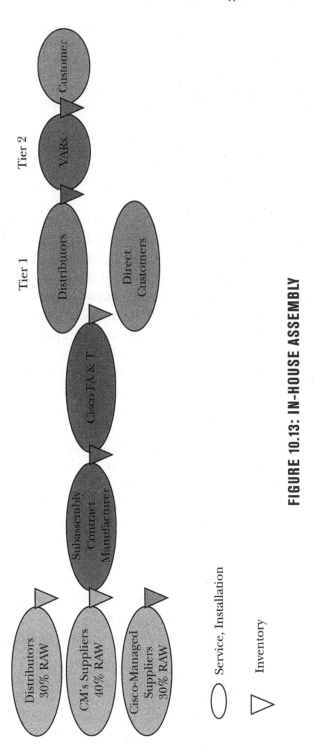

FIGURE 10.13: IN-HOUSE ASSEMBLY

ducing the cash-to-cash cycle (CCC), which is the interval between the time when Dell pays its suppliers and the time when it is paid by its customer. In Figure 10.14, note the "shipper merge in transit" box, implying that the final configuration of customer order (such as matching the PC and the monitor with customer order) is performed by shippers such as UPS. FedEx recently announced that it would reengineer itself to provide a more elaborate merge-in-transit service for Cisco Systems (Blackman 1999). As shown in Figure 10.15, Dell has cut its assembly time of a desktop PC to five hours. These innovations let Dell minimize its CCC. The relationship among the factors defining CCC is shown in Figure 10.16.

In the garment industry, Li and Fung of Hong Kong (Magretta 1998b) has found a very effective way of quickly dissecting the value chain of a product and allocating different chunks of the chain to globally dispersed companies. Thus it is possible that a company not owned by Li and Fung could be participating in the value chains of several products controlled by Li and Fung. Li and Fung retains direct control on the front-end processes (design, engineering, and production planning) and back-end processes (quality control, testing, and logistics) of the value chain. In the garment industry, the front- and back-end processes tend to be very knowledge-intensive. Li and Fung also coordinates all handoffs among participating companies and acts as a reservoir of technical and business knowledge for all partners. It allocates the middle stages of the value chain, which tend to be more labor-intensive (raw material and component sourcing, and managing production operations), to partners. All business partners are responsible for negotiating business terms individually.

Reengineering the value chain also entails reengineering the interfaces among the resulting entities. In Chapter 2 we discussed the issues in managing the interface between two domains within a single traditional enterprise. The central point was that each domain would attempt to maximize the total system value, deriving solutions from its own vantage point. The domains would bid against each other to have their solutions adopted. The decision makers at the corporate level had the authority to accept or reject individual bids and get the accepted solution implemented.

In a modular supply chain, however, there may not exist a dominant central authority. There is no direct way of enforcing the implementation of a solution if the entities do not belong to a single enterprise. It must be done through cooperation, which is not easy to achieve within a value

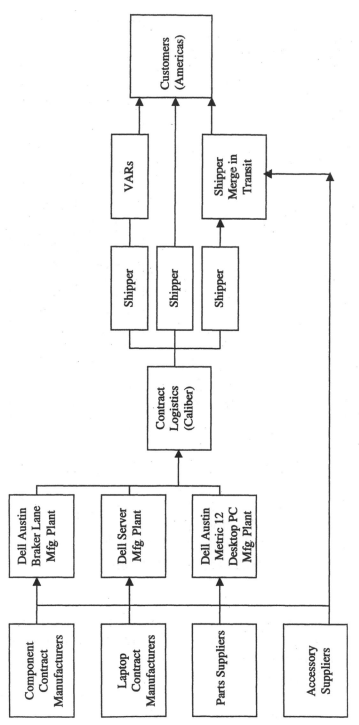

FIGURE 10.14: DELL'S DIRECT BUSINESS MODEL

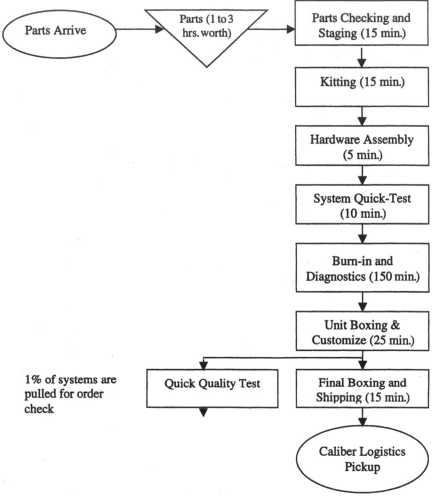

Total manufacturing time: 5 hr. avg.
Time from order received to loaded on truck: 36 hrs.

FIGURE 10.15: MANUFACTURING AT DELL

chain. A lead company may orchestrate the best solution for the extended enterprise, but this is not usually possible. There must therefore be agreement on the principles of just behavior and on the procedures that enable all parties to verify compliance by others.

The principles for regulating behavior will, obviously, vary with the nature of the products and business processes, but at its core there must exist

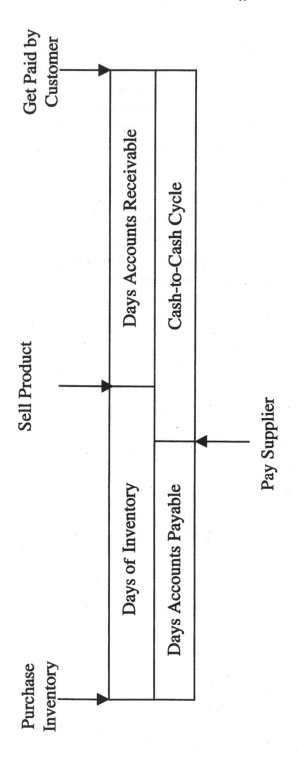

Cash-to-Cash Cycle = Days of Inventory + Days Accounts Receivable – Days Accounts Payable

FIGURE 10.16: CASH-TO-CASH CYCLE

a general agreement on a few specific metrics. Such metrics in a supply chain would include inventory turns, cycle time, process costs, product availability on retail shelves, fill rates, demand accuracy, and profitability. It would also include categories of information that would be shared, such as data for measuring success against targets, data for identifying exceptions in sales and orders, data for supporting decisions such as sales promotions, point-of-sale data, historical shipment information, and inventory status. Methods of data sharing, such as EDI or groupware, must also be stipulated.

Companies must agree on performance requirements for individual activities, verification procedures, and reward assessments. Nissan, for example, has created an interface agreement with all its suppliers. It has defined clear roles for its suppliers, specified joint examination of the ways costs could be reduced, and outlined a commitment to help improve the processes of suppliers that did not measure up. To enable verification, Nissan requires all business processes to be transparent. Access to each other's databases and joint audits are some of the ways of providing transparency.

Cost and Value Implications

Consider a value chain linking three entities, a manufacturer, a distributor, and customers. We analyze four different business models for this value chain. The flows of material corresponding to the four business models are shown in Figure 10.17. Business model A is the traditional one, where all sales to customers take place through a distributor. In model B, the distributor generates customer orders for the manufacturer, who supplies the customers directly. That is, the distributor does not take possession of goods at any point. In model C, as in B, the distributor generates orders; customers can be supplied by both the distributor and the manufacturer. Finally, in model D, the distributor is totally eliminated from the supply chain.

Model A

This is the traditional model wherein a manufacturer's products are sold to customers through distributors. The implications for flows of information, funds, and material are examined below.

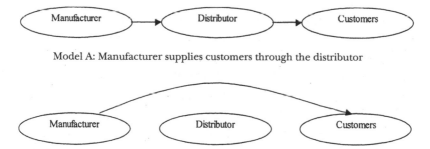

Model A: Manufacturer supplies customers through the distributor

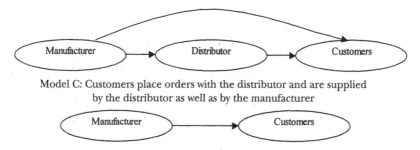

Model B: Manufacturer supplies directly to customers, who place orders with the distributor

Model C: Customers place orders with the distributor and are supplied
by the distributor as well as by the manufacturer

Model D: Customers place orders with the manufacturer and are supplied by the manufacturer

FIGURE 10.17: MATERIAL FLOWS IN A VALUE CHAIN

Information Flow

A distributor generates sales orders from prospective customers and relays them to the manufacturer, either after aggregating and batching them by item category or directly, so that the manufacturer may aggregate them at the end of the day. Note that there are different order-processing cost implications in these two modes of information flow. In the first mode the distributor incurs a higher cost of order processing, but it can withhold customer-specific information from the manufacturer. In the second mode, as the orders are being batched, variance of demand experienced by the manufacturer would increase.

Payment

The distributor can take possession of goods and pay for them either when the goods arrive at the distributor's site or when they are at the manufacturer's site. If the distributor takes possession at the manufacturer's site, it would pay a lower unit price for the products and can have them shipped directly to the customer's site, saving the warehousing and handling cost at its own site. It would, however, have to incur the additional cost of logistics.

Sales

The distributor may increase sales revenue by increasing its marketing budget or by reducing the unit price. Both measures may increase sales (flow of goods) in the channel. The manufacturer would gain more if its margin remained fixed, but the distributor would expect the manufacturer to share its gain with it. The manufacturer can share channel gain by reducing the unit price it charges to the distributor. Alternatively, the manufacturer can make a side payment to the distributor (Chakravarty and Martin 1991). Note that the manufacturer's gain is increased by economies of scale, which are enhanced as sales increase.

Inventory

Cycle and buffer inventories may be carried by both the manufacturer and the distributor, whereas customers would carry only buffer inventory. The size of the cycle inventory held by the distributor will depend on the frequency of delivery from the manufacturer, the distributor's profit margin, the setup cost, and the mean demand for products. The cycle inventory at the manufacturer's site will be different, as it will be determined by the manufacturer's own decisions, such as frequency of delivery and profit margin, its setup cost (different from that of the distributor), and mean demand. The determinants of buffer inventory, such as forecast error, mean demand, and service level, will be common to both the manufacturer and the distributor. Delivery lead time (manufacturer to distributor) will impact the distributor's buffer inventory, whereas production lead time will impact the manufacturer's buffer inventory. Note, however, that if the distributor takes possession of the goods at the manufacturer's site, the manufacturer would save all its inventory-related costs.

The inventory position of the manufacturer will depend on echelon stock (inventory at the manufacturer plus inventory in the pipeline all the way to the customer), customer orders outstanding, and planned production, up to a certain time period.

Logistics

The logistics function can be organized in four different ways. The first option would be for the manufacturer to incur the logistics cost of delivery to the distributor, and the distributor would incur the cost of delivering to customers. In the second option, the manufacturer would incur the logistics cost of delivery to the distributor and from the distributor to customers. The third option is similar to option 2, but now the distributor incurs the entire cost of logistics. In option 4, the whole logistics operations is outsourced to a third party.

Delayed Differentiation

The distributor can act as a VAR. It can do so in several ways, such as repackaging and customizing the products for individual market segments. Delayed differentiation may add value to the chain in several ways. If products are delivered to the customer from the distributor rather than the manufacturer, delivery lead time may be reduced. Lead time reduction decreases the customer's buffer inventory and so increases the customer's profit margin. It also enables the manufacturer to cut back on the number of item categories it must handle, and permits it to ship semi-finished products, in bulk, to the distributor. This reduces the logistics cost to the manufacturer. Delayed differentiation also reduces the manufacturer's value added, and hence its inventory and logistics costs. The manufacturer's share of channel profit, however, is reduced, and the supplier now must incur the additional cost of providing customization at its site.

Model B

The distributor controls the transmission of customer orders to the manufacturer, but it does not take physical possession of the goods. Therefore, the distributor would have nothing to gain by batching (lot-sizing) orders transmitted to the manufacturer. Orders generated by the distributor would, of course, depend on the pricing structure it can negotiate with the manufacturer, its own fixed costs, and the elasticity of demand for the manufacturer's product.

The flow of funds in the supply chain can take two possible routes. It can flow from the customer to the manufacturer on delivery of goods, and from the manufacturer to the distributor as payment for order generation. Alternatively, it can flow from the customer to the distributor on receipt of goods, and from the distributor to the manufacturer after adjusting for the distributor's margin. Note that these two modes of payments would induce different behaviors from the manufacturer and the distributor, as they optimize their individual profits. As in model A, the distributor may offer price discounts and/or increase its marketing budget to increase customer orders.

There would, obviously, be no inventory carried by the distributor. The cycle and buffer inventories at the manufacturer would now depend on customers' demand characteristics (mean demand and forecast errors), production lead time, setup cost, and cost of capital. The size of the customer's buffer inventory would depend on demand characteristics, delivery lead time, and cost of capital. Logistics service can be provided by a third party or the manufacturer.

In a slight variation, the distributor may take ownership of goods at a manufacturer's site but use a third-party logistics provider to move the products from the manufacturer to the customer. In this case the distributor will have to bear the cost of inventory (at the manufacturer's site) and logistics. Note that the customer does not interact with the manufacturer and will make payment to the distributor when it receives the goods. The distributor will have to pay the manufacturer on assuming ownership of products.

Model C

This is a combination of models A and B. This is similar to the dynamic replenishment model of Cisco, discussed earlier. The customers can be supplied by manufacturer or distributor. Unlike the Cisco model, however, model C channels all customer orders through the distributor. As orders arrive, the distributor would satisfy them from its own inventory. If the distributor does not have sufficient quantities in its inventory, it asks the manufacturer to supply directly to customers. For orders with a long lead time, the manufacturer may supply directly to customers even if the items are inventoried with the distributor. Both the manufacturer and the distributor may incur inventory costs. Customers pay whoever supplies products to them. The distributor and manufacturer need to work out a policy regarding direct supplies from manufacturer and the minimum inventory that the distributor must hold.

Model D

This model represents the option of not using a distributor. Orders from customers are expected to be lower than in the other three models. However, since the distributor's margin is eliminated, the unit price charged to customers is expected to be lower than in other cases, which in turn may increase sales. The manufacturer will carry all inventories—cycle and buffer—and customers would hold some buffer inventory. Logistics can be done by the manufacturer or by a third party.

10.6 DIGITAL BUSINESS COMMUNITY

A digital business community (DBC) is a set of companies with shared interests who together seek market dominance within industry environment. It relies on its capability to be agile and to respond quickly to customer requirements (Tenenbaum 1997). It does so by making sure that the mix of capabilities necessary to satisfy key customers can be satisfied by members of the DBC and that value added by partners in a community is seamlessly integrated to form the total value chain. It is, of course, not enough for the community members to be electronically connected. They must have their business processes aligned as well, and there should be sharing of information and knowledge throughout the value chain. It is clear that for such a DBC to be effective, it must have clear objectives in terms of performance targets, industry context, and its life cycle. Clearly visible and well-understood rules of collaboration among member companies would be required for this purpose.

Baldwin and Clark (1997) outline the foundations of a business community: architecture, an interface, and standards. Architecture defines the types of capabilities and the degree of process alignment and system transparency in a community. Interface determines protocols of sharing information and decision rules for resource sharing and task management. Standards facilitate intra-community communication and performance evaluation of individual companies. The architecture and interface together determine the boundaries of operations of community members. Individual companies can be innovative provided they stay within the architectural boundaries, which are also called the visible rules.

The visible rules may have elements that are common to most partners, but some rules may be specific to pairs of partners interacting with each other. For example, the rules of interfacing a supplier and a manufacturer may be very different from the rules of interfacing a distributor and a reseller. In the former case the emphasis is on component quality, JIT delivery, and BOM structure. In the latter case the rules may emphasize packaging, branding, labeling, and pricing. Thus rules need to be worked out for each pair of interacting partners first, and from these an inventory of master rules will emerge. Each of these rules must then be identified with specific sets of stakeholders (suppliers, manufacturers, distributors, etc.), and a system for managing these rules must be formulated.

The business community used by Cisco Systems is shown in Figure 10.18 (Cisco Systems 1998). The rules of interfacing Flextronics with LSI Logic, on the supply side, and the rules of interfacing Ingram Micro with Gold, on the distribution side, may have some common elements. The question is to what extent Cisco, as a community leader, may influence the shaping of these rules. How much control (of operations) should LSI Logic be required to forgo? Would it be the same for other suppliers such as Arrow? How would such decisions be made? What would be Cisco's role in the rule-making process? A rule is good only if compliance can be ensured. However, the more restrictive a rule, the higher the risk of noncompliance. Rule making, therefore, must incorporate various trade-offs in establishing optimal levels of rule restrictiveness.

Effective rules of interfacing cannot be created overnight. They evolve gradually as two interacting partners try out different ways of managing their operations jointly. Consider vendor-managed inventory. To implement VMI, the replenishment processes must be altered significantly, but other processes such as order receiving, invoice reconciliation, and shipping may remain unchanged. In VMI the vendor ships to the customer periodically, based on the customer's usage pattern. The customer may report its stock status periodically, which can be used by the vendor to determine the replenishment quantity. Alternatively, the vendor may directly monitor the customer's inventory. The customer is relieved of a burden, as the vendor takes over the job of forecasting and triggering replenishment orders in time to avoid stock-outs.

Thomas and Betts (T&B), a supplier of electrical components, worked with a distributor to gradually fine-tune the visible rules in the context of VMI. As a first step, T&B and the distributor cleared out any slow-moving T&B products. If a competing line was being replaced, T&B allowed the distributor to blend the two lines, gradually replacing the old products. T&B then took over the job of deciding when and how much of the product to ship to the distributor. Its planners evaluated distributor sales figures and forecasts and designed replenishment controls that automatically replenish distributor inventory at the right time. No limits were placed on the size of replenishments. For small distributors inventory was replen-

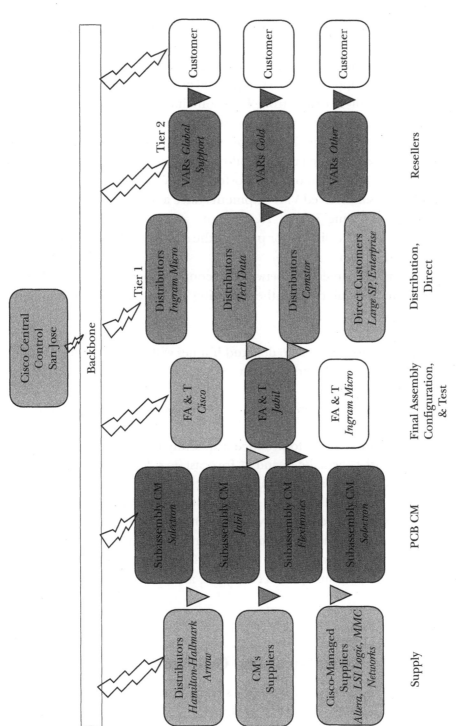

FIGURE 10.18: BUSINESS COMMUNITY

ished once a week, and for large distributors it was twice a week. T&B paid for the cost of logistics on all shipments. T&B designated approximately 1,000 items as superstock, which was guaranteed to be delivered within 48 hours of receiving orders. It did not offer guarantees for fill rate, but it had achieved 99.7% shipping accuracy; it packed and labeled each product individually. A shipment carton from T&B contained an assortment of products. Each carton was received at a distributor warehouse, where it was opened and unpacked and the contents were put away. If there was a shortage or error, T&B accepted responsibility and made adjustments to the invoice. Invoice reconciliation was posted in the distributor's accounts payable, and T&B invoiced VMI shipments once a month. The distributor committed to carrying 75% of T&B's items, and stock balances were reported every day. The distributor paid for the software that linked the VMI operations to its own.

An interface agreement between a contract manufacturer and on OEM may, in general, contain the following stipulations, with an agreed value of ϵ.

- Manufacturer delivers to demand if it lies within $\pm \epsilon$% of forecast
- Manufacturer delivers forecast $+ \epsilon$% if actual demand exceeds forecast $+ \epsilon$%
- Manufacturer delivers forecast $- \epsilon$% if actual demand is less than forecast $- \epsilon$%
- Manufacturer plans to produce a quantity that equals the forecast
- If forecast $<$ actual demand $<$ forecast $+ \epsilon$%, manufacturer produces an additional amount equaling demand $-$ forecast at a premium unit price c_1 ($c_1 > c$). If forecast $- \epsilon$% $<$ actual demand $<$ forecast, manufacturer delivers the forecast quantity at unit price c_1 and charges an additional price c_2 per unit of undelivered quantity (forecast $-$ demand). The price c_2 reflects the cost of salvaging inventory. For a multi-period control, c_1 and c_2 are set as a function of time.

Operationalizing Collaboration

A business community must possess a strong foundation of collaborative behavior, as it leads to cost-effective, efficient, and customer-focused solutions. The benefits of collaboration accrue through operational practices targeted to inter-enterprise synergy.

The so-called trading partner framework may be used for inducing collaboration. A set of trading partners may have different competencies, different views of the marketplace, and different information sources. Collaboration therefore needs to be structured for different scenarios of partner capabilities. The business partners must agree to a plan that clearly describes what is going to be sold, how it would be merchandised and promoted, what markets it would be sold in, and the time frame in which sales would take place. The plan is implemented through partner companies' operational systems, which are accessed using approved communications standards. Any partner can adjust its plans within established parameters. The characteristics of a generic collaboration plan are a process model, front-end agreements, information sharing, common metrics, and a set of rules. The process model defines how and where collaboration fits into supply chain processes; front-end agreements form the basis for measuring and evaluating performance, accountability, contingencies, and changes; information sharing schemes define the data elements to be shared, frequency of updates, and data security; common metrics are the set of measures all partners use for evaluation and benchmarking activities; and the set of rules defines how partner activities will be controlled.

The front-end agreement is perhaps the most critical element of collaboration. It is a mission statement that outlines the shared understanding and objectives of collaboration, confidentialities and empowerment. It defines goals in terms of specific opportunities, means of measurement to be used, business processes to be used, and assessment of impacts on each member's business practices. It determines the adequacy of each party's competencies, resources, and control systems in terms of their capability to contribute to the overall goals. It traces collaboration agreements to trading-partner competencies and identifies business functions that would be key executors of the collaboration plan. It also determines the outline of information to be shared (such as decision support tools and data), frequency of updates, forecasts, technology, and recovery and response times. It determines the services and parameters for triggering orders. The staffing levels and time commitments are estimated, and resources are assigned to processes and process management. The agreement also establishes ground rules for handling conflicts between partners. Ongoing evaluation and benchmarking are also included in the collaboration agreement. An example of a front-end agreement between two companies is shown in the Appendix to this chapter.

Organizational Alignment

For effective collaboration, organizations need to evolve to a market-driven structure. Collaboration plans must be integrated with complementary processes such as product design, product development, and market development. For example, the planning process will become more customer-centric in the way it links sales, marketing, finance, logistics, production, and supplier management. If financial objectives become difficult to meet, it would trigger adjustments and communicate the incorporation of these adjustments into joint business plans across the value chain. The need for establishing trust will increase, and so the importance of customer relationship management is enhanced manyfold. At the tactical (and operational) level, where customers and suppliers are interfaced, partners must know each other's processes well enough to leverage their complementary competencies. Multifunctional teams must be constituted so that a merchandise or category manager can work with inventory management, store operations, and logistics. Similarly, customer services, sales, marketing, distribution, and production should be the responsibility of another multifunctional team.

Measuring collaboration requires the generation of both process control and process achievement metrics. Such metrics must help manage process variability and process constraints. Some of the common metrics of interest would be percent out of stock on the store shelf, inventory turns, total cycle time, forecast accuracy, profitability, return on assets, order fill rate, process cost, process yield, and cash-to-cash cycle.

Security Considerations

Any collaboration scheme, unless secured adequately, will result in failure. The key attributes of a secure system are authentication, authorization, integrity, confidentiality, auditing, and non-repudiation. Authentication ensures that remote users connected to a system are who they are supposed to be. All downloaded information must be authenticated so as not to be compromised. Authorization includes application privileges and the ability to assign rights to others. A digital signature attests to the integrity of the contents of messages. To obtain confidentiality at a system level, all attributes of a connection, including time, frequency, and data queries, must be kept private. Needs of confidentiality are best met through some form of encryption. The security system must also provide for auditing of access and potential breaches using an audit trail. Finally, it must ensure that all parties involved in initiating transactions within the application are tied to those transactions.

RosettaNet

RosettaNet is a digital business community for the IT supply chain that has brought together companies such as Cisco, Compaq, Hewlett-Packard, Intel, Microsoft, Oracle, SAP, ABB, American Express, EDS, CompUSA, Ingram Micro, FedEx, and UPS. The need for common electronic business interfaces as determined by this DBC is depicted in Figure 10.19, showing the parallel between a human-to-human business exchange and a server-to-server electronic business exchange. In order to communicate in a human-to-human business exchange, humans must be able to produce and hear sound. Further, they must then agree on common phonemes, used to create individual words in a language. Grammatical rules are then applied to the words to create a dialog. That dialog forms the business process, which is conducted (or transmitted) through a device such as a telephone.

The fundamental system of exchanging sounds in a human-to-human business exchange can be compared to the Internet, which enables two servers to exchange information during a server-to-server electronic business exchange. HTML/XML functions as the language of this electronic exchange. And, presently, e-commerce applications serve as the instrument by which electronic business process is transmitted.

RosettaNet is focusing on building a master dictionary to define properties for products, partners, and business transactions. This master dictionary is used to support the e-business dialog known as the Partner Interface Process (PIP). RosettaNet PIPs create new areas of alignment within the overall IT supply chains, allowing IT supply chains partners

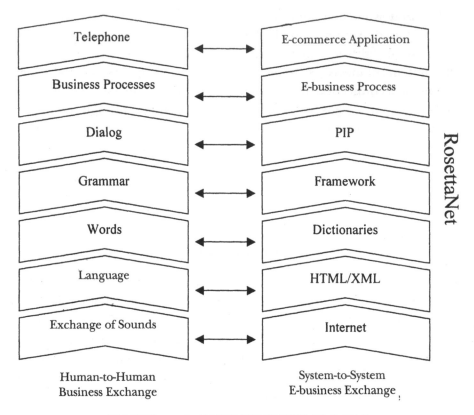

FIGURE 10.19: COMMON NEEDS IN A DBC

to leverage e-commerce applications and the Internet as a business-to-business commerce tool.

Business process modeling is used to identify and quantify the individual elements of a business process, creating a clearly defined model of the supply chain partner interfaces as they exist today. This is called as-is modeling, and it is shown as a constituent of business process architecture in Figure 10.20. This model reflects the results of extensive research at every level of the supply chain, which is then analyzed to identify any misalignments or inefficiencies. Through the analysis of the detailed as-is model, a generic to-be process emerges, showing the opportunities for realignment in the form of a PIP target list and estimating the business impact of implementing the resulting PIPs (savings as a function of time and money). The purpose of each PIP is to provide common business/data models

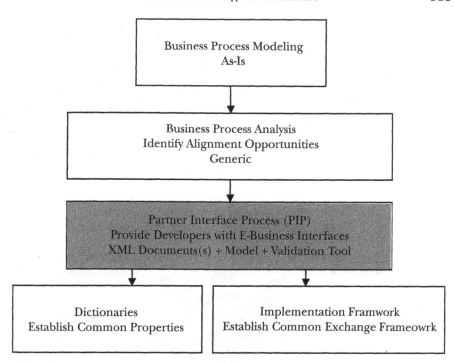

FIGURE 10.20: BUSINESS PROCESS ARCHITECTURE

and documents, enabling system developers to implement RosettaNet
e-business interfaces.

As part of RosettaNet's projects, two data dictionaries are being developed to provide a common set of properties required by PIPs. The first is
a technical properties dictionary (technical specifications for all product
categories), and the second is a business properties dictionary, which includes catalog properties, partner properties (attributes used to describe
supply chain partner companies), and business transaction properties.
These dictionaries, coupled with the RosettaNet implementation framework (exchange protocol) (Figure 10.19) form the basis for each Rosetta-
Net PIP.

Supply Chain Partnership

RosettaNet requires the formation of an industry-wide partnership to carry
out the stated mission of this initiative. The RosettaNet managing board is
responsible for defining the interface development projects and setting

the initiative's priorities. Furthermore, members of this board hold the primary responsibility for promoting and implementing these interfaces. The RosettaNet managing board consists of 28 individuals representing global members of the IT supply chain, including hardware manufacturers, software publishers, distributors, resellers, system integrators, end users, technology providers, financial institutions, and shippers. In Figure 10.21 the companies on the RosettaNet managing board are shown.

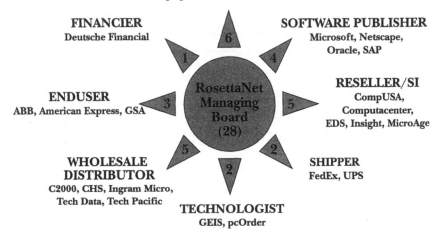

MANUFACTURER
Cisco, Compaq Hewlett-Packard, IBM, Intel, Toshiba

FINANCIER
Deutsche Financial

SOFTWARE PUBLISHER
Microsoft, Netscape, Oracle, SAP

ENDUSER
ABB, American Express, GSA

RESELLER/SI
CompUSA, Computacenter, EDS, Insight, MicroAge

RosettaNet Managing Board (28)

WHOLESALE DISTRIBUTOR
C2000, CHS, Ingram Micro, Tech Data, Tech Pacific

SHIPPER
FedEx, UPS

TECHNOLOGIST
GEIS, pcOrder

FIGURE 10.21: ROSETTA BUSINESS COMMUNITY

(Source: www.rosettanet.org.)

10.7 VIRTUAL ENTERPRISE

A virtual enterprise is an architecture that permits temporary relationships among participating companies in a value chain. A VE adds value through (1) rapid reconfiguration of the supply chain to match customers, products, and markets, (2) joint operations among partners leading to high return on assets, increased market access, and new technology, (3) seamless interfacing of suppliers, manufacturers, distributors, logistics providers, and other entities, and (4) reduction of cost in the supply chain and improvements in order cycle, on-time delivery, and stock-outs. The four major issues in a VE are creation of the VE, definition of business processes, coordination of processes, and termination of the VE.

There are several scenarios in which a VE can be especially useful. For example, for products with short life cycles, it can facilitate frequent reengineering of supply chains. That is, new suppliers with different capabilities can be added and/or an existing supplier in the VE can be purged. If customers wish quicker delivery, several options can be considered: holding finished goods inventory, incorporating more agile partners, and using logistics providers for direct or emergency shipments. If capacity becomes a problem in the short term, new (temporary) partners can be added. If superior service providers emerge (due to improvements in product and process technology or management practices), the VE would be well positioned to induce them to join the alliance. Finally, for the strategy of a total customer solution, discussed in Chapter 9, it may become necessary to widen the net to include suppliers who, along with the current members of the VE, can deliver complete solutions. The expanded VE would ensure single-point ordering for the customer.

To create a VE, the value chain is organized as groups of tasks, and a business partner is identified with each group of tasks. All partners must agree to rules of collaboration, as in the trading partners framework discussed earlier. Partners must commit to a conflict resolution procedure and agree to address changes in collaboration rules as new partners are added or deleted. The resources made available (and the constraints imposed) by each partner because of such reengineering must be revealed and agreed upon by all.

Business Process Coordination

In a VE, the business processes of each partner are carefully mapped in terms of inputs to the process, controlling elements and decisions, and outputs of a process (NIIIP 1998), as discussed in Chapters 1 and 3. Typical processes would be order management, logistics management, purchasing, production management, product flow, invoice routing, and inventory control. Inputs to a process relate to materials, equipment, worker knowledge, and funds. Process controls define how a process is to be performed, yield control parameters, resource requirements, and links to other processes. Decision elements determine the optimal rate of output and resource utilization. Outputs of the process identify the products and services to be transferred to the next process. Customers of each process (internal or external) are identified, and the status of all processes in terms of input, output, controls, and decision are made transparent industry-wide. The three major flows linking the system—material, money, and information—are clearly identified. Note that the material flow chain need not be identical with the money flow chain, and so the information flow chain must be designed to support the material and money flows and, at the same time, must maintain consistency among them. The principles of collaboration planning, discussed earlier, and Internet-based ordering, as in Figure 10.6, are promoted. The terms of business contracts, including transfer prices, delivery schedule, risks, inventories, and penalties, are designed for individual processes or groups of processes (participating companies).

Coordination in a VE requires moving resources to target locations for work to begin, accounting of resource sharing among partners, controlling the execution of tasks, monitoring work progress at each process, and sequencing of tasks of related processes that share critical resources such as machines, equipment, and skilled workers. The information architecture to implement coordination could be based on EDI, groupware, the Internet, or extranets. The architecture may be denoted as star (web) or delta, depending on the nature of linkages among the participating companies.

Star (web) architecture, as shown in Figure 10.22, provides an automated means of exchanging information between partners. However, it also presents unique challenges. Integrating an entire enterprise in this point-to-point manner may require developing specific interfaces from each entity to every other entity in the web. Finding information within a star architecture may not be easy. Because the interface environment provides no comprehensive directory service, users have no way of knowing where desired information is located. They may not even know that it exists. Controlling information in this environment is another challenge, as it offers no protocol for information flows across the VE. It is difficult to control the time when information is updated, who does it, and what impact the changes would have on related information. Point-to-point architecture limits a company's ability to integrate forward with customers and backward with suppliers. Since suppliers and customers may also develop other systems to solve their specific problems, integration of suppliers and customers using point-to-point connectivity may become infeasible.

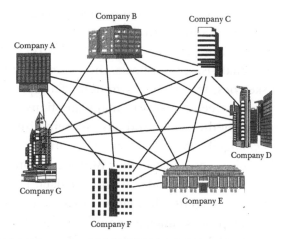

FIGURE 10.22: STAR (WEB) ARCHITECTURE

(Source: NIIIP 1998. Copyright www.niiip.org.)

Delta architecture, shown in Figure 10.23, uses a central hub to which all entities are connected. Note that addition and deletion of suppliers is now easier. However, a certain degree of centralization of decision-making (in the hub) becomes unavoidable.

The digital environment places increased emphasis on managed business processes and the applications and data that support these processes.

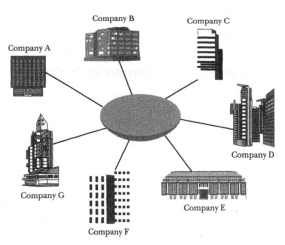

FIGURE 10.23: DELTA ARCHITECTURE

(Source: NIIIP 1998. Copyright www.niiip.org.)

As part of a business process, information needs to be captured (explicitly and implicitly), shared, and controlled. Information, both formal and informal, is a valuable asset that includes verbal instructions, applications, computer model analysis, spreadsheets, three-dimensional models, ideas, drawings and sketches, test data, and multimedia representations. To do their jobs, engineers need answers to many questions, such as what their current tasks are, whether new products are being introduced, whether the design has changed, which CAD system controls the model, where it may be, whether it is accessible, whether BOMs are available, and whether there are existing shop orders and cost estimate data. The objective should be development of intelligent information models, discussed in Chapter 3.

Information for process management must also be addressed at the virtual enterprise level. Vertical organizations shaped around specific business infrastructures are less likely to exist in the future. For example, in concurrent engineering, organizational boundaries are less important; it is multifunctional teams that count. This can be accomplished if tools, applications, and infrastructure support increased parallelism with less formal boundaries. Therefore, business processes and workflow to support these processes must be well defined and understood by everyone involved. In an integrated process, data and applications that previously operated in stand-alone islands now need to be shared. The need to improve the business process places heavy demands on information technology: parallel activities, accurate status information, people working as a team, and data sharing.

Example 1: Defense Industry

Two small commercial businesses joined forces temporarily to create a VE. In the process, they invited a group (General Dynamics and Lockheed) to join the enterprise so that the VE could benefit from their technical expertise. In addition, significant interaction took place between the VE and other external groups, including government standards groups and potential customers of the enterprise's product.

The VE concept was formulated through meetings and information exchange with partners. Once the VE was established, software that supported the needed protocols and facilitated the VE was installed. Connections were established between member companies for creating linkages supporting VE operations. Contract terms and conditions that determined the involvement, contributions, and expectations of member companies were established. Workflow models for new products and services were developed. Security measures needed to insulate the VE from nonparticipating portions of the member companies were established. The VE put into place components that performed product design. It searched for new members that could provide needed skills and technologies using intelligent agents. It established new relationship defining terms, conditions, and security to bring new member company into the VE. It started collaborative design by defining needed data (via product data model) and facilitated sharing of design, and design verification data. The VE consulted with government agencies and others on manufacturing practices for the new product. It manufactured products using the shared component data (availability, quality, etc.). Finally, it designed a system to coordinate manufacturing, assembly, and shipment between VE members and to share financial benefits with the VE team.

Example 2: Telecommunications

Huber & Suhner (H&S) is a Swiss-German manufacturer that produces data transmission cables, coaxial connectors, and other telecommunications equipment (Riggers 1998). Its customers include Nokia, Ericsson, and Motorola. Because of short lead times, the company was faced with the problem of having to assemble several thousand parts from over 20,000 components at short notice. Huber & Suhner's objectives were to transform its supply chain from its current "octopus" structure to a network structure, shown in Figure 10.24. It wanted to create a community of suppliers, designers, assemblers, and logistics providers. It also wanted to be faster than its competition in entering new markets and accessing new resources.

The company, in a pilot project, identified four major business processes on the network that comprised 3 suppliers, 4 manufacturing sites, 2 distributors, and up to 20 large customers that generate 80% of Huber and Suhner's revenue (shown in Figure 10.25). The partner companies, including logistics providers, are independent enterprises. The companies included in the network were suppliers such as Carle Leipold, Aeschli (Germany and England); manufacturers such as H&S (Switzerland, United States, England); distributors such as H&S and SME (France, Germany); and customers such as Motorola, Hewlett-Packard, Nokia, Ericsson, and Siemens.

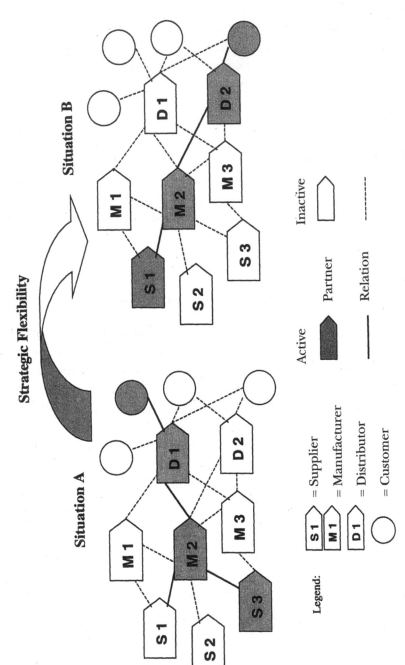

FIGURE 10.24: HUBER AND SUHNER'S CONCEPT OF VE

(Source: Teleflow, http://www.item.unisg.ch.)

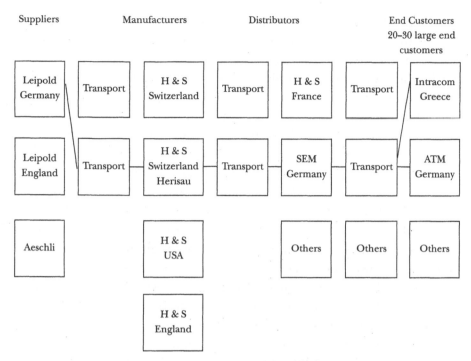

| Suppliers | | Manufacturers | | Distributors | | End Customers
20–30 large end
customers |

FIGURE 10.25: HUBER & SUHNER VIRTUAL ENTERPRISE

(Source: Teleflow 1998.)

10.8 INTELLIGENT AGENTS

We have seen how reengineering the value chain for different business models, with different architectures of ownership, collaboration, and information accessibility, can create tremendous value in the system. Such decentralized organizations, with quick reconfiguration capabilities, require a multitude of decisions for real-time actions. These decisions require a large variety of data, which keep changing at a fast pace. Human decision makers, good at strategy formulation and conceptualization, are not known to be effective at absorbing and evaluating a large mass of data quickly. Software programs, designed for mass processing of data, are of only limited help. In today's dynamic environment, such as in a VE, decision making may involve continuous monitoring, matching, and multi-attribute negotiations. The source of information is not static; data must be pulled from different entities at different times (see Chapter 3). Moreover, as new business models and alliances evolve and new forms of collaborative arrangements take shape, the nature of decisions as well as the categories of data required may also change drastically.

Intelligent agents (IA), unlike the traditional software programs, may prove to be very useful in such dynamic environments. Such agents behave like human assistants in that they can perform tasks based on individual goals. The goals are set by humans, but the agents decide, on their own, how, where, and when to get the jobs done. A traditional procurement program, for example, can trigger the purchase of a certain raw material if its price is at a desired level. An intelligent agent, on the other hand, will be responsible for scanning the market, searching for new suppliers, and making deals on terms that would be most favorable to the buyer. For maximum effectiveness such agents should be autonomous, should be able to communicate with other agents and humans, should be focused on a specialized set of tasks, and should be modular so that they can be created, deployed and terminated, with minimal effort. In digital chains between suppliers and customers, agents are beginning to be used in consumer-to-business transactions such as price negotiations and auctions of finished goods. In a business-to-business environment (e-business), where businesses buy/sell components and/or other resources from/to other businesses, agent prototypes are being created for use in the near future. Ford and Oracle have announced a joint venture called AutoXchange that would act as a deal maker between manufacturers and suppliers of auto parts (Johnston 2000). It would earn money from four sources: transaction fees, trading commissions charged to companies that use the exchange to sell off excess supplies, supply chain management fees, and advertisements (Kumar and Feldman 1999). Maes, Guttman, and Moukas (1999) have suggested six types of roles for intelligent agents: need identification, product brokering, merchant brokering, negotiation, purchasing, and service evaluation. They identify capabilities of some currently available software in terms of these six business functions.

Agents can help customers identify their needs on the basis of tastes revealed in their past buying behavior. Amazon.com uses a notification agent that scans publishers' catalogs and notifies the customer when a new book by a certain author or in a certain category becomes available. Agents may analyze customers' buying behavior to infer needs that customers themselves might not be aware of. They may suggest products that match these needs or alert customers about the potential of new products.

The need identification agent usually generates a set of attributes that match several customer needs. The product brokering agent enables customers to narrow down the list of products to those that best meet their needs by guiding them through a large product attribute space. Based on constraints specified by customers, the agent filters out unwanted products. Agents may also identify customer peers on the basis of their product ratings and suggest to a customer products that have been rated highly by the customer's peers. This is known as collaborative filtering.

The merchant brokering agent identifies suppliers who can supply the products needed by a customer and who match closely in terms of other customer-desired attributes such as price, quality, and delivery. Virtual database technology has been reported to be helpful in merchant brokering. Technologies such as XML and mobile agents (Java-based) may prove to be more flexible, open-ended, and easy to implement.

A buyer or seller wanting to negotiate with his/her counterpart can create an agent and activate it in a centralized marketplace such as that run by Kasbah (Kasbah 1999). The agents seek out potential buyers or sellers and negotiate with them to reconcile bid prices and acceptable prices. The agents are endowed with negotiation strategies such as a transaction completion deadline and variation in acceptable price with time. The buying agent offers a bid to the selling agent, which may respond with a binding yes or no (Kasbah) or a counterproposal (Tete-a-tete, http://ecommerce.media. mit.edu/tete-a-tete). Kasbah provides three functions (linear, quadratic, exponential) to the buyers for increasing bid price for a product over time. Kasbah also incorporates a "better business bureau" that compiles peer ratings of how well buyers and sellers have managed their part of the deal. Buyers and sellers below a certain user-specified reputation threshold may be barred from participating in negotiation with that user.

Tete-a-tete agents cooperatively negotiate across multiple facets of a transaction, including warranties, delivery times, service contracts, return policies, loan options, gift services, and other value-added services. The agents evaluate XML-based proposals and order them based on criteria set by the agent's owner. Proposals are modified based on critiques of earlier proposals. Both product and supplier features are considered in a negotiation process. Thus a buyer may settle for a less-than-ideal product from a good supplier in preference to an ideal product that is available only from a less reputable supplier.

Bundle-Based Negotiations

Fan, Stallert, and Whinston (1998) describe a prototype of a market-clearing agent, where buyers and sellers trade in bundles of resources. A resource bundle specifies the quantities of resource types being traded, and the bundle is sold as a whole for a single price. Thus a bundle (b_1, b_2, b_3) at price p implies that a buyer is willing to pay price p for a bundle of resources containing b_1 units of resource 1, b_2 units of resource 2, and b_3 units of resource 3. If p and b_i are negative, it implies that a seller is willing to sell the bundle of resources at price p. Note that $b_i = 0$ implies that the bundle does not contain resource type i.

Assume that at the time of market clearing, there are n orders, n being the sum of buy and sell orders. The jth order specifies the number of jth bundles to be bought or sold. The market-clearing agent will match the resource quantities from the buy orders with the resource quantities of the sell order. Assume that the agent selects l ($l \leq n$) orders to create an optimal match. Let x_j equal the number of jth bundles used in the optimal matching set.

Since the market is financed through the trading surplus, the agent will attempt to maximize this surplus. The trading surplus from the matching process can be written as

$$\text{Trade surplus} = \sum_{j=1}^{n} x_j p_j,$$

where p_j is the surplus from a single bundle of type j.
Hence the optimal matching problem can be stated as

$$\text{Maximize} \sum_{j=1}^{n} x_j p_j \qquad (10.1)$$

subject to

$$\sum_{j=1}^{n} b_{ij} x_j = 0 \text{ for } i = 1 \text{ to } m \qquad (10.2)$$

$$x_j \leq N_j, \text{ for } j = 1 \text{ to } n \qquad (10.3)$$

$$x_j \geq 0$$

where b_{ij} is the units of resources of type i present in a bundle of order j, m is the number of types of resources, and N_j is the number of bundles of type j available for matching.

Note that since $N_j \geq 0$ and $b_{ij} \geq 0$, constraint 10.2 can be written, without loss of generality, as

$$\sum_{j=1}^{n} b_{ij}x_j \leq 0, i = 1 \text{ to } m \qquad (10.4)$$

Let y_i $(i = 1$ to $m)$ and u_j $(j = 1$ to $n)$ be the variables in the dual formulation of the problem corresponding to constraints 10.4 and 10.3, respectively. The dual of the order matching problem can be expressed as

$$\text{Minimize } \sum_{j=1}^{n} N_j u_j$$

subject to

$$\sum_{i=1}^{m} b_{ij}y_i + u_j \geq p_j, j = 1 \text{ to } n$$

$$y_i \geq 0, i = 1 \text{ to } m$$

$$u_j \geq 0, j = 1 \text{ to } n$$

Note that y_i can be interpreted as the cost of one unit of resource of type i, irrespective of the order type. Hence the theoretical transaction price T_j for the jth bundle can be written as

$$T_j = \sum_{i=1}^{m} b_{ij}y_i$$

The actual transaction price T_j^* will be established through negotiation such that

$$T_j \leq T_j^* \leq p_j$$

Observe that since the orders are matched at the resource level, there may not be a one-to-one match between buyers' and sellers' orders. Several sellers may supply to a single buyer, and several buyers may buy from a single seller. To minimize the number of transactions between buyers and sellers, another linear program can be set up by defining z_{irq} as the number of units of resource i sold by seller q to buyer r. A transaction $v_{rq} = 1$ if $\sum_i z_{irq} > 0$. Hence the objective would be to

$$\text{Minimize } \sum_q \sum_r v_{rq}$$

subject to

$$\sum_r z_{irq} = b_{iq}x_q, \text{ for all } i \text{ and } q$$

$$v_{rq} \geq \sum_i z_{irq}/M$$

where M is a large number, v_{rq} is a binary variable $(0,1)$, and x_q is the number of bundles of type q in the optimal matched set.

Shaw and Sikora (1998) discuss a similar intelligent agent in the context of manufacturing resource management.

REFERENCES

Baldwin, Y., and K. Clark (1997), "Managing in an Age of Modularity," *Harvard Business Review*, September–October, pp. 84–93.

Blackman, D. (1999), "FedEx and Cisco," *Wall Street Journal*, November 4.

Booz, Allen, and Hamilton, and Economist Intelligence Unit (1999), "Competing in the Digital Age: How the Internet Will Transform Business," (http://www.bah.com/greatideas/pptdata/index.htm).

Callahan, C., and B. Pasternack (1999), "Corporate Strategy in the Digital Age," *Strategy and Business*, quarter 2, issue 15, pp. 10–18.

Chakravarty, A., and G. Martin (1991), "Operational Economies of a Process Positioning Determinant," Computers and Operations Research, vol. 18, no. 6, pp. 515–530.

Cisco Systems (1998), "Networked Manufacturing for the 21st Century," Keiser Associates, Vienna, Va.

Fabias, P. (1997), "EC Riders," http://www.cio.com/archive/061597_commerce_print.htm.

Fan, M., J. Stallert, and A. Whinston (1998), "Mechanism and Process Design for Supply Chain Agent Organization Based on Bundle Markets," working paper, school of business, University of Texas, Austin.

Hoy, T. (1999), "Manufacturing Assembly Pilot," Auto Industry Action Group, http://www.aiag.org/map/main2.html.

Johnston, M. (2000), "Cisco Joins Ford-Oracle Auto Xchange Venture," February, http://www.idg.net/crd_idgsearch_137899.html.

Kalogeridis, C. (2000), "Internet Generation Next," http://www.anxo. com/downloads/actionline1.pdf.

Kasbah (1999), http://kasbah.media.mit.edu.

Kumar, M., and S. Feldman (1999), "Business Negotiation on the Internet," IBM Research Division, Thomas J. Watson Research Center, Yorktown Heights, N.Y.

Lohr, S. (1999), "Microsoft On-Line Alliance with Ford Expected," *New York Times,* September 20.

Maes, P., R. Guttman, and A. Moukas (1999), "Agents that Buy and Sell: Transforming Commerce as We Know It," Massachusetts Institute of Technology Media Lab, Cambridge, Mass, http://ecommerce.media.mit.edu/papers/caem98.pdf.

Malone, T., J. Yates, and R. Benjamin (1987), "Electronic Markets and Electronic Hierarchy," *Communications of ACM,* vol. 30, no. 6, pp. 484–497.

Margherio, L. (1998), "Emerging Digital Economy," U.S. Department of Commerce, Washington, D.C., http://www.ecommerce.gov/emerging.htm.

Magretta, J. (1998a), "The Power of Virtual Integration: An Interview with Dell Computer's Michael Dell," *Harvard Business Review,* March–April, pp. 73–84.

Margretta, J. (1998b), "Fast, Global, and Entrepreneurial Supply Chain Management: Hong Kong Style," *Harvard Business Review,* September–October, pp. 103–114.

NIIIP (National Industrial Information Infrastructure Protocols) (1998), NIIIP Reference Architecture, http://www.niiip.org/public-forum/index-ref-arch.html.

Riggers, B. (1998), "Agile Manufacturing Strategies for Networks," University of St. Gallen, Institute for Technology Management, St. Gallen, Switzerland.

Rocks, D. (1999), "Why Office Depot Loves the Net," *Business Week,* September 27, pp. 66–68.

Shaw, M., and R. Sikora (1998), "A Multi-Agent Framework for the Coordination and Integration of Information Systems," *Management Science,* vol. 44, no. 11, pp. 565–578.

Teleflow (1999), "Project Management for VS Integration," http://www.item.unisg.ch/Homepages%20projekte/Teleflow.

Tenenbaum, J. (1997), "Electronic Commerce: ATP Program Ideas and Technologies," Comerc One, Inc., Walnut Creek, Calif.

Ticol, D., A. Lowy, and R. Kalakota (1998), "Joined at the Bit," working paper, Georgia State University, school of business, Atlanta, Georgia.

TPN (Trading Process Network) (1999), http://www.tpn.geis.com/index.html.

VICS (Voluntary Interdisciplinary Commerce Standards) (1998), Collaborative Planning Forecasting and Replenishment, http://www.cpfr.org/Guidelines.html.

Trading Partners: Corner Store, Inc., and ReliaSupplier, Inc.

I. Agreement & Statement

    A. Purpose

        Corner Store and ReliaSupplier agree to collaborate in key supply chain processes using standards. Our goal is to increase mutual efficiencies and delight the end consumer through dynamic information sharing, focus on common goals and measures, and commitment to collaborative processes. We recognize that there are many business process, technological, and organizational changes required by this collaboration, and we commit to apply resources to make these changes in order to make our collaboration effective and meet our mutual goals.

    B. Confidentiality

        All communication will be governed by antitrust regulations. Both trading partners commit here to absolute confidentiality in the use of information shared.

II. Goals and Objectives

    A. Opportunity

        Corner Store and ReliaSupplier will seek to reduce stock-outs, increase sales, reduce business transaction costs, improve the use of capital (especially that involved in inventory), and facilitate trading partner relationships.

    B. Measurement of Success

        Corner Store and ReliaSupplier agree to focus on key results-oriented measures: retail in-stock, inventory turns (at retail), and forecast accuracy (measured when the forecast can impact production, eight weeks prior). Goals for specific products are attached, but the overall goal is 96% retail in-stock, six turns at retail, <15% sales forecast error (eight weeks out), and <20% order forecast error (eight weeks out). We also agree to maintain several measures involving performance of specific parts of the process. Our performance against all of these measures will be the basis of our quarterly face-to-face reviews. Details on these measures (scope, data source, responsibility for maintaining, frequency of measure, frequency of reporting, construction of the algorithm, unit of measure) will be attached to this agreement. Supporting processes are detailed in the process model (to be attached).

III. Discussion of Competencies, Resources, and Systems

Based on our earlier discussion of the competencies, resources, and systems that each party brings to the partnership, we agree to follow the CPFR scenario, where the retailer has ultimate responsibility for the sales forecast, the manufacturer has ultimate responsibility for the order forecast, and the manufacturer has ultimate responsibility for order generation.

IV. Definition of Collaboration Points and Responsible Business Functions

A. Collaboration Points

Collaboration points include the joint business plan, the sales forecast, and the order forecast. Collaboration on the sales and order forecast will be driven by the following item-level exception criteria and values:

1. Sales Forecast Exception Criteria

Retail in-stock <95%, sales forecast error >20%, sales forecast differs from same week prior year >10%, change in promotional calendar or number of active stores.

2. Order Forecast Exception Criteria

Retail in-stock <95%, order forecast error >20%, annualized retail turns < goal (as noted on item management profile table), entry of new events that impact inventory/orders, emergency orders requested >5% of weekly forecast.

B. Responsible Business Functions

The following business functional units are impacted by and

responsible for the success of collaborative planning, forecasting, and replenishment:

1. Retailer
   Merchandising/buying, forecasting, inventory management.
2. Manufacturer
   Sales team, planning/forecasting, distribution.

V. Information Sharing Needs

Information sharing will be open and routine as needed to support collaboration processes. No information on competitor activity will be shared. As it is our expectation that communication will be timely, the collaboration information cycle time will be measured.

A. Areas of Information Sharing

Shared information includes data necessary to measure success (common metrics), such as retail in-stock percentage, inventory, and forecast accuracy; data necessary to identify exceptions in the sales and order forecast, such as retail in-stock percentage, inventory turns and levels (retail, in transit, warehouse), sales and order forecast accuracy, vendor order fill rate, and so on; data necessary to support decisions about exception items, such as promotions, planned inventory actions and other events that impact the forecast, point-of-sale data, historical shipments, sales and order forecasts, current item retail in-stock percentage, current inventory turns, number of valid stores, and so on; and item management profiles, including item identifiers and logistics rules (rounding rules, order minimum and multiples, configurations, etc.). Complete details are listed in the data model (to be attached).

B. Frequency of Updates

Forecasts will be created and shared on a weekly basis. Exceptions, supporting data, and item management data will be shared daily, and metrics will be calculated and shared monthly.

C. Method of Data Sharing

Where possible, data sharing will be accomplished using standard data formats such as EDI transaction sets.

D. Recovery and Response Times

Response time for collaboration system used by Corner Store and ReliaSupplier should be no more than 30 seconds. Steps have been taken to limit recovery time following a system fault to 12 hours.

VI.   **Service and Ordering Commitments**
One of the ways both companies expect to benefit from collaboration is through commitments to supply and to consume through orders (subject to a range of deviation). The range of deviation will be reevaluated periodically. ReliaSupplier agrees to support the agreed-to forecast (at a frozen period of seven days) with timely shipments within 3% of the forecast (measured weekly). In return, Corner Store agrees to consume the forecast through orders within 3% (measured weekly). In the event that these limits are exceeded, the respective party will notify the other as soon as possible and will determine a resolution plan. In recognition of the spirit of this agreement, both parties agree to commit the resources and systems necessary to maintain the upstream planning processes, which will in turn allow the timely identification and resolution of potential issues.

VII.   **Resolution of Disagreements**
In the event of a disagreement, the ultimate process owners will have final say over the sales forecast, order forecast, and order generation if intermediate efforts at resolution are not successful. All other disagreements will be handled by a meeting of the leaders of the affected functional areas, and ultimately the agreement owners.

VIII. **Agreements Review Cycle**
This agreement will be reviewed each year in January, and the undersigned will reaffirm the effectiveness of the process by renewing the agreement.

11

Supply Chain Models

11.1 SCOPE OF QUANTITATIVE MODELING

In a supply chain with a large number of participants, a multitude of decisions must be made about investments, coordination and cooperation, customer service, and profit maximization. Many such decisions have system-wide implications and become extremely complex given the uncertainties and large number of decision variables in the system. In such scenarios quantitative modeling can be useful in arriving at optimal decisions and managerial insights. Such decision scenarios include how to configure a supply chain, which products to target to which destinations in the chain, how to maximize cooperation between customers and producers in order to minimize waste in the system, how to design flexible contracts among parties, and how to share inventories and profits in a network of suppliers.

By its very nature, quantitative modeling can best be applied to scenarios that are well structured. Consequently, in most supply chain models, it is assumed that a single organization has total ownership of the supply chain. Multi-echelon inventories that link supplier, manufacturer, and distributor, all owned by the same company, have been modeled extensively since the late seventies. Quantitative models have also been used for supplier evaluation, capacity allocation, and coordination in an MRP environment. Treatment of modular supply chains, on the other hand, has begun only in the 1990s (and includes issues such as incentives for coordination and contract structuring. In this chapter, therefore, our discussion is focused on two issues: how to configure a supply chain in terms of choice of products, choice of suppliers, and choice of plant locations, and how to maximize decentralized coordination.

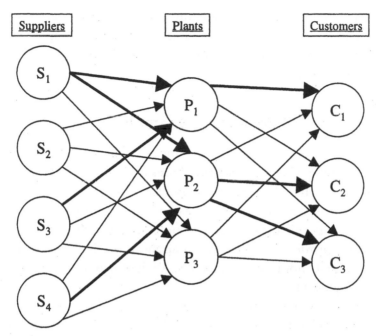

FIGURE 11.1: SUPPLY CHAIN CONFIGURATION

11.2 SUPPLY CHAIN CONFIGURATION

Consider the supply chain scenario shown in Figure 11.1, involving suppliers, plants, and customers, each of which may be at a different geographical location. All three customers C_1, C_2, and C_3 buy the same product, which comprises components 1 and 2. Suppliers S_1 and S_2 supply component 1, while suppliers S_3 and S_4 supply component 2. For the network shown, note that we may build one, two, or three plants (i.e., seven possibilities). For any chosen plant, possible supplier sets would be $\{(S_1,S_3), (S_1,S_4), (S_2,S_3), \text{ and } (S_2,S_4)\}$. Finally, a plant may be used to supply one, two, or all three customers (i.e., seven possibilities).

Since there may be uncertainties in customer demand and delivery lead times, both component and finished-product inventories may have to be held at the plants. All shipping between suppliers and plants, and between plants and customers, can be undertaken by the firm itself, or outsourced to third-party logistics providers. Finally, if the locations involve several different countries, there may be differentials in tariffs, taxes, local-content requirements, and labor costs. It is therefore clear that there are several supply chain choices, of which one is shown with bold arrows in Figure 11.1. This solution implies that suppliers S_1 and S_3 supply plant P_1, which satisfies customer C_1, while suppliers S_1 and S_4 supply plant P_2, which satisfies customers C_2 and C_3.

An Industry Example

A company in the telecommunications industry has modeled such problems by generating all possible supply chains and doing a complete financial analysis for each such chain. It chooses the supply chain that yields the most profit or best return on assets (or some other metric).

Consider a specific product of the company called Fiber-Optic Access System, which provides access to video applications, high-speed Internet access, and high-bandwidth applications. A bill of material of the product is shown in Figure 11.2.

The major customer of the product is in Japan. Closure material 1 is acquired from Belgium, closure material 2 from the Netherlands, and the cable material from Japan; all other material can be purchased in the United States. The company needed to decide where to locate the intermediate assembly plants for the paddle board, closure assembly, ONU-MUX, and power supply, and where to locate the final assembly plant. The company also considered joint ventures and licensing arrangements for such operations. The labor rates for assembly operations, tariffs, and taxes varied from country to country. The cost of transportation of raw material and intermediate assemblies depended on plant location decisions.

The company considered several manufacturing and subcontracting options. In terms of manufacturing, they considered expanding their cur-

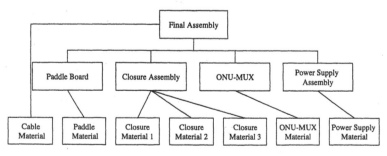

FIGURE 11.2: FIBER-OPTIC ACCESS SYSTEM BOM

rent plant in the northeastern United States, adding a new factory in the United States, and adding a new factory in Japan. They considered four subcontracting options: subcontract the intermediate or final assembly in the United States or in Japan, and subcontract the final assembly in China or Taiwan. This generated a large number of manufacturing and subcontracting combinations, out of which the company chose to consider the options shown in Table 11.1 for detailed financial analysis. Option 1 is explained further, in a form suitable for financial analysis, in Figure 11.3. A detailed income statement for each option, such as the one shown in Figure 11.3, was prepared. Net present values (NPVs) for all options were computed. The option with the highest NPV turned out to be option 8— use a joint venture in China for all manufacturing and assembly operation. This solution required transportation of all raw material (from Europe, the United States, and Japan) to China, and transportation of the final product from China to Japan.

Configuration with Pricing Decisions

In the above industry example, the supply chain was designed for a single product. It was also assumed that the unit price of the product was independent of sales quantity. If that is not the case, the allocation of plant financing cost to products will be nontrivial. In a real-world problem the

Options	Intermediate Assembly (Manufacture or Acquisition)	Final Assembly (Manufacture or Acquisition)
1.	Current plant in Boston	Current plant in Boston
2.	New plant in Japan	New plant in Japan
3.	New plant in the United States	New plant in the United States
4.	Add capacity in Boston	Add capacity in Boston
5.	Use current plant in Boston	Subcontract in Japan
6.	Subcontract in Japan	Subcontract in Japan
7.	Subcontract in China	Subcontract in China
8.	Use joint venture in China	Use joint venture in China
9.	Subcontract in the United States	Subcontract in the United States
10.	Subcontract in the United States	Use plant in Boston
11.	Subcontract in the United States	Use new plant in the United States

Table 11.1: Supply Chain Options

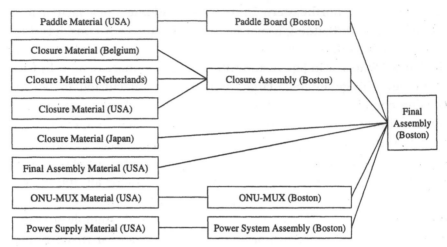

FIGURE 11.3: SUPPLY CHAIN (OPTION 1)

margins would depend on how sensitive sales are to unit price and how much of the financing cost is absorbed by a product. Consider a simple analysis with two products produced in a single plant (in country 1), and sold in countries 1 and 2. Let the sales quantity q be expressed as $q = ap^n$, $n < 0$, where p is the unit price and a is a scaling constant that captures product and country characteristics. Simon (1989) discusses many application scenarios where empirical evidence of such demand curves have been found. Henderson and Quandt (1980) suggest that for industries other than utilities, $n < -1$. In our example we assume $n = -2$. We denote the unit manufacturing costs of products as v_1 and v_2 (both produced in country 1). It is obvious that if the products are sold in country 1, neither tariff nor transportation cost will be incurred. The tariff rates of products sold in country 2 are assumed to be d_{12} and d_{22}, respectively. Similarly, transportation costs per unit of products are γ_1 and γ_2. We also assume that all costs and revenues are expressed in country 1's currency. For product i, let $x_i = v_i +$ allocated plant financing cost. It is therefore clear that the tariff charged per unit of product i in country 2 will be $d_{iz}x_i$.

The profit function, letting k denote country, would be

$$z = \sum_{i=1}^{2} \sum_{k=1}^{2} \left\{ \left(\frac{q_{ik}}{a_{ik}} \right)^{-1/2} - (1 + d_{ik})x_i - \gamma_k \right\} q_{ik} \qquad (11.1)$$

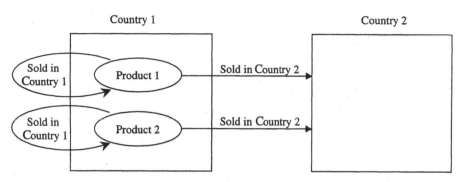

FIGURE 11.4: GLOBAL PLANT AND MARKETS

Note that $d_{11} = d_{21} = 0$, and $\gamma_1 = 0$.

Since the plant financing cost F must be completely recovered through the allocated costs, we must also have

$$\sum_{i=1}^{2} \sum_{k=1}^{2} (x_i - v_i)q_{ik} = F \tag{11.2}$$

To maximize z subject to constraint 11.2, we form the Lagrangian equation

$$L = z + \lambda \left\{ \sum_{i=1}^{2} \sum_{k=1}^{2} (x_i - v_i)q_{ik} - F \right\}$$

Equating $\frac{\partial L}{\partial x_i} = 0$, and $\frac{\partial L}{\partial q_{ik}} = 0$, we have the following equations:

$$\sum_{k} (\lambda - 1 - d_{ik})q_{ik} = 0, \, i = 1,2 \tag{11.3}$$

$$\frac{1}{2}\left(\frac{a_{ik}}{q_{ik}}\right)^{\frac{1}{2}} - x_i(1 + d_{ik} - \lambda) - \gamma_k - \lambda v_i = 0, \, i = 1,2; \, k = 1,2 \tag{11.4}$$

Equations 11.2, 11.3, and 11.4 together generate seven equations to solve for seven unknowns: four q_{ik}, two x_i, and λ. The q_{ik} can be eliminated easily, so that we have three equations in x_1, x_2, and λ to solve, as shown below.

$$\frac{a_{12} d_{12} (x_1 - v_1)}{(\lambda - 1)\{x_1(1 + d_{12} - \lambda) - \lambda v_1 - \gamma_2\}^2} +$$

$$\frac{a_{22} d_{22} (x_2 - v_2)}{(\lambda - 1)\{x_2(1 + d_{22} - \lambda) - \lambda v_2 - \gamma_2\}^2} = 4F \quad (11.5)$$

$$x_1 d_{12} - \gamma_2 = \left\{ \frac{a_{12}}{a_{11}} \times \frac{1 + d_{12} - \lambda}{\lambda - 1} \right\}^{1/2} \quad (11.6)$$

$$x_2 d_{22} - \gamma_2 = \left\{ \frac{a_{22}}{a_{21}} \times \frac{1 + d_{22} - \lambda}{\lambda - 1} \right\}^{1/2} \quad (11.7)$$

Chakravarty (2000a) has analyzed a generalized version of this problem by permitting several plants distributed among many countries. He also adds another set of constraints to ensure adequate capacity at each plant. His analysis reveals that tariffs behave like market imperfections, setting up arbitrage-like opportunities, with overhead (plant financing) allocation as the instrument of arbitrage. He establishes three categories of products helpful in generating managerial insights: those with no overhead absorption, those with overhead absorption in certain countries, and those with overhead absorption in all countries.

Cost of Financing

Hodder (1984) and Hodder and Dincer (1986) analyze a similar problem with a single product, with emphasis on financing cost. With a single product, allocation of finance cost is a non-issue. Hence they include finance cost in the objective function. They assume that plants in country j come in a fixed size M_j, requiring a fixed investment of F_j. The model structure for any two countries j and k is shown in Figure 11.5.

Cost of sales in country k from country $j = q_{jk}\{v_j(1 + d_k) + T_{jk}\}$

After-tax profit $= q_{jk}\{p_k - v_j(1 + d_k) - T_{jk}\}(1 - \eta_k)$

Financing cost in country $j = F_j r_j$

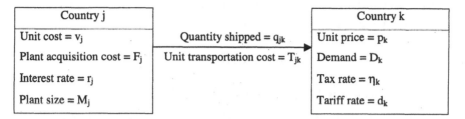

Country j	Country k
Unit cost = v_j	Unit price = p_k
Plant acquisition cost = F_j	Demand = D_k
Interest rate = r_j	Tax rate = η_k
Plant size = M_j	Tariff rate = d_k

Quantity shipped = q_{jk}
Unit transportation cost = T_{jk}

FIGURE 11.5: SUPPLY PLANNING WITH UNCERTAINTIES

Therefore, profit Π is written as

$$\Pi = \sum_k \left[\sum_j q_{jk}\{p_k - v_j(1 + d_k) - T_{jk}\} - F_j r_j \right](1 - \eta_k)$$

Note, however, that the financing cost in country j will be zero if no plant is established in that country. A binary variable x_j is used to model the plant opening decision. That is, x_j equals 1 if a plant is opened in country j, 0 otherwise. The profit expression with constraints would be

$$\Pi = \sum_k \left[\sum_j q_{jk}\{p_k - v_j(1 + d_k) - T_{jk}\} - F_j r_j x_j \right](1 - \eta_k) \quad (11.8)$$

The demand and supply constraints would be

$$\sum_j q_{jk} \geq D_k \quad (11.9)$$

$$\sum_k q_{jk} \leq M_j x_j \quad (11.10)$$

Hodder and Dincer do not relate price p_k to demand D_k. Instead, they treat price p_k, tariff cost $v_j(1+d_k)$, and transportation cost as random variables. The optimization can now be framed as,

Maximize $E(\Pi)$

subject to

$$V(\Pi) \leq u \quad (11.11)$$

where $E(\Pi)$ and $V(\Pi)$ are the expected value and variance of profit Π, respectively. We may restate the above as,

$$\text{Maximize } E(\Pi) - \lambda V(\Pi)$$

subject to constraints 11.9, 11.10, and 11.11.

Cohen and Lee (1989) have applied a mathematical programming model to analyze Apple Computer's global supply chain. It is a four-echelon model involving suppliers, manufacturing plants, distribution centers, and markets. The decision variables include choice of suppliers, products, and production quantities at plants, and supply links between plants and distribution centers and between distribution centers and markets. Arntzen et al. (1995) analyze Digital Equipment Company's supply chain using a model that minimizes a weighted sum of total cost and production (and transportation) days. They explicitly incorporate tariff drawbacks as decision variables in their model. Other decision variables include plant location, capacity, manufacturing technology, product mix, shipping modes, and shipping quantities. They report that Digital has used their model for new product decisions and supply strategies for components.

11.3 SUPPLY CHAIN PLATFORM

We saw in the industry example of Fiber-Optic Access System, that the supply chain was designed for a single product. The emphasis in that case was on product and production economics. If, on the other hand, customer satisfaction is the primary objective, supply chains will be targeted to specific markets or specific customers. In most cases, however, supply chain design is driven by both production economics and customer satisfaction. This implies that a supply chain should be able to support multiple products from multiple plants in multiple markets. It does not, however, mean that it would be optimal to sell in all markets, purchase components from all vendors, and manufacture products in all possible locations. It should be obvious that the optimal configuration may select a subset of markets, a subset of products, a subset of possible plant locations, and a subset of vendors. We call such a configuration a supply chain platform. Note that in Figure 11.1, the configuration shown in bold lines may qualify as a platform. Plant F_3, supplier B_1, and all links associated with them do not belong to the platform.

The decisions a company must make in configuring a platform include what products (including components and assembly) to support on the platform, which suppliers to involve in long-term relationships, where to manufacture, and which markets to serve. AT&T in the early 1980s, for example, had a plant in The Hague, the Netherlands, producing 5E, and another in Huizen (China) producing ISM. Since the products had many common components, the company saved about $10 million by relocating the plants to a common location in Brazil. Motorola builds "converged" factories where pagers, cellular phones, and two-way radios share common facilities.

There are many issues that must be examined before establishing a supply chain platform. Product structure is important, as it determines the necessary components and commonalities among them. Supplier capability is important, as the platform should not be supporting too many suppliers. Supplier, plant, and market locations are important not only for local economies but also for shipping costs.

There are many uncertainties that impact a supply chain design, of which we consider two: uncertainty about being able to deliver the product to customers on time, and uncertainty about product demand. If demand exceeds capacity, the excess demand is lost. If delivery is delayed beyond the agreed date, the customer may decline to buy it or he/she may buy it only at a discounted unit price.

The Model

Let $I(i \in I)$, $M(m \in M)$, $K(k \in K)$, $J(j \in J)$, and $U(u \in U)$ be the sets of products, markets, components, plant locations, and suppliers, respectively. We define a product choice $e = (i,m)$ as product i sold in market m. Let $\hat{E}(e \in \hat{E})$ be the set of all such product choices, and let E be a subset of $\hat{E}$. We characterize product demand by its mean μ and variance σ^2. The unit price is denoted by p, and the discounted price for late delivery by p'. Component unit costs are denoted by β and the total cost of components in a product by B. Shipping cost per unit is denoted as α for components and γ for end products. Financing cost per unit capacity is C, and the cost of assembling a product is A. We let the agreed delivery lead time be T and the lost sales due to capacity limitations be L. The tariff and tax rates are assumed to be d and η, respectively.

Let $f(x)$ = pdf of demand, and $g(t)$ = pdf of lead time. Assume that the planned production capacity is Q. The lost sales can then be written as

$$L = \int_Q^\infty (x - Q) f(x) dx$$

The proportion of the products delivered before time T can be written as

$$w = \int_0^T g(t) dt$$

It is clear that the interactions between demand and lead time uncertainties would lead to four different scenarios, depending on whether or not demand exceeds capacity Q and actual delivery time exceeds T. The expected revenues in each of these scenarios can be expressed as shown in Table 11.2.

Aggregating over the four scenarios, we can write the expected revenue as

$$E(Rev) = \{pw + p'(1 - w)\} \left\{ \int_{-\infty}^Q x f(x) dx + Q \int_Q^\infty f(x) dx \right\}$$

Demand x	Lead Time t	Revenue
$x \leq Q$	$t \leq T$	$pw \int_0^Q xf(x)dx$
$x \leq Q$	$t > T$	$p'(1 - w) \int_0^Q xf(x)dx$
$x > Q$	$t \leq T$	$pwQ \int_Q^\infty f(x)dx$
$x > Q$	$t > T$	$p'(1 - w)Q \int_Q^\infty f(x)dx$

Table 11.2: Demand and Lead Time Scenarios

We can rewrite $E(\text{Rev})$ as,

$$E(Rev) = \{pw + p'(1 - w)\}\left\{\int_{-\infty}^{\infty} xf(x)dx - \int_{Q}^{\infty} (x - Q)f(x)dx\right\}$$
$$= R(\mu - L) \qquad\qquad (11.12)$$

where

$$R = pw + p'(1 - w)$$
$$L = \int_{Q}^{\infty} (x - Q)f(x)dx$$

The elements of cost are the unit cost of components (β), the cost of shipping components to assembly plants (α), the labor cost of assembly (A), and the cost of shipping the final product to markets (γ). In addition, a tariff cost at the rate d is incurred in the country where the product is sold. The total component cost (including shipping of components) per unit of product i manufactured in country j can be written as

$$B_{ij} = \sum_{k \in K_i} \underset{u \in U}{\text{Minimum}}\, (\beta_{ku} + \alpha_{kuj})$$

Hence the total landed cost per unit of product i in country m can be written as

$$\text{Unit cost} = B_{ij} + A_{ij} + \gamma_{ij.}$$

The unit cost in country m after paying the tariff would be

$$\text{Unit cost} = (B_{ij} + A_{ij} + \gamma_{ij})(1 + d_m) \qquad\qquad (11.13)$$

Next, to model the capacity cost, note that we can express capacity Q as

$$Q = \mu + z\sigma$$

where z is obtained from the standard normal distribution corresponding to the service level being satisfied. The value of reserve capacity will be equal to $z\sigma$. Thus the cost of reserve capacity for a single product in location j would be $C_j z\sigma$. For a group of products sharing plant capacity at location j, this can be written as

$$\text{Capacity cost} = C_j z \left(\sum_i \sigma_i^2 \right)^{1/2}$$

Next consider the profit associated with a product choice $e = (i,m)$ in set E. From equation 11.13 we write

$$H_e = (B_{ij} + A_{ij} + \gamma_{ij})(1 + d_m)$$

Therefore the expected cost of products would be

$$E(\text{Cost}) = H_e(\mu_e - L_e)$$

From equation 11.12, we have expected revenue as

$$E(\text{Rev}) = R_e(\mu_e - L_e)$$

Therefore

$$\text{Operating profit} = (R_e - H_e)(\mu_e - L_e)$$

Incorporating the cost of capacity, the expected profit before tax would be

$$\sum_{e \in E} (R_e - H_e)(\mu_e - L_e) - C_j \left\{ \sum_{e \in E} \mu_e + z \left(\sum_{e \in E} \sigma_e^2 \right)^{1/2} \right\}$$

Net profit after tax would be

$$\Pi_E = \left[\sum_{e \in E} (R_e - H_e)(\mu_e - L_e) - C_j \left\{ \sum_{e \in E} \mu_e + z \left(\sum_{e \in E} \sigma_e^2 \right)^{1/2} \right\} \right](1 - \eta_m)$$

$$(11.14)$$

It is clear that z would be set as

$$z = \text{Maximum}\,(z_c, z^*)$$

where z_c corresponds to the service level stipulated by the customer and z^* is the optimal value of z obtained by equating $\frac{\partial \Pi_E}{\partial z}$ to zero. After simplifying $\frac{\partial \Pi_E}{\partial z} = 0$, the equation for z^* can be written as

$$\sum_{e \in E} [(R_e - H_e)\sigma_e F(\mu_e + z^*\sigma_e)] = \sum_{e \in E} (R_e - H_e)\sigma_e - C_j \left(\sum_{e \in E} \sigma_e^2 \right)^{1/2}$$

(11.15)

where

$$F(y) = \int_{-\infty}^{y} f(x) dx$$

Equation 11.15 can be solved iteratively for z^*. Observe that Π_E in equation 11.14 can be rewritten as

$$\Pi(E, j, U) = \left\{ \sum_{e \in E} x_{ejU} - zC_j \left(\sum_{e \in E} y_e \right)^{1/2} \right\} (1 - \eta_m)$$

where

$$x_{ejU} = \{(\mu_e - L_e)(R_e - H_e) - \mu_e C_j\}$$

$$y_e = \sigma_e^2$$

Next, let

$$X_{E,j,U} = \sum_{e \in E} x_{ejU}$$

$$Y_E = \sum_{e \in E} y_e$$

Then the objective function is written as

$$\Pi(E, j, U) = \{X_{E,j,U} - zC_j Y_E^{1/2}\}(1 - \eta_m)$$

(11.16)

The problem now transfers to maximization of function 11.16 subject to $E \subset \hat{E}$ and $U \subset W$, where W is the set of all possible supplier groups. Observe that function 11.16 is convex with respect to $X_{E,j,U}$ and Y_E. As in

Chapter 7, it can be shown that for the optimal group E, with $e \in E$ and $g \notin E$,

$$\frac{x_{ejU}}{y_e} > \frac{x_{gjU}}{y_g}$$

The procedure for establishing the supply chain platform can therefore be summarized in the following five steps.

- For each supplier group $U = \{u_1, u_2 \ldots\}$, determine the set of products that can be completely assembled.
- Arrange this set in descending order of the ratio x_{ejU}/y_e.
- Assign products from the top of the list to the set E and recompute optimal z for E.
- Recompute $\Pi(E,j,U)$ with the new value of z.
- Stop assigning products to E, when $\Pi(E,j,U)$ starts to decrease.

Chakravarty and Balakrishnan (2000) have solved this problem for different problem scenarios. With ten products, ten markets, five suppliers, and no lead time constraint, their solution is shown in Figure 11.6. Only two suppliers (1 and 3) were used.

With the inclusion of a lead time constraint so that $w = 0.95$, it was

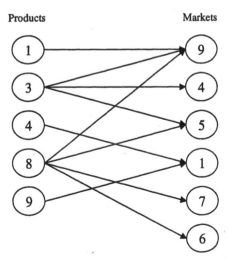

FIGURE 11.6: PRODUCT AND MARKET CHOICES

optimal not to serve market 4. Most of product 3 sold in market 4 was rerouted to market 1. Total profit decreased from $10,562,956 to $10,485,412 as a result of the lead time constraint.

11.4 COORDINATION IN A SUPPLY CHAIN

In Chapter 2 we discussed the notion of integration in a decentralized system (i.e., with business processes not aligned) in the context of manufacturing-marketing interface creation. The supply chain was assumed to be integral (not fragmented) in terms of ownership. It was shown how a local solution in one domain can be improved by the other domain, and how such improvements could continue until solutions from both domains converge. In Chapter 9 we briefly discussed the coordination issues in a modular supply chain with fragmented ownership. It was pointed out that both information and financial incentives would have major roles in such coordination. We next discuss how financial incentives can be used to influence the business practices of partners in a supply chain.

Consider a three-echelon system comprising a manufacturer, a retailer, and customers, as shown in Figure 11.7.

To decouple the manufacturer from the retailer, it is assumed that the manufacturing cycle T' is different from the retailer's ordering cycle T, and that the unit retail price s is different from the unit wholesale price P. Demand D for the product is assumed to be a function of the price charged to the consumers, and is deterministic. Setup costs are incurred by both manufacturer and retailer. The retailer incurs a setup cost A in processing customers' orders before placing its order with the manufacturer (every T period). The manufacturer, on the other hand, will incur

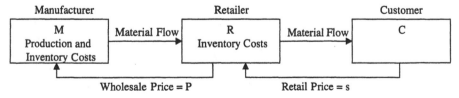

FIGURE 11.7: SYSTEM WITH MANUFACTURER, RETAILER, AND CUSTOMER

two types of setup costs: the cost K of processing the retailer's orders, incurred every time delivery is made (every T period) to the retailer, and a machine setup cost A_0, incurred once every manufacturing cycle. As in Chakravarty and Martin 1991 and Lee and Rosenblatt 1986, we assume $T' = mT$, where m is a positive integer. The production rate with the current production equipment is assumed to be ρ units per period ($\rho > D$). The manufacturer's unit production cost is assumed to be c ($c < P$). The discounted wholesale price, used for coordinating with the retailer, is p, so that $c < p < P < s$. The inventory carrying cost rate is $r\$/\$$. Demand is assumed to decrease linearly with price, so that $D = a - bs$.

Retail Operations

For a given value of wholesale price P, the retailer will maximize its profit by optimally setting the values of retail price s and its order cycle T. Allowing for setup cost and inventory carrying cost, it can be verified that the retailer's profit will be written as

$$R(P,s,T) = sD - PD - \frac{A}{T} - \frac{T\,Pr\,D}{2}$$

$$= \left(s - P - \frac{T\,Pr}{2}\right)(a - bs) - \frac{A}{T} \qquad (11.17)$$

where $D = a - bs$.

Maximizing $R(P, s, T)$ with respect to s and T, we have

$$T^* = \left\{\frac{2A}{Pr(a - bs)}\right\}^{1/2}$$

$$s^* = \frac{1}{2}\left[P + \frac{a}{b} + \left\{\frac{A\,Pr}{2(a - bs^*)}\right\}^{1/2}\right]$$

$$R^*(P,s^*,T^*) = \left(s^* - P - \frac{T^*\,Pr}{2}\right)(a - bs^*) - \frac{A}{T^*} \qquad (11.18)$$

For $A = \$100$, $D = 7{,}000 - 200s$, $r = 0.2$, and $P = \$25$, it can be verified that

$$s* = \$29.96$$

$$T* = 0.1953 \text{ years}$$

$$R(P, s*, T*) = \$3,964.28$$

Manufacturing Operations

Since $\rho > D$, it is clear that the machine will be used for production for only a part of the manufacturing cycle. The manufacturer will accumulate inventory during the time manufacturing is in progress, even though DT units will be withdrawn (to supply the retailer) every T periods. The inventory accumulation diagram will be as in Figure 11.8.

Chakravarty and Martin (1991) have shown that the average inventory level of the manufacturer can be written as

$$I_a = \frac{DT}{2}\left\{(m - 1) - \frac{D}{\rho}(m - 2)\right\}$$

It is also apparent that the manufacturer's setup cost per period will equal $(A_0 + mK)/mT$. Therefore we may write the manufacturer's profit as

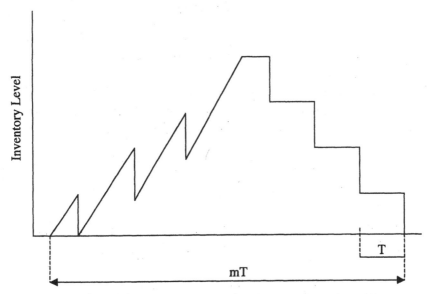

FIGURE 11.8: MANUFACTURER'S INVENTORY ACCUMULATION

$$M(P,T,s,m) = (P - c)D - \frac{A_0 + mK}{mT} - \frac{DT}{2} cr \left\{ (m - 1) - \frac{D}{\rho}(m - 2) \right\}$$

(11.19)

where $D = a - bs$. Since the retailer fixes the values of s, and T, so the manufacturer will maximize its profit by appropriately choosing m.

Since m is an integer, it can be shown that the optimal value for m will be written as

$$m^*(m^* - 1) \le \frac{2A_0}{T^2 Dcr \left(1 - \frac{D}{\rho}\right)} \le m^*(m^* + 1)$$

Continuing with the same numerical example, if $A_0 = \$400$, $K = \$100$, $\rho = 300$, and $c = \$18$, then

$$m^* = 3$$

$$M^* = M(P,T^*,s^*,m^*) = \$5,536.42$$

Price Incentive

In terms of supply chain coordination, the relevant question now is whether both manufacturer and retailer would gain if the manufacturer's wholesale price is reduced.

We may rewrite equations 11.17 and 11.19 by substituting the discounted wholesale price p for P to express manufacturer's gain (MG) and retailer's gain (RG) due to price discounting as

$$RG = R(p,T,s) - R(P,T^*,s^*)$$

(11.20)

$$MG = M(p,T,s,m) - M(P,T^*,s^*m^*)$$

(11.21)

Also, by adding $R(p,T,s)$ and $M(p,T,s,m)$ we obtain the joint profit in the system as

$$J(p,T,s,m) = (s - c)D - \frac{A_0 + m(A + K)}{mT} - \frac{DT}{2} r(p + c\mu)$$

(11.22)

where

$$\mu = (m - 1) - \frac{D}{\rho}(m - 2)$$

Therefore, the objective would be to maximize equation 11.22 subject to $RG = \alpha MG$, where RG and MG are as in equations 11.20 and 11.21, respectively. The value of α determines how the system gain would be shared between the retailer and the manufacturer. A value of 1 for α would imply that the two partners share the gains equally.

By eliminating p from equations 11.20, 11.21, and 11.22, the optimization of 11.22 can be restated as

$$\text{Maximize } J(s,T,m,\alpha) = (s - c)D - \frac{A_0 + mA + mK}{mT} - \frac{T}{2}crD\mu -$$

$$\frac{\frac{T}{2}r\left\{(M^* - \alpha R^*) + \dfrac{A_0 - m\alpha A + mK}{mT} - \dfrac{T}{2}crD\mu + (\alpha s + c)D\right\}}{\dfrac{T}{2}\alpha r + (1 + \alpha)} \tag{11.23}$$

subject to

$$\frac{A_0 + mK}{mT} + \frac{T}{2}crD\mu + M^* + cD \leq \frac{sD - \left(\dfrac{A}{T} + R^*\right)}{1 + \dfrac{Tr}{2}} \tag{11.24}$$

where $D = a - bs$. The constraint 11.24 ensures that $MG \geq 0$ and $RG \geq 0$.

Observe that $\frac{\partial J}{\partial s} = 0$ will find an expression for s in terms of T. A one-dimensional grid search over the values of T can now be devised that would find the optimal value of equation 11.23 subject to constraint 11.24.

Continuing with the numerical example, and with $\alpha = 1$, the optimal value of p is found as $p = \$23.15$. The revised values of s, T (written as s', T', etc.) would be

$$s' = \$26.95$$
$$T' = 0.247 \text{ years}$$
$$m' = 2$$
$$R' = \$4,791$$
$$M' = \$6,363$$

Thus the manufacturer would reduce the unit wholesale price by $1.85 (7.4%) and the retailer would cut the retail price by $2.79 (9.4%). Both manufacturer's and retailer's profits would increase by about $826 each (since $\alpha = 1$), implying a system gain of $1,652. Note that the system gain increases if α is decreased from 1.0, but in that case most of the profit would accrue to the retailer. For example, when $\alpha = 0.1$, the retailer would gain $1,518, and the manufacturer $152, with a total system gain of $1,670.

Several earlier versions of this problem with stricter assumptions have been studied by Monahan (1984), Banerjee (1986), and Chakravarty and Martin (1988, 1989).

11.5 IMPACT OF UNCERTAINTY REDUCTION

We have seen how cooperation in product pricing between the manufacturer and retailer can increase the profits of both parties. In this analysis it was assumed that the variance of customer demand was not high, so demand could be approximated by its average value. In the apparel industry, where the lead time between order placement by the retailer and delivery by the manufacturer can be very high, the impact of demand variance can be critical. Skinner (1992) reports that these lead times can vary between eight months and five months. Reducing forecast error, therefore, assumes a high priority. A common industrial approach (Skinner 1992) is to observe actual orders for a period t and use this information to place an order, with a revised forecast to be delivered $T - t$ periods later (T is the total lead time available whether or not demand forecast is revised).

Iyer and Bergen (1997) have used stochastic modeling techniques to analyze whether forecast revision increases profits of the retailer or the manufacturer or both. They assume a normal demand distribution with known variance σ^2. To study forecast revisions, however, they assume that the mean demand θ is not known with certainty. Specifically, they assume that θ is normally distributed, with mean μ and variance τ_2. Thus the distribution of demand x, described as $f(x)$, can be defined by $N(\mu, \sigma^2 + \tau^2)$.

To obtain a revised forecast at time t, they first revise the distribution $g(\theta)$ of θ based on the observed demand up to time t. They use the result of Berger 1985 to obtain a posterior distribution of $g(\theta)$ and hence $f(x)$, based on the observed demand d during t, as

$$g(\theta|d) = N\left(\mu(d), \frac{1}{\rho^2}\right)$$

where

$$\mu(d) = \frac{\sigma^2 \mu}{\sigma^2 + \tau^2} + \frac{\tau^2 d}{\sigma^2 + \tau^2}$$

$$\rho^2 = \frac{1}{\sigma^2} + \frac{1}{\tau^2}$$

The posterior distribution of $f(x)$ can thus be written as

$$f(x|d) = N\left(\mu(d), \sigma^2 + \frac{1}{\rho^2}\right)$$

Using the above results, and assuming a single-period (news vendor) model, expressions for expected profit and other characteristics can be established for the two scenarios: without and with forecast revision.

Retailer Operations

As before, we assume that the retail unit price is s, manufacturer's unit cost is c, the manufacturer charges the retailer p per unit and the cost of carrying inventory is $r\$/\$$. We let the shortage cost (cost of goodwill) and the number of units stocked for the season to be v (per unit per period) and Q, respectively.

Note that the traditional problem (without forecast revision) is a standard news vendor problem, so that the expected retailer's profit for the season is written as

$$\Pi(R) = s \int_{-\infty}^{Q} f(x)dx + s \int_{Q}^{\infty} Q f(x)dx - pr \int_{-\infty}^{Q} (Q - x) f(x)dx -$$

$$v \int_{Q}^{\infty} (x - Q) f(x)dx - pQ$$

$$(11.25)$$

The optimal inventory level Q^* can be shown to be

$$Q^* (R) = \mu + z\sqrt{\sigma^2 + \tau^2} \qquad (11.26)$$

where z is the value of the standard normal variate corresponding to a given service level. The optimal service level in the news vendor model equals $(s + v - p)/(s + v + pr)$. The expected quantity sold, similar to expression 11.12, can be written as

$$S = \mu - L$$

where

$$L = \int_{Q}^{\infty} (x - Q) f(x)dx$$

is the expected lost sales, and can also be expressed as

$$L = B(z)\sqrt{\sigma^2 + \tau^2}$$

where $B(z)$ is the right-hand linear loss function of a standard normal distribution, corresponding to the value of z.

The expected maximum profit can be written as

$$\Pi^* (R) = s(\mu - L) - pr\{Q^* - (\mu - L)\} - vL - pQ^*$$

Substituting for Q^* and L, the above can be simplified to

$$\Pi^* (R) = (s - p)\mu - p(1 + r)z\sqrt{\sigma^2 + \tau^2} - (s + v + pr)B(z)\sqrt{\sigma^2 + \tau^2}$$

$$(11.27)$$

It is clear that the corresponding profit for the manufacturer will be expressed as

$$\Pi^*(M) = (p - c)(\mu + z\sqrt{\sigma^2 + \tau^2}) \tag{11.28}$$

Next consider the retailer's profit $\Pi_F(R)$ and manufacturer's profit $\Pi_F(M)$ when a forecast revision is undertaken at time t. It is clear from equation 11.26 that with an observed demand of d, we may express $Q_F^*(R)$ as

$$Q_F^*(R) = \mu(d) + z\sqrt{\sigma^2 + \tfrac{1}{\rho^2}}$$

However, since the observed demand d is also expected to have a normal distribution, the expected value of $Q_F^*(R)$ would be

$$EQ_F^*(R) = \int_{-\infty}^{\infty} Q_F^*(R)f(d)\,dd = \int_{-\infty}^{\infty} \mu(d)f(d)\,dd + z\sqrt{\sigma^2 + \tfrac{1}{\rho^2}}$$

which simplifies to

$$EQ_F^*(R) = \mu + z\sqrt{\sigma^2 + \tfrac{1}{\rho^2}} \tag{11.29}$$

The retailer's profit corresponding to equation 11.27 can now be written as

$$\Pi_F^*(R) = (s - p)\mu(d) - p(1 + r)z\sqrt{\sigma^2 + \tfrac{1}{\rho^2}} - (s + v + pr)B(z)\sqrt{\sigma^2 + \tfrac{1}{\rho^2}}$$

Integrating over all d, we can write

$$E\Pi_F^*(R) = (s - p)\mu - p(1 + r)z\sqrt{\sigma^2 + \tfrac{1}{\rho^2}} - (s + v + pr)B(z)\sqrt{\sigma^2 + \tfrac{1}{\rho^2}} \tag{11.30}$$

Similarly, the manufacturer's profit would be

$$E\Pi_F^*(M) = (p - c)(\mu + z\sqrt{\sigma^2 + \tfrac{1}{\rho^2}}) \tag{11.31}$$

From equations 11.27 and 11.30 it follows that

$$E\Pi_F^*(R) - \Pi^*(R) = \{zp(1 + r) + B(z)(s + v + pr)\}$$
$$\left\{(\sigma^2 + \tau^2)^{1/2} - \left(\sigma^2 + \frac{1}{\rho^2}\right)^{1/2}\right\}$$

It can be easily verified that

$$\tau^2 - \frac{1}{\rho^2} = \frac{\tau^4}{\sigma^2 + \tau^2} > 0$$

implying

$$\tau^2 > \frac{1}{\rho^2}$$

Hence

$$E\Pi_F^*(R) > \Pi^*(R), \text{ if } z > 0 \qquad (11.32)$$

That is, forecast revision will increase the retailer's profit if the retailer plans to satisfy more than 50% of customer demand.

From equations 11.28 and 11.31 we have, similarly,

$$E\Pi_F^*(M) = \Pi^*(M) = (p - c)z\left\{\left(\sigma^2 + \frac{1}{\rho^2}\right)^{1/2} - (\sigma^2 + \tau^2)^{1/2}\right\}$$

It is clear that

$$E\Pi_F^*(M) < \Pi^*(M), \text{ if } z > 0 \qquad (11.33)$$

Since customers expect a service level exceeding 90%, it follows that while the retailer would benefit from forecast revision, it will not be in the manufacturer's interest. Also note that the total system profit can be written as

$$E\Pi_F^*(R) + E\Pi_F^*(M) =$$
$$(s - c)\mu - \{z(c + pr) + B(z)(s + v + pr)\}\sigma\left(1 + \frac{1}{1 + \left(\frac{\sigma}{\tau}\right)^2}\right)^{1/2} \qquad (11.34)$$

It is clear that as $\frac{\sigma}{\tau}$ decreases, the total system profit in equation 11.34 as well as the retailer's profit in 11.30 will increase. The manufacturer's profit equation 11.31 would, however, decrease. Generally, the ratio $\frac{\sigma}{\tau}$ will be low for high-fashion goods (t being very high). Hence the manufacturer will be even more reluctant to participate in forecast revision for high-fashion goods.

Iyer and Bergen (1997) suggest several schemes that ensure that the manufacturer can increase its profit by participating in forecast revision. We briefly review a few of these suggestions.

Consignment Inventory

With consignment inventory, the retailer does not place any order. The manufacturer has access to customer demand and is responsible for replenishing the retailer's inventory with a consignment of goods so that a certain customer service level is satisfied. Since the retailer does not own the consignment, any units left unused at the end of the season belong to the manufacturer. It is clear that for a service level specified by the value of z, the manufacturer's consignment will equal $Q^*(R)$. Thus the manufacturer's cost would equal $c \cdot Q^*(R)$. The retailer will pay the manufacturer only for the units sold, which equals $\mu - L$. Hence the manufacturer's profit would be written as

$$\Pi^*(M) = p(\mu - L) - cQ^*(R)$$

Substituting from equation 11.26 and simplifying,

$$\Pi^*(M) = (p - c)\mu - cz\sqrt{\sigma^2 + \tau^2} - pB(z)\sqrt{\sigma^2 + \tau^2} \quad (11.35)$$

For the forecast revision case, the corresponding profit would be

$$\Pi_F^*(M) = (p - c)\mu - cz\sqrt{\sigma^2 + \frac{1}{\rho^2}} - pB(z)\sqrt{\sigma^2 + \frac{1}{\rho^2}} \quad (11.36)$$

From equations 11.35 and 11.36 we have

$$\Pi_F^*(M) - \Pi^*(M) = \{cz + pB(z)\}\left\{(\sigma^2 + \tau^2)^{1/2} - \left(\sigma^2 + \frac{1}{\rho^2}\right)^{1/2}\right\}$$

That is,

$$\Pi_F^* (M) > \Pi^* (M), \text{ if } z > 0$$

Note that the retailer does not pay for inventory carrying as it never owns the inventory, and it pays the manufacturer only for the units sold (not Q_R^*). Hence

$$\Pi^* (R) = s(\mu - L) - vL - p(\mu - L)$$

which simplifies to

$$\Pi^* (R) = (s - p)\mu - (s - p + v)B(z)(\sigma^2 + \tau^2)^{1/2}$$

Thus we can write

$$\Pi_F^* (R) - \Pi^* (R) = (s - p + v)B(z)\left\{(\sigma^2 + \tau^2)^{1/2} - \left(\sigma^2 + \frac{1}{\rho^2}\right)^{1/2}\right\}$$

That is,

$$\Pi_F^* (R) > \Pi^* (R), \text{ if } z > 1/2.$$

Both retailer and manufacturer improve their respective profits by revising forecasts.

Price Discount

Consider a price discount scheme with $p_1 \geq p$. The manufacturer charges p_1 per unit if the retailer's order size is less than $Q^* (R)$ and p per unit if the order size exceeds or is equal to $Q^* (R)$. Iyer and Bergen prove that such a price discount scheme would improve both the manufacturer's and retailer's profits when a forecast revision is implemented.

Volume Commitment

In a multi-product system, the retailer makes a commitment at time $t = 0$ to purchase a total of I units across M products. The exact order quantity of each product, on the other hand, is decided at time t after observing actual demand. Iyer and Bergen show that if $I = M\mu + Mz(\sigma^2 + \tau^2)^{1/2}$, the profit with revised forecasts at time t can be written as

$$E\Pi_F^* = (s + pr)M\mu - (p + pr)I -$$

$$(s + v + pr)M\left(\sigma^2 + \frac{1}{\rho^2}\right)^{1/2} \int_{-\infty}^{\infty} \phi_z(Z)B(Z)dZ$$

where $\phi_z(Z)$ is normally distributed with mean μ_z and variance σ_Z^2, where

$$\mu_z = \frac{I - M\mu}{M\left(\sigma^2 + \frac{1}{\rho^2}\right)^{1/2}} = \frac{z(\sigma^2 + \tau^2)^{1/2}}{\left(\sigma^2 + \frac{1}{\rho^2}\right)^{1/2}}$$

$$\sigma_Z^2 = \frac{\tau^4}{M(\sigma^2 + \tau^2)\left(\sigma^2 + \frac{1}{\rho^2}\right)}$$

The authors go on to establish that under the above conditions

$$E\Pi_F^*(R) > \Pi^*(R)$$

$$E\Pi_F^*(M) > \Pi^*(M)$$

That is, both manufacturer and retailer increase their respective profits with revised forecasts, given a commitment to purchase a specified total quantity of M products.

11.6 SHARING INVENTORY IN A NETWORK

For a typical inventory system the inventory level can be shown to rise very rapidly with customer service (Silver, Pyke and Peterson 1998). Thus there is a good chance of not being able to satisfy customer demand, no matter how high the inventory level happens to be. It is clear that by pooling their inventories together, retailers can reduce their inventory costs without affecting their service levels. Narus and Anderson (1996) point out that in modular supply chains, retailers have started to make cooperative arrangements to share inventory and other customer services. They predict that such arrangements would decrease costs by up to 20%.

Okuma Corporation has approached this problem by requiring each of its 46 distributors in North and South America to carry only a minimal number of machine tools in its inventory. However, the company tries to ensure that nearly all Okuma machine tool categories and parts are in stock at all times—either in its central warehouse or somewhere in its supply chain. Distributors can locate any part through Okumalink, which is a shared information system, and obtain its status. If a distributor does not possess an item required by a customer, it tracks down the distributor that is closest to the customer and has the item, and arranges for it to be delivered directly to the customer's facility.

Grahovac and Chakravarty (2000), based on the analyses of Sherbrooke (1967) and Axsater (1990), study inventory sharing in a supply chain with low-demand items. The demand is described by a discrete probability distribution, such as Poisson, with arrival rate λ. They consider a supply chain of N retailers and a central distribution center. When a customer demand occurs at a retailer and the item is in stock, the demand is filled immediately and, at the same time, a replacement order is placed with the distribution center. If the item is not in stock, an emergency order is placed in the system. The distribution center has the first crack at satisfying the emergency order. If it is out of stock, any one of the retailers that has the item may service the emergency order. If all retailers are out of stock, the emergency order is backlogged at the distribution center. The distribution center services the backlogged emergency orders before regular orders. The shipping time from the manufacturer to the distribution center is L. Shipping time for regular orders from distribution center to retailers is T, and for emergency orders (from distributors or from other retailers) it is F. Retailer i's stocking level is S_i.

The authors show that the steady-state probability $\Pi_i(S_i - k)$ of there being $S_i - k$ units of inventory at retailer i can be written as

$$\Pi_i(S_i - k) = \Pi_i^0 \frac{S_i! \mu_i^{k-S_i}}{k!(\lambda_i + \delta_i)^{k-S_i}}, \; k = 1, 2 \ldots S_i - 1 \qquad (11.37)$$

$$\Pi_i(S_i - k) = \Pi_i^0 \frac{S_i! \lambda_i^{(k-S_i)}}{k! \eta_i^{(k-S_i)}}, \; k = S_i + 1, S_i + 2 \ldots \qquad (11.38)$$

$$\frac{1}{\Pi_i(0)} = \sum_{k=0}^{S_i} \frac{S_i! \mu_i^{(S_i-k)}}{k!(\lambda_i + \delta_i)^{(S_i-k)}} + \sum_{k=S_i+1}^{\infty} \frac{S_i! \lambda_i^{(k-S_i)}}{k! \eta_i^{(k-S_i)}} \tag{11.39}$$

To solve for $\Pi_i(S_i - k)$ we would need to evaluate expressions 11.40 to 11.45 first.

$$\delta_i = \sum_{r \neq i} \frac{P_r(1 - \beta_r)\lambda_r}{1 + \sum\limits_{r \neq i} \beta_r} \tag{11.40}$$

where δ_i is the rate of transshipment request at i, and β_i the rate of regular shipment to i.

$$P_r = \frac{\sum\limits_{\text{Max}(S_0, S_k - M) \leq j \leq S - M} e^{-\lambda T} \cdot (\lambda L)^j / j!}{\sum\limits_{j \geq S_k - M} e^{-\lambda T} \cdot (\lambda L)^j / j!} \tag{11.41}$$

where P_r is the probability that an order at retailer i is transshipped from another retailer given that retailer i is out of stock. $S = \sum\limits_{i=0}^{N} S_i$.

$$\frac{1}{\mu_i} = T + \frac{B_0}{\lambda_0} \tag{11.42}$$

where $\frac{1}{\mu_i}$ is the average lead time for regular orders.

$$\frac{1}{\eta_i} = F + \frac{1}{\lambda_0} \sum_{j > S - M} (j - S_0)e^{-\lambda L} (\lambda L)^j / j! \tag{11.43}$$

where $\frac{1}{\eta_i}$ is the average lead time for emergency orders and S_0 is the inventory with the distribution center.

$$B_0 = \sum_{j=0}^{\infty} (j - S_0)e^{-\lambda L} (\lambda L)^j / j! \tag{11.44}$$

where B_0 is the average back order at the distributor.

$$M = \sum_{i=1}^{N} \lambda_i \{(\alpha_i + \beta_i)T + (1 - \beta_i)F\} \tag{11.45}$$

where M is the average total number of items in transit to retailers, and α_i is the rate of emergency shipment to i.

We cannot still solve for $\Pi_i(S_i - k)$, since the values of β_i and α_i are not known. We can express β_i in terms of $\Pi_i(j)$ as

$$\beta_i = \sum_{k=1}^{S_i} \Pi_i(k) \tag{11.46}$$

and

$$\alpha_i = P_i(1 - \beta_i) \tag{11.47}$$

We first solve for $\Pi_i(S_i - k)$ with assumed values of α_i and β_i. We then update α_i and β_i using equations 11.46 and 11.47 and repeat the computation of $\Pi_i(S_i - k)$. This iteration is continued until values of $\Pi_i(S_i - k)$ converge.

Next, we express the total cost in a centralized system, comprising transportation, inventory, and customer waiting, as

$$TC = \sum_{i=0}^{N} h_i S_i + \sum_{i=1}^{N} \lambda_i \{\alpha_i c_T + (1 - \beta_i)(c_F - c_T)\} + \sum_{i=1}^{N} c_w B_i \tag{11.48}$$

where B_i, the average backlog at the retailer, is computed as

$$B_i = \sum_{k=S_i}^{\infty} (k - S_i)\Pi_i(k)$$

c_T, c_F, and c_w are the cost per unit of regular shipment, the cost per unit of emergency shipment, and the cost of customer waiting per unit time, respectively. Inventory carrying cost is h_i per unit per period.

In a decentralized system, we assume that the distributor and retailers absorb their respective holding costs as well as a portion of the customer's waiting cost. We let R ($0 \leq R \leq 1$) denote the relative share of customer waiting and emergency shipping cost, absorbed by retailers. Thus cost minimization can be written as

$$\text{Minimize } C_0 = h_0 S_0 + (1 - R) c_w \sum_{i=1}^{N} B_i +$$

$$(1 - R) \sum_{i=1}^{n} \lambda_i \{ \alpha_i c_T + (1 - \beta_i)(c_F - c_T) \} \quad (11.49)$$

where S_i is the value that maximizes C_i where

$$C_i = h_i S_i + R c_w B_i + R \lambda_i \{ \alpha_i c_T + (1 - \beta_i)(c_F - c_T) \}, \ i = 1, 2, \ldots N \ (11.50)$$

Grahovac and Chakravarty (2000), based on a large number of computational runs, conclude that sharing inventory in the network can reduce total system cost by 20%, as predicted by Narus and Anderson (1996). Their results show that overall system inventories do not necessarily decrease. When inventories do decrease, the distributor happens to be the major beneficiary, sometimes at the expense of the retailers. This is counter to uncertainty reduction and price discounting for supply chain coordination, where retailers benefited most. This may suggest that integrating inventory sharing with uncertainty reduction and/or price discounting may have interesting implications. Note that consignment inventory, considered in the context of uncertainty reduction, is an extreme case of inventory sharing, where $R = 0$.

It is clear that the implied control policy in the Grahovac and Chakravarty model is $(S-1, S)$. While this is satisfactory for low-demand items such as repair parts, for the general case it may not be optimal to order one unit at a time. Anupindi and Bassok (1999) study inventory sharing for high-demand items in the context of centralization of inventory. They consider a two-retailer single-period model to investigate under what conditions it would be better to pool retailer's inventory at a central location. They also study a decentralized system where only a fraction of customers who do not get their orders satisfied at one retailer consider buying from other retailers. Unlike Grahovac and Chakravarty (2000), Anupindi and Basok's model does not incorporate shipping costs.

We next discuss a system like that of Okuma Corporation, where a virtual centralized warehouse has access to physical stocks held by individual retailers. A customer order can be satisfied by any of N retailers in the network. Recall that this is similar to the business-to-business e-commerce model used by Cisco Systems (called dynamic replenishment), discussed

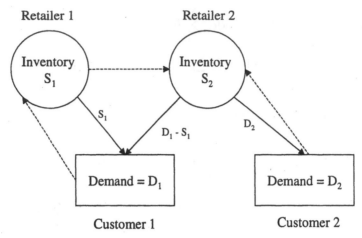

Retailer 1   Retailer 2

FIGURE 11.9: INVENTORY SHARING

in Chapter 10. To keep the exposition tractable, however, we consider a two-retailer single-period model, as in Anupindi and Bassok 1999.

The retailers form an extended enterprise as shown in Figure 11.9, where the solid lines represent the physical flow of goods and the dotted lines represent the transmission of order information. In the example in Figure 11.9, retailer 2 shares its inventory with retailer 1 by satisfying a part of the order of retailer 1's customer (customer 1). Retailer 2 also satisfies its own customer. It is assumed in Figure 11.9 that $D_1 > S_1$ and $D_1 + D_2 \leq S_1 + S_2$. Note that since retailers will locate themselves close to their own customers, there will be an additional shipping cost (u_{21} per unit) for the amount $D_1 - S_1$ from retailer 2 to customer 1.

The demand scenario shown in Figure 11.9 is one of six possible. The complete list of scenarios and their revenue implication are shown in Table 11.3.

In Table 11.3, s_i denotes unit surplus (unit price − unit cost) from customer i, and u_{ji} is the unit transportation cost from retailer j to customer i. Ignoring inventory holding and shortage costs, the expected surplus revenue corresponding to each scenario can be written as

$$\Pi_1 = \int_{D_1=-\infty}^{S_1} \int_{D_2=-\infty}^{S_2} (s_1 D_1 + s_2 D_2) f(D_1) f(D_2) dD_1 dD_2$$

	Scenario	Revenue
1	$D_1 \leq S_1, D_2 \leq S_2$	$s_1D_1 + s_2D_2$
2	$D_1 \leq S_1, D_2 \geq S_2$ $D_1 + D_2 \leq S_1 + S_2$	$s_1D_1 + s_2D_2 - u_{12}(D_2 - S_2)$
3	$D_1 \geq S_1, D_2 \leq S_2$ $D_1 + D_2 \leq S_1 + S_2$	$s_1D_1 + s_2D_2 - u_{21}(D_1 - S_1)$
4	$D_1 \leq S_1, D_2 \geq S_2$ $D_1 + D_2 \geq S_1 + S_2$	$s_1D_1 + s_2(S_1+S_2 - D_1) - u_{12}(S_1 - D_1)$
5	$D_1 \geq S_1, D_2 \leq S_2$ $D_1 + D_2 \geq S_1 + S_2$	$s_1(S_1 + S_2 - D_2) + s_2D_2 - u_{21}(S_2 - D_2)$
6	$D_1 \geq S_1, D_2 \geq S_2$	$s_1S_1 + s_2S_2$

Table 11.3: Demand Scenarios

$$\Pi_2 = \int_{D_1=-\infty}^{S_1} \int_{D_2=S_2}^{S_1+S_2-D_1} \{s_1D_1 + s_2D_2 - u_{12}(D_2 - S_2)\} f(D_1) f(D_2)dD_1dD_2$$

$$\Pi_3 = \int_{D_2=-\infty}^{S_2} \int_{D_1=S_1}^{S_1+S_2-D_2} \{s_1D_1 + s_2D_2 - u_{21}(D_1 - S_1)\} f(D_2) f(D_1)dD_2dD_1$$

$$\Pi_4 = \int_{D_1=-\infty}^{S_1} \int_{D_2=S_1+S_2-D_1}^{\infty} \{s_1D_1 + s_2(S_1 + S_2 - D_1) -$$
$$u_{12}(S_1 - D_1)\} f(D_1) f(D_2)dD_1dD_2$$

$$\Pi_5 = \int_{D_2=-\infty}^{S_2} \int_{D_1=S_1+S_2-D_2}^{\infty} \{s_1s_1(S_1 + S_2 - D_1) + s_2D_2 -$$
$$u_{21}(S_2 - D_2)\} f(D_1) f(D_2)dD_2dD_1$$

$$\Pi_6 = \int_{D_1=S_1}^{\infty} \int_{D_2=S_2}^{\infty} (s_1S_1 + s_2S_2) f(D_1) f(D_2)dD_1dD_2$$

where $f(D_i)$ is the pdf of customer i's demand. Thus the total expected surplus would be

$$\Pi = \Pi_1 + \Pi_2 + \Pi_3 + \Pi_4 + \Pi_5 + \Pi_6$$

An optimal solution can be obtained by equating $\frac{\partial \Pi}{\partial S_1}$ and $\frac{\partial \Pi}{\partial S_2}$ to zero. The differential coefficients $\frac{\partial \Pi}{\partial S_i}$ become complex. For example,

$$\frac{\partial \Pi_1}{\partial S_1} = s_1 F(S_1)\{S_1 + S_1 f(S_1)\} + s_2 S_2 f(S_1) - s_1 F(S_1)F(S_2) - s_2 f(S_1)G(S_2)$$

where

$$G(x) = \int F(x)dx$$

$$\frac{\partial \Pi_2}{\partial S_1} = s_1 F(S_1)F(S_2) + \frac{1}{2}(s_1 + s_2 - u_{12})F(S_1)F(S_2) -$$

$$(s_1 S_1 + s_2 S_2)f(S_1)F(S_2) - (s_1 + s_2 - u_{12})F(S_2)$$

In a similar way, $\frac{\partial \Pi_3}{\partial S_1}$ to $\frac{\partial \Pi_6}{\partial S_1}$ can be determined to solve for

$$\sum_{k=1}^{6} \frac{\partial \Pi_k}{\partial S_1} = 0$$

$$\sum_{k=1}^{6} \frac{\partial \Pi_k}{\partial S_2} = 0$$

11.7 SUPPLY CONTRACT

A contract is an agreement between two parties that sets out both what is expected of each in a business transaction and measures for dealing with deviations from the expected. The subject of contracts is not a new field of study, and it is a major consideration in the legal profession. The economics of contracts has also been amply studied and represented in the economics literature (e.g., Katz 1989). There is now a renewed interest in studying supply contracts in the context of the extended enterprise, information flow, the speed of information processing, and flexibility, as discussed in Chapter 10.

In a modular supply chain the decentralized entities arrive at their own optimal decisions based on local information, whereas in an integral system centralized decisions are based on global information. Thus the total system profit in an integral supply chain, $\Pi(I)$, tends to exceed the profit in a modular supply chain, $\Pi(M)$. A contract, in this context, is a means of bringing $\Pi(M)$ closer to $\Pi(I)$, not just by coordination but by creating appropriate incentives for the entities involved. Contracts can also be structured for facilitating long-term partnerships in a supply chain. Such contracts may not necessarily be beneficial in the short term, but in the long term both parties may benefit by gradually aligning their systems to be compatible with each other.

Tsay, Nahmias, and Agrawal (1999) suggest a typology of supply chain contracts and list eight types of contracts. This includes contracts related to specification of decision rights, pricing, minimum purchase commitments, quantity flexibility, buy-back policies, allocation rules, lead time, and quality. Many of the above categories do not represent typical contract scenarios, however. For example, one form of decision rights specification (Lee and Whang 1999) may involve a consignment policy for redistributing inventory carrying costs among the entities, but it does not specifically state measures for controlling deviations from this policy. The so-called minimum-purchase-quantity contract is a form of price discount based on purchase quantity, and we have discussed this in section 11.5. Similarly, use of allocation rules may not be a contract. A buyer may use allocation rules to decide how to split its purchase requirement among several competing suppliers. The contracts of real interest in today's economy are the so-called quantity-flexible contracts and lead-time contracts. By allowing buyers and sellers to deviate a little from their commitments so as to obtain a better price, an incentive is created in the system that helps in making all-around improvements in the system. Such flexibility allows participants to better avail themselves of opportunities presented to them. Similarly, in today's e-commerce economy there is a high premium on shortening lead times. Innovative contracts that reduce lead times are therefore called for.

Contractual Threshold

Consider a contractual situation between a retailer and manufacturer. Let the retailer's forecast be defined by mean μ and a forecast error of σ^2. Assume that the manufacturer negotiates a quantity flexibility of $\pm\ x$ from the mean μ, where x is a decision variable. Let the actual demand from retailer be a random variable D. The contractual terms can be stipulated in terms of delivery quantity Q (Chakravarty 2000b).

For the contractual terms,

$$Q = D, \text{ if } \mu - x \le D \le \mu + x$$

$$Q = \mu + x, \text{ if } D > \mu + x$$

$$Q = \mu - x, \text{ if } D < \mu - x$$

In regard to the supplier's actions, the supplier produces μ units for the period. If $\mu < D \le \mu + x$, the supplier produces an emergency batch or purchases a quantity $q = D - \mu$ at a premium unit price of c_1. If $\mu - x \le D \le \mu$, supplier disposes of the quantity $q = \mu - D$ at some cost c_2 (inventory holding and salvage). In a multi-period case, c_1 and c_2 will vary with time.

The retailer's revenue can be written as

$$\text{Revenue} = \int_{\mu-x}^{\mu+x} sDf(D)dD + \int_{\mu+x}^{\infty} s(\mu + x)f(D)dD + \int_{-\infty}^{\mu-x} s(\mu - x)f(D)dD$$

where s is the unit retail price. The retailer's cost would be

$$\text{Cost} = \int_{\mu-x}^{\mu+x} pDf(D)dD + \int_{\mu+x}^{\infty} p(\mu + x)f(D)dD +$$

$$\int_{-\infty}^{\mu-x} p(\mu - x)f(D)dD + \int_{-\infty}^{\mu-x} h(\mu - x)f(D)dD$$

where p the unit wholesale price and h is the inventory holding cost per unit per period. Thus the retailer's profit (revenue minus cost), after integrating out the terms, can be written as

$$\Pi(R) = (s - p)(\mu + x) - s(\mu - x)F(\mu - x) + s\int_{-\infty}^{\mu-x} DF(D)dD -$$

$$(s - p) \int\limits_{\mu - x}^{\mu + x} F(D)dD - h(\mu - x)F(\mu - x) \quad (11.51)$$

In a similar way, revenue and cost for the supplier would be expressed as

$$\text{Revenue} = \int\limits_{\mu - x}^{\mu + x} pDf(D)dD + \int\limits_{-\infty}^{\mu - x} p(\mu - x)f(D)dD + \int\limits_{\mu + x}^{\infty} p(\mu + x)f(D)dD$$

$$\text{Cost} = \int\limits_{\mu - x}^{\mu + x} cDf(D)dD + \int\limits_{\mu + x}^{\infty} c(\mu + x)f(D)dD + \int\limits_{-\infty}^{\mu - x} c(\mu - x)f(D)dD +$$

$$\int\limits_{\mu}^{\mu + x} c_1 (D - \mu)f(D)dD + \int\limits_{\mu + x}^{\infty} c_1 x f(D)dD +$$

$$\int\limits_{\mu - x}^{\mu} c_2(\mu - D)f(D)dD + \int\limits_{-\infty}^{\mu - x} c_2 x f(D)dD$$

where c is the unit manufacturing cost and c_1 and c_2 are the premiums as stipulated in the contract.

Therefore, the supplier's profit, after integrating out terms, can be written as

$$\Pi(S) = (p - c)(\mu + x) + c_1 x + (c_1 + c - p) \int\limits_{\mu}^{\mu + x} F(D)dD +$$

$$(c_2 + c - p) \int\limits_{\mu - x}^{\mu} F(D)dD \quad (11.52)$$

The retailer would like to maximize $\Pi(R)$. However, the supplier will not participate in the contract if it cannot guarantee a certain profit Y for itself. Thus the optimization problem can be written as

$$\text{Maximize } \Pi(R)$$

subject to

$$\Pi(S) \geq Y$$

Forming a Lagrangian function, we can write

$$\text{Maximize } L = \Pi(R) + \lambda\{\Pi(S) - Y\},$$

and determine λ such that $\Pi(S) \geq Y$.

Since $\Pi(R)$ and $\Pi(S)$ are concave in x, at the optimal value we would have

$$\frac{\partial L}{\partial x} = \frac{\partial \Pi(R)}{\partial x} + \lambda \frac{\partial \Pi(S)}{\partial x} = 0 \tag{11.53}$$

Differentiating equations 11.51 and 11.52, we can rewrite equation 11.53 as

$$A + \lambda B = 0$$

where

$$A = (s - p) + s(\mu - x)F(\mu - x) - (s - p)F(\mu - x) - $$
$$(s - p)F(\mu + x) + (h + s)F(\mu - x) + (h + s)(\mu - x)f(\mu - x)$$

$$B = (p - c - c_1) + c_1 F(\mu + x) - c_2 F(\mu - x) - $$
$$(p - c)\{F(\mu + x) + F(\mu - x)\}$$

The equation $A + \lambda B = 0$ can now be solved for x for values of λ, until $\Pi(S) = Y$.

Perhaps a simpler approach would be to evaluate equations 11.51 and 11.52 for various values of x, and choose the x for which $\Pi(S) = Y$.

Multiple Products with Capacity Constraints

Chakravarty (2000c) has studied quantity flexibility in the apparel industry, where retailers revise their orders closer to the selling season. Firms such as Liz Claiborne regularly outsource items from many suppliers (Flaherty 1993). Apparel designers and buyers estimate demand for fashion lines for the coming season, and allocate production quotas (i.e., pre-production orders) to the suppliers. The suppliers use this information to free up production capacity for Liz Claiborne items and make commitments for raw materials and production equipment with their own suppliers. Actual orders from Liz Claiborne's retailers arrive at the beginning of the season. If these orders are different from the pre-production orders, Liz Claiborne must consider adjusting its order quantities. Although Liz Claiborne attempts to coordinate changes in its orders with suppliers, the impact on suppliers of such changes can be severe if they delay acquisition of production capacity until after receiving firm orders.

A multi-tier pricing contract is one way of managing the production and supply operations in such a situation. Such a contract may be structured as follows: (1) a unit price of μ for a quantity x agreed to in advance, (2) a surcharge Γ per unit increase in the order from x to z, and (3) surcharge γ per unit decrease in the order from x to z. In addition, the company would incur a manufacturing cost β per unit of quantity x, an upward capacity adjustment cost of Δ per unit, and a downward capacity adjustment cost of δ per unit. The pre-production order size of x is determined in anticipation of the customer's order, but the quantity z is determined in real time after firm orders from retailers are known. Both x and z are determined by the company, given the cost parameters. The value of x is communicated to the supplier well before the start of the selling season so that the supplier can be proactive in reserving x units of capacity for the company. The supplier also needs to make commitments of raw material and/or tooling, equivalent to x units of supply, to its own suppliers.

In the apparel industry most suppliers are overseas (usually in East Asia). To save transportation costs, the products are shipped in bulk to the buyer's facilities in the United States. The company completes finishing operations such as final assembly, labeling, applying the trademark, and packaging in its own facilities before shipping the product to its retailers. In a multi-product situation, it is clear that the company will have a capacity-balancing problem as well, for if z exceeds x for one product, x *must* exceed z for some other product so that the total capacity used remains unchanged. We assume that the company's production facility can be made flexible or partially flexible. As flexibility of the facility increases, values of both Δ and δ will decrease. However, since total production capacity is limited and the required capacity per unit of production may not be identical for all products, the company's decision-making process must allow for multiple trade-offs in determining the values of x (original order) and z (revised order) for each product.

In preparing for the next period's production, both the company and its suppliers must make advance commitments for raw materials, components, tooling, and workers corresponding to an order quantity of x units. Sometime before the start of that period, actual demand D becomes known. Production resources mobilized for quantity x will need to be adjusted up or down depending on the value of D. It is clear that if x exceeds D, there will be a downward adjustment in resource configuration, since carrying of inventory is not allowed. If D exceeds x, an upward adjustment will be made to satisfy demand. The supplier will recoup its losses from such adjustments by appropriately fixing the surcharge parameters Γ and γ. Thus, corresponding to the company's problem where x and z are optimized, there exists a dual problem in which the supplier determines the optimal premiums to charge. We study only the company's problem. That is, for given values of all costs related to order revision, the company must determine x before D is known, and z after D is known. We also assume that the company's sales price has been determined by competitive forces in the market and is not affected by the values of cost parameters.

The problem can be solved by using a scenario approach (as in Eppen, Martin, and Schrage 1989) to model the uncertainties in demand. We let D_{is} denote anticipated demand for product i in scenario s, and D_{iR} observed demand in real time. Similarly, we denote the planned scenario-specific production quantity by z_{is} and the actual production quantity, determined in real time, by z_{iR}. The upward and downward adjustments in production quantity in scenario s are denoted as R_{is}, and r_{is}, respectively. The unit price to the customer is denoted as α_i. The company's profit can be written as

$$\Pi_i = (\alpha_i - \beta_i - \mu_i)z_i - Bx_i - (\Delta_i + \Gamma_i)(z_i - x_i)$$

Letting $w_i = \alpha_i - \beta_i - \mu_i$, $A_i = \Delta_i + \Gamma_i$, and $a_i = \delta_i + \gamma_i$, we rewrite Π_i as

$$\Pi_i = w_i z_i - A_i(z_i - x_i) - B_i x_i$$

Denoting the probability of scenario s by p_s and the maximum quantity of i that can be produced in the facility by m_i, the expected profit (net of adjustment and capacity costs) can now be written as

$$\Pi = \sum_s \sum_{i \in I} \{w_i z_{is} - (A_i R_{is} + a_i r_{is})\} p_s - \sum_{i \in I} B_i x_i \qquad (11.54)$$

where

$$\sum_s p_s = 1 \qquad (11.55)$$

Since the policy is not to hold inventory,

$$z_{is} \le D_{is}, \ \forall_i \text{ and } s \qquad (11.56)$$

The capacity constraint of the facility is expressed as

$$\sum_i \frac{z_{is}}{m_i} \le 1, \ \forall_s \qquad (11.57)$$

$$\sum_i \frac{x_i}{m_i} \le 1 \qquad (11.58)$$

The upward and downward adjustments in production quantity can be defined as

$$R_{is} \ge z_{is} - x_i, \ \forall_i \text{ and } s \qquad (11.59)$$

$$r_{is} \ge x_{is} - z_i, \ \forall_i \text{ and } s \qquad (11.60)$$

Since Π is to be maximized, it is clear from equations 11.54, 11.59, and 11.60 that if $R_{is} > 0$, $r_{is} = 0$ and vice versa. Observe that scenario s is a random variable, but D_{is} are predetermined. To determine p_s we create scenarios of demand variation according to the following scheme:

$$D > \mu + k\sigma \Rightarrow \text{High}$$

$$\mu + k\sigma \ge D \ge \mu - k\sigma \Rightarrow \text{Medium}$$

$$D > \mu - k\sigma \Rightarrow \text{Low}$$

Using the above ranges of product demand, the probability p_s is computed for a known demand distribution.

In real time, after the actual demand D_{iR} is known, a decision must be made whether or not to increase (or decrease) the production order size, which is set at respective x_i values. If there are no capacity constraints, the revised values would equal D_{iR} or x_i, depending on the values of x_i, w_i, A_i, and a_i. On the other hand, if there is a capacity constraint, the revised values of z_{is} ($= z_{iR}$) may be equal to x_i, D_{iR}, or a value between x_i and D_{iR} if $D_{iR} > x_i$. Note that for the case $D_{iR} < x_i$, x_i will be set equal to D_{iR}, as discussed earlier, at a cost of $a_i (x_i - D_{iR})$, before any other real-time adjustments are made.

Chakravarty (2000c) suggests a ranking scheme of products,

$$\alpha_i = (w_i - A_i) - \underset{j \neq 1}{\text{Min}} \left\{ \frac{m_j}{m_i} (w_j + a_j) \right\}$$

and shows that it is optimal to pick products from this ranked list for capacity allocation when there is a capacity shortage.

Quantity Flexibility in a Multi-Period Environment

The only model for schedule revision in multiple periods is by Tsay and Lovejoy (2000). However, unlike Chakravarty 2000b discussed earlier, in their model the threshold of quantity flexibility w_j and a_j in period j are not decision variables. Specifically, at period t the retailer provides the supplier with a set of information in several categories, written as a vector

$$\{r(t)\} = [r_0(t), r_1(t) \ldots r_j(t) \ldots]$$

where $r_0(t)$ is the actual purchase in period t and $r_j(t)$ is the forecast purchase in period $t + j$, the forecast being made in period t. The thresholds α_j and w_j are expressed as $\alpha = [\alpha_1, \alpha_2 \ldots]$ and $w = [w_1, w_2 \ldots]$. The flexibility that the buyer has in revising $\{r(t)\}$ is expressed as

$$[1 - w_j] r_j(t) \leq r_{j-1}(t + 1) \leq [1 + \alpha_j] r_j(t)$$

That is, the quantity $r_j(t)$ is revised to $r_{j-1}(t+1)$ in period $t+1$.

Tsay and Lovejoy minimize a convex cost function subject to the above flexibility constraints and show that optimal $r_j(t)$ can be expressed as

$$r_j t = \text{Max} \{T_j(t), (1 - w_{j+1})r_{j+1}(t - 1)\}, j = 0, 1, 2 \ldots$$

where

$$T_j(t) = \frac{(1 + A_j)f_j(t) - l_j(t)}{1 + A_j}$$

$$l_j(t) = \begin{cases} I(t - 1), \text{ for } j = 0 \\ [l_{j-1}(t) + (1 - \Omega_{j-1})r_{j-1}(t) - (1 + A_{j-1}(t)]^t, j \geq 1 \end{cases}$$

and where $f_j(t)$ is the quantity sold to customer and $I(t)$ is the inventory at a supply chain node at the end of period t. A and Ω are expressed as

$$1 + A_j = \prod_{q=1}^{j} (1 + \alpha_q)$$

$$1 + \Omega_j = \prod_{q=1}^{j} (1 + w_q)$$

11.8 SUPPLY CHAIN ACCOUNTING

In Chapters 9 and 10 we have discussed at length the problems of motivating the entities in a modular supply chain to work toward a common goal. For example, when manufacturers use distributors for a product, the distributor may not be as concerned as the manufacturer about stock-outs. In a slightly different scenario, a retailer may experience stock-outs and loss of customer goodwill because of the unreliability of its suppliers. The cost of stock-outs is borne by the retailer with no recourse. A mechanism for holding the suppliers accountable for their actions would go a long way toward mitigating these problems.

Lee and Whang (1999) propose an accounting procedure that ensures that the total system cost is fully allocated to site managers in each period. All costs are traced to individual entities so as to eliminate free riders. Such a procedure can also be shown to generate the right kind of incentives for aligning business processes of firms in a supply chain.

Consider a retailer and manufacturer denoted by subscripts r and m, respectively. The simple idea behind this incentive scheme is that the manufacturer will pay for inventories held in the system. In return, the retailer will pay the manufacturer for any shortages at its facility. The accounting procedure is based on the notion of echelon inventory and echelon holding cost of inventory.

For the manufacturer and retailer shown in Figure 11.10, echelon inventory x_m, at the manufacturer, is the sum of the on-hand inventories at the manufacturer and retailer and the amount in transit from the manufacturer to the retailer, minus the backorders at the retailer. Thus we can write

$$x_m = I_m + I_r + w_m - B_r \qquad (11.61)$$

$$x_r = I_r + w_r - B_r, w_r = 0 \qquad (11.62)$$

where I represents on-hand inventory, B the backorders, and w the in-transit inventory (w_m is in transit from the manufacturer to the retailer). Echelon holding costs (rate) at manufacturer and retailers are $\bar{h}_m$, and $\bar{h}_r$, where $\bar{h}_m = h_m$ and $\bar{h}_r = h_r - h_m$. Note that $\bar{h}_r$ represents the holding cost due to value added at the retailer. We also define two functions, $x^+ = \text{Max}(x, 0)$, and $x^- = \text{Max}(-x, 0)$. Next, we trace the payments from the manufacturer to the retailer and vice versa.

The manufacturer must pay for its inventory I_m, the retailer's inventory I_r, and the in-transit inventory w_m, corresponding to the value added by it, h_m. Therefore the manufacturer's cost is written as

$$C(M) = h_m(I_m + I_r + w_m)$$

Using equation 11.61, we rewrite $C(M)$ as

$$C(M) = h_m(x_m + B_r)$$

Verify that we can express B_r as

Manufacturer $\longrightarrow$ Retailer $\longrightarrow$

FIGURE 11.10: SUPPLY CHAIN ACCOUNTING

$$B_r = x_r^+ - x_r$$

Hence $C(M) = h_m\{x_m + x_r^+ - x_r\}$

The retailer must pay for any backlog for two reasons: loss of goodwill at a cost of v per unit, and a payment to the manufacturer at a cost of h_m per unit. The retailer must also pay for its own inventory (the value-added part) and the cost of goodwill:

$$\text{Inventory cost} = (h_r - h_m) x_r^+$$

$$\text{Goodwill cost} = v x_r^-$$

Retailer and Manufacturer's Costs

Hence the total cost to the manufacturer would be

$$TC(M) = C(M) - h_m x_r^-$$

We rewrite $TC(M)$ as,

$$TC(M) = h_m(x_m - x_r) + h_m\{x_r^+ - x_r^-\} = h_m x_m$$

Similarly, the retailer's total cost would be,

$$TC(R) = (h_r - h_m) x_r^+ + (h_m + v) x_r^-$$

By adding and subtracting the term $(h_r - h_m) x_r^-$, we simplify $TC(R)$ to

$$TC(R) = (h_r - h_m) x_r + (h_r + v) x_r^-$$

Total System Cost

The total system cost can be written as

$$TC(M) + TC(R) = h_m x_m + (h_r - h_m) x_r + (h_r + v) x_r^-$$

which can be rewritten as

$$TC(M) + TC(R) = h_m(x_m - x_r) + h_r x_r + h_r x_r^- + v x_r^-$$

Observing that $h_r x_r = h_r\{x_r^+ - x_r^-\}$, we have

$$TC(M) + TC(R) = h_m(x_m - x_r) + h_r x_r^+ + v x_r^-$$

Next, since

$$x_m = I_m + I_r + w_m - B_r$$

and

$$x_r = I_r - B_r$$

we have

$$x_m - x_r = I_m + w_m$$

Hence the system cost is written as

$$TC(M) + TC(R) = \{h_m(I_m + w_m)\} + \{h_r x_r^+ + v x_r^-\}$$

The total cost has two clear components, one for the manufacturer (with subscript m) and the other for the retailer (with subscript r). The cost component $h_m(I_m + w_m)$ is fully controlled by the manufacturer, and the cost component $h_r x_r^+ + v x_r^-$ is fully controlled by the retailer. Thus the two parties incur cost for what they control.

Cachon and Zipkin (1999) have studied a generalized version of the above problem where the backordering cost is shared by both parties instead of being charged wholly to the manufacturer.

REFERENCES

Anupindi, R., and Y. Bassok (1999), "Centralization of Stock: Retailers vs. Manufacturers," *Management Science,* vol. 45, pp. 178–191.

Arntzen, B., G. Brown, T. Harrison, and L. Trafton (1995), "Global Supply

Chain Management at Digital Equipment Corporation," *Interfaces*, vol. 25, pp. 69–93.

Axsater, S. (1990), "Modeling Emergency Lateral Shipments in Inventory Systems," *Management Science*, vol. 36, pp. 1329–1338.

Banerjee, A. (1986), "On a Quantity Discount Pricing Model to Increase Vendor Profits," *Management Science*, vol. 32, pp. 1513–1517.

Berger, J. (1985), *Statistical Decision Theory and Bayesian Analysis*, 2nd ed., Springer-Verlag, New York.

Cachon, G., and P. Zipkin (1999), "Competitive and Cooperative Inventory Policies in a Two-Stage Supply Chain," *Management Science*, vol. 45, pp. 936–953.

Chakravarty, A. (2000a), "Product Variety, Capacity and Pricing Decisions in a Global Manufacturing Environment," working paper, A. B. Freeman School of Business, Tulane University, New Orleans.

Chakravarty, A. (2000b), "Optimal Contractual Thresholds," working paper, A. B. Freeman School of Business, Tulane University, New Orleans.

Chakravarty, A. (2000c), "Revising Production and Supply Quantities in Real Time with Capacity Constraints and Supplier Surcharge: A Single Period Model," working paper, A. B. Freeman School of Business, Tulane University, New Orleans.

Chakravarty, A., and N. Balakrishnan (2000), "Impact of Global Supply Chain Uncertainties on Product Definition," working paper, A. B. Freeman School of Business, Tulane University, New Orleans.

Chakravarty, A., and G. Martin (1988), "An Optimal Joint Buyer-Seller Discount Pricing Model," *Computers and Operations Research*, vol. 15, no. 3, pp. 271–281.

Chakravarty, A., and G. Martin (1989), "Discount Pricing Policies for Inventories Subject to Declining Demand," *Naval Research Logistics*, vol. 36, pp. 89–102.

Chakravarty, A., and G. Martin (1991), "Operational Economies of a Process Positioning Determinant," *Computers and Operations Research*, vol. 18, no. 6, pp. 515–530.

Cohen, M., and H. Lee (1989), "Resource Deployment Analysis of Global Manufacturing and Distribution Networks," *Journal of Manufacturing and Operations Management*, vol. 2, pp. 81–104.

Eppen, G., K. Martin, and L. Schrage (1989), "A Scenario Approach to Capacity Planning," Operations Research, vol. 37, no. 4, pp. 517–527.

Flaherty, T. (1993), "Liz Claiborne Inc. and Reuntex Industries," Harvard Business School Case 9-693-098, Boston, Mass.

Grahovac, J., and A. Chakravarty (2000), "Sharing and Lateral Transshipment of Inventory in a Supply Chain with Expensive Low Demand Items," working paper, A. B. Freeman School of Business, Tulane University, New Orleans, (will appear in Management Science).

Henderson, J., and R. Quandt (1980), *Microeconomic Theory: A Mathematical Approach*, McGraw-Hill, New York.

Hodder, J. (1984), "Financial Market Approaches to Facility Under Uncertainty," *Operations Research*, vol. 32, pp. 1374–1380.

Hodder, J., and Dincer (1986), "A Multifactor Model for International Plant Location and Financing Under Uncertainty," *Computers and Operations Research,* vol. 13, pp. 601–609.

Iyer, A., and M. Bergen (1997), "Quick Response in Manufacturer Retailer Channels," *Management Science,* vol. 43, pp. 559–570.

Katz, M. (1989), "Vertical Contractual Relations," in *Handbook of Industrial Organization,* vol. 1, R. Schmalensee and R. Willig (eds.), Elsevier, New York.

Lee, H., and M. Rosenblatt (1986), "A Generalized Quantity Discount Pricing Model to Increase Supplier's Profits," *Management Science,* vol. 30, pp. 1524–1539.

Lee, H., and S. Whang (1999), "Decentralized Multi-Echelon Supply Chains: Incentives and Information," *Management Science,* vol. 45, pp. 633–640.

Monahan, J. (1984), "A Quantity Discount Pricing Model to Increase Vendor Profits," *Management Science,* vol. 30, pp. 720–726.

Narus, J., and J. Anderson (1996), "Rethinking Distribution," *Harvard Business Review,* July–August, pp. 112–120.

Sherbrooke, C. (1968), "METRIC: A Multiechelon Technique for Recoverable Item Control," *Operations Research,* vol. 16, pp. 122–141.

Silver, E., D. Pyke, and R. Paterson (1998), Inventory Management and Production Planning and Scheduling, 3rd ed., John Wiley and Sons, New York.

Simon, H. (1989), *Price Management,* North Holland, New York.

Skinner, R. (1992), "Fashion Forecasting at Oxford Shirtings," Proceedings of Quick Response Conference, University of Chicago, Chicago, March 17–18, pp. 90–107.

Tsay, A., and W. Lovejoy (2000), "Quantity Flexibility Contracts and Supply Chain Performance," *Manufacturing and Service Operations Management,* vol. 1, pp. 89–111.

Tsay, A., S. Nahmias, and N. Agrawal (1999), "Modeling Supply Chain Contracts: A Review," in *Quantitative Models of Supply Chain Management,* S. Tayur, M. Magazine, and R. Ganeshan (eds.), Kluwer Academic Publishers, Boston, Mass.

12

Responsive Manufacturing

12.1 PARADIGM SHIFT

Today more than ever, a manufacturer must operate as one of the entities in a modular supply chain. The (inevitable) switch of power to customers in a supply chain has been accelerated. For a manufacturer to be competitive, therefore, rapid and appropriate response to customers' shifting needs is a must. A response that is determined in real time after observing the changes in customers' needs is seldom cost-effective. Although it has the allure of simplicity, it can be very expensive and/or off target. As Hayes, Wheelwright, and Clark (1984) point out, good manufacturing systems are like organic entities. They evolve gradually as the knowledge base of the system is enhanced, effecting improvements in manufacturing capabilities. Therefore, a reactive response through quick fixes in the system can be counterproductive. Real-time responses cannot always be avoided, and in that case, they must be incorporated in the planning process.

For example, it is well known that a flexible system is ideal for responding quickly to changes in market demands. Flexibility, however, must be built over time through investment in technology, process reengineering, and worker training. Quality improvements similarly require changes in attitude in addition to investment in technology and worker training. While manufacturing systems must be proactive, uncertainties in customer needs set a limit to the degree of proactivity possible. The Allis Chalmers Company in Milwaukee (Chapter 1) which invested heavily in flexible manufacturing in the 1970s is a case in point. Jaikumar's (1986) further attests to this, finding that investment in flexible technology in the United States has not been market driven.

The answer to what the appropriate level of proactivity should be emerges from a company's competitive strategy and, more specifically, its manufacturing strategy. The resource-based view (Wernerfelt 1984) is to develop and exploit a set of resources, such as technology and organizational skills, for specific products and market positioning. In the context of manufacturing strategy, this view can be captured by a performance frontier, as shown in Figure 12.1.

The points *A* and *B* in Figure 12.1 correspond to the positioning of two competing companies. According to Porter 1980, *B* is a low-cost competitor and *A* competes by product differentiation. *A* would provide larger product variety through its flexible processes, whereas *B* would employ a focused facility that generates economies of scale from less-flexible manufacturing equipment. But this process view does not explain why Japanese companies in the 1980s were able to offer a wider product variety at a lower cost. Teece and Pisano (1994) provide an explanation through the so-called dynamic capabilities approach. This approach posits that firms can differentiate themselves not only through the bundles of human and organizational capabilities they possess, but also through knowledge (and programs) that help create new capabilities. Examples of such new capabilities would be a more precise process control for improving quality, reducing the number of steps involved in shaping a part, and lean manu-

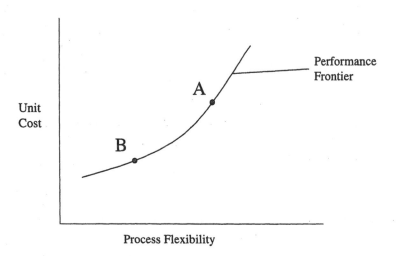

FIGURE 12.1: PERFORMANCE FRONTIER

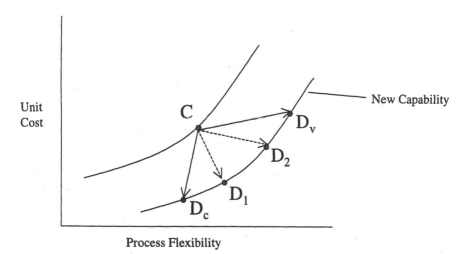

Unit Cost

New Capability

C

D_V

D_2

D_1

D_c

Process Flexibility

FIGURE 12.2: INCORPORATING NEW CAPABILITIES

facturing (JIT with TQM). The performance frontier associated with the new capability is shown in Figure 12.2.

A company positioned at point C with inferior capability may wish to move to any point between D_C and D_V on the new performance frontier. Moving to D_c implies that the company is using a cost-reduction strategy in adopting the new capability. Similarly, a move to point D_v implies a differentiation strategy. The points D_1 and D_2, on the other hand, imply improvements in both cost and differentiation and explain how Japanese companies gained competitive advantage in the 1980s.

Observe that productivity (and the associated strategy) discussed so far is targeted to the economies of the product and its in-house position. This implicitly assumes that a customer would buy company A's product even if it solves only a part of the customer's problem. That is, the customer may need to buy other components and/or products from different suppliers to be able to use A's product. In the new economy of today, this assumption may not hold, and company A may have to create alliances with other suppliers to be able to provide a complete solution to customers. A graphical representation of such an alliance will be confusing. A mathematical approach will be more precise.

Let R be the product range that can be supplied by an alliance with n partners and R_i the product range provided by the ith partner, so that

$R = \sum_{i=1}^{n} R_i$. Let $c_j(R_j)$ represent the performance frontier (as in Figure 12.1) of the jth partner, and $v(R)$ represent the value associated with product range R. Thus the profit in the alliance would be

$$\Pi(R) = v(R) - \sum_{j=1}^{n} c_j(R_j) - A$$

where A is the company's share of the cost of forming and running the alliance. Consider the case when $c_j(R_j) = R_j^2/a_j$. $\text{Min} \sum_j c_j(R_j)$ will be obtained at $R_j = a_j\left(\frac{R}{b}\right)$, where $b = \sum_{i=1}^{n} a_i$, and $\text{Min} \sum_{j=1}^{n} c_j(R_j) = R^2/b$.

Thus,

$$\Pi(R) = v(R) - \frac{R^2}{b} - A$$

The optimal value of R can be found by solving $\frac{2R}{b} - \frac{\partial v(R)}{\partial R} = 0$. The optimal profit will then equal $\Pi(R)$, with $R = R^*$. Thus the optimal product range, and hence process flexibility, will be determined by the characteristics of the partners in the alliance.

To analyze the impact of different capabilities, let subscript k denote a capability level. Note that each value of k will generate a performance frontier as in Figure 12.1. Thus the optimal profit for a given k, corresponding to $\Pi(R^*)$, can be written as

$$\Pi^k = \Pi(R^k) - G^k$$

where G^k is the cost of developing the kth capability. Hence

$$\Pi^* = \underset{k}{\text{Maximum}} (\Pi^k) = \underset{k}{\text{Maximum}} \{\Pi(R^k) - G^k\}$$

While we used a simple quadratic form for $c(R)$ for the sake of exposition, its form in a real-world application may be very complex. This will mainly be determined by the nature of interactions among products and processes. It is clear that $c(R)$ will be impacted by economies of scale, equipment flexibility, investment in new technology, investment in setup cost reduction, and the cost of outsourcing, in addition to the cost of labor and material.

12.2 PRODUCT-PROCESS INTERACTIONS

To understand this interaction between products and processes, we start from customer needs in terms of products attributes and product performance level and track it all the way to realized sales in Figures 12.3 to 12.7. The impact of customer needs on realized product quality, manufacturing lead time, and manufacturing cost is shown in Figure 12.3. Demand for the product, based on product quality and other factors, is determined as shown in Figure 12.4. In Figure 12.5, production quantity and capacity requirements are established. In Figure 12.6, the impact of scheduling uncertainties on actual output is established. Finally, in Figure 12.7, the notion of inventory balance is explained. We first describe the charts in Figures 12.3 to 12.7 individually before discussing how they can be used together (Chakravarty and Ghose 1993).

Note in Figure 12.3 that design complexity of a product increases if the desired product performance level and/or the number and type of desired product attributes increases. Design complexity, in turn, increases

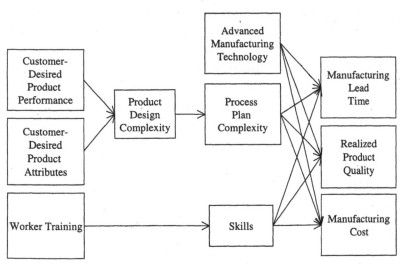

FIGURE 12.3: PRODUCT DESIGN AND MANUFACTURING

FIGURE 12.4: PRODUCT DEMAND

the complexity of the process plan. The manufacturing lead time increases with process complexity and decreases with workers' skill level and/or acquisition of advanced manufacturing technology. It is also clear that the realized product quality decreases with process complexity and increases with the skill level of workers and/or advanced technology. Manufacturing cost, as shown in Figure 12.3, increases with all three factors: process complexity, skill level, and advanced technology.

Realized product quality and manufacturing lead time (waiting time of customers) are next tracked to customers' demand for the product, as shown in Figure 12.4. It is clear that the quality level of a product, as perceived by customers, increases with the realized (actual) quality level, warranty budget, after-sale service budget, and advertising budget. It may also increase with price. Demand for the product would increase with perceived product quality and sales promotion budget. Demand decreases, however, with manufacturing lead time (customer waiting time) and the backlog (shortage) of the product.

Next, product demand and process plan complexity (from Figure 12.3) are tracked to production quantity schedule and production capacity schedule, as shown in Figure 12.5. Determination of the best production-

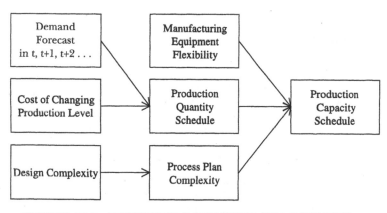

FIGURE 12.5: PRODUCTION QUANTITY AND CAPACITY

quantity schedule, given demand forecasts and relevant costs, is amply discussed in the operations management literature (Krajewski and Ritzman 1992). A process plan determines the number of processing steps and amount of time needed to complete the step, the need for expensive equipment, and the risk of errors by workers. Process plans also specify whether a particular process is unique (shared by a few products) or common (shared by many products). Observe next that required production capacity increases with the number of processing steps. The duration for which a machine is blocked by a process can increase if these are errors by workers. Note that a set of unique processes may require only a few flexible machines or a large number of dedicated machines. Finally, the required production capacity increases with the quantity to be produced in a period.

The quantity and capacity schedules are then tracked to realized production output in real time, as shown in Figure 12.6. Competitive pressures and capacity shortfalls may mandate schedule changes in real time. If the competition quotes a shorter delivery lead time, the company may have to follow suit by cutting manufacturing lead time on certain products. Similarly, if new products must be introduced at short notice to fend off a competitor, the production schedules of existing products must be revised. The company may want to switch to new processes proactively to move to a different performance frontier (see Figure 12.2), requiring changes in production schedules. Finally, machine breakdowns may necessitate delaying certain products. The revised quantity schedule along with the capacity schedule and work-in-process inventory determine which

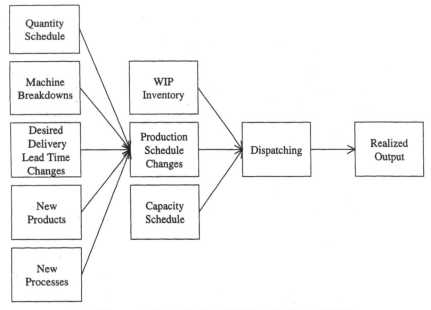

FIGURE 12.6: REALIZED OUTPUT IN REAL TIME

products should be dispatched to which machine for processing. Since products may be kept waiting between processing steps, dispatching rules substantially impact the output rate of completed products.

The realized quantities and inventories or backlogs in period $t-1$, together with demand in period t (from Figure 12.4), are then tracked to realized sales, inventories, and backlogs in period t in Figure 12.7. If the total available quantity exceeds demand, inventory will have to be carried in period t. If not, there will be a backlog unless demand equals available quan-

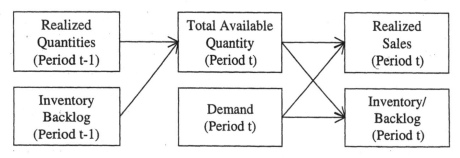

FIGURE 12.7: REALIZED SALES AND INVENTORY/BACKLOG

tity. Available quantity equals realized production quantity plus inventory, or realized production quantity minus backlog, all in the previous period.

Consider now the use of the charts in Figures 12.3 to 12.7. Let us assume the company is considering broadening the attribute bundle of some of its products. From Figure 12.3 we see that design complexity will increase, increasing process complexity. This would decrease the realized (actual) quality of the product (i.e., the rate of defective items may increase) and increase manufacturing lead time. The company may wish to recover all or part of the quality and lead time by investing in worker training, advanced manufacturing equipment, or both. A broader attribute bundle in a product will increase the level of perceived quality. The company will be justified in increasing the unit price for its offering, which in turn would further increase the level of perceived quality. The company would naturally increase advertising for the new, more appealing product and strengthen warranty and after-sale service to go with it. As a result, the perceived quality may even exceed what it was before despite a possible drop in realized quality. The higher perceived quality together with increased sales promotion may increase product demand in spite of longer lead times.

Next, note in Figure 12.5 that the increased process complexity will definitely increase the cost of capacity. If product demand increases as well, capacity cost would increase as the production rate increases. If demand variability increases (whether or not demand increases), investment in flexible machines may pay dividends.

Observe in Figure 12.6 that with a sufficient investment in production capacity the realized output in real time may increase. This, together with the assumption that demand increases, would increase actual sales.

Thus, whether or not it is profitable to broaden the attribute bundle would depend on whether the investments in worker training, advanced technology, warranty, after-sale service, advertising, sales promotion, and production capacity can be more than offset by increased revenue from a combination of increased unit price and possible increase in sales. Also, note that tracking the impact through the charts reveals a number of different investment options available to the company to support the attribute-bundle-broadening strategy. The company may invest in all or only a few of the available options. For example, it may decide to invest in worker training and advertising and nothing else. Next, through tracking the interactions we also see that there exists a large number of chains linking the customer-desired product attribute bundle (Figure 12.3) to realized sales

(Figure 12.7). The company may focus its action strategy on only one such chain, or it may decide to act on several chains simultaneously. Thus investing in worker training, advertising, and machine flexibility will correspond to focusing on a single chain. Investing in worker training as well as advanced technology will be tantamount to acting on at least two chains simultaneously. There are thus two related sets of questions: how much to invest where, and how much information to retrieve. A single-chain strategy may require less information retrieval than a multiple-chain strategy. The information retrieval problem can be solved by using the relationship chain management (RCM) techniques discussed in detail in Chapter 3. The investment optimization problem can be solved by framing it as a mathematical program.

12.3 ACQUISITION OF ADVANCED MANUFACTURING TECHNOLOGY

As we have seen in Figure 12.3, acquisition of advanced manufacturing technology (AMT) is one of the ways of improving the competitive capability of a company. As Clark (1996) explains, Searle Medical Instruments Group could move to an improved performance frontier by investing in AMT (Schemener 1977). It does not, however, imply that AMT would improve the competitiveness of all companies in all industries (Jaikumar 1986). Boyer et al. (1997) point out that the capabilities of AMT can be fully realized when companies also invest in upgrading the skills of their workforce. The complete picture may be much broader. As we have seen in Figures 12.3 to 12.6, AMT interacts not only with manufacturing variables but also with marketing and strategic variables such as advertising, warranty, differentiation, and price. In the manufacturing domain, the impact of AMT is seen in terms of savings in inventory, throughput time, labor, scrap, and machine time. In the marketing domain the benefits are seen through product quality, product variety, and ability to respond quickly to changes in product demand and product design.

Acquisition of AMT has been treated as a capital expenditure item in most companies. As such, it is required to pass a "hurdle rate" test, suitable for stand-alone traditional technologies. Since AMT requires significant capital investment, to managers the perceived risk in AMT acquisition is very great. For this reason the hurdle rate is kept artificially high. Discounted cash flow (DCF) is still the most-used technique for making a decision on AMT. Lefly and Sarkis (1997) in a survey found that more than 80 percent of the companies in the United States use the payback-period procedure (using DCF) for evaluating AMT. Finance-based criteria such as ROI and payback period are likely to fail in the context of AMT because its long-term strategic benefits, such as quality, flexibility, delivery reliability, and response to market needs, are not easily quantifiable. There is also a synergistic effect among technologies such as CAD/CAM, FMS, and automated storage and retrieval system (AS/RS) that the financial techniques are unable to capture.

The analytic hierarchy process (AHP) proposed by Saaty (1980) is a weighted-factor scoring model that has been used in many studies for evaluating large, complex systems. Participants are asked to do a pair-wise comparison of all attributes of the system (Warbalickis 1988). AHP uses these comparisons to arrive at a ranking (or relative dominance) of all attributes. Though it provides a structure, the number of pair-wise comparisons can be very large. A more severe weakness is that in their comparisons, the participants are not asked to factor in strategic and customer-related implications of their evaluations.

Thus the available evaluation approaches include quantitative (financial) methods, subjective (judgmental) scoring techniques, AHP, optimization approaches, and simulation. Each of these techniques addresses a limited aspect of a very complex problem. A broader perspective, similar to interaction tracking, would therefore seem to suggest itself.

Naik and Chakravarty (1992) have proposed a three-level hierarchical approach to assess the strategic and operational fit of AMT. They assume that the AMT comprises a machining (processing) system, a materials handling system, a tool handling system, an inspection system, and system integration software. Strategic evaluation takes place at the top level of the hierarchy, operational evaluation is conducted at the middle level, and the bottom level is meant for financial evaluation. Strategic evaluation is split into four hierarchical stages of its own. Each stage is focused at a set of attributes related to a specific aspect of business. As shown in Figure 12.8, the attributes in stage 1 are the competitive strategy options available to the company. Stage 2 attributes are customer needs; in stage 3 they are manufacturing capabilities; and in stage 4 the attributes define several AMT configurations (in terms of machining, materials handling, tool handling, inspection, and system integration).

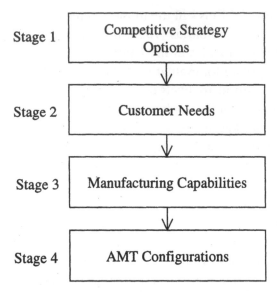

FIGURE 12.8: STRATEGIC EVALUATION

The hierarchy of stages used for operational evaluation is shown in Figure 12.9. The hierarchy of stages for financial evaluation is shown in Figure 12.10.

Observe in all three diagrams, Figures 12.8 to 12.10, that the hierarchies are similar to, but more straightforward than, those used for product-process interaction tracking in Figures 12.3 to 12.7. Because of the simpler structure we may analyze each level of the hierarchy using an approach similar to the house of quality, discussed in Chapter 6.

Consider strategic evaluation. A house of quality may be used for analyzing the interactions between attributes in each pair of adjacent stages. Assume that the competitive strategy options in stage 1 are innovation, customization, low cost, and a hybrid of innovation and customization. Similarly, assume that the attributes of customer needs are price offered, attractiveness of product features, performance and reliability, delivery speed and reliability, product line breadth, and speed of new product introduction. These attributes may be linked by a house of quality, as shown in Figure 12.11.

Entries in the cells are importance ratings—N (nil), L (low), M (medium), H (high), and V (very high). These entries are made by asking questions such as "How important is high performance and reliability to

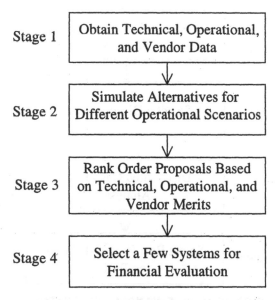

Stage 1	Obtain Technical, Operational, and Vendor Data
Stage 2	Simulate Alternatives for Different Operational Scenarios
Stage 3	Rank Order Proposals Based on Technical, Operational, and Vendor Merits
Stage 4	Select a Few Systems for Financial Evaluation

FIGURE 12.9: OPERATIONAL EVALUATION

customization strategy?" For example, in Figure 12.11 the entry in the cell corresponding to performance and reliability vs. customization is H, implying that the importance of product performance to customization is high.

Customer needs, in turn, are related to the attributes of manufacturing capability, as in Figure 12.12.

Finally, the manufacturing capabilities are related to AMT configurations as shown in Figure 12.13. Note that the cell entries in Figure 12.13 measure how suitable each of the manufacturing capabilities is to each of the customer needs.

Our objective is to track the strategy options all the way to AMT configurations. We first track them to customer needs by merging Figures 12.11 and 12.12. To do so, let us assume that we wish to determine the importance of high throughput to customization. We may track customization to throughput by using the second row of Figure 12.11 and the first column of Figure 12.12, as shown in Figure 12.14.

Note that there are six paths linking customization to throughput, and any one of these six paths can be used to track one to the other. Obviously,

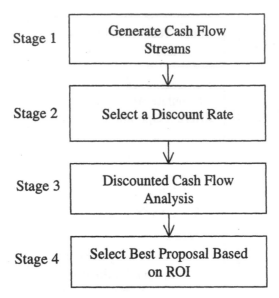

Stage 1	Generate Cash Flow Streams
Stage 2	Select a Discount Rate
Stage 3	Discounted Cash Flow Analysis
Stage 4	Select Best Proposal Based on ROI

FIGURE 12.10: FINANCIAL EVALUATION

the path chosen will have the best overall importance rating. To determine a path's importance rating, consider the path through the rectangle labeled as low price. Since low price is not consistent with customization, high throughput together with low price cannot be important to customization, however high the importance of high throughput to low price may be. Therefore, if u_{ij} is the importance of the ith manufacturing capa-

Strategy Options	Low Price	High Feature Attractiveness	High Performance and Reliability	High Delivery Speed and Reliability	Broad Product Line	High Speed of New Product Intro
Innovation	N	V	H	N	N	V
Customization	N	M	H	M	N	N
Low Cost	V	L	M	M	H	N
Innovation and Customization	N	H	H	L	N	H

FIGURE 12.11: TRACKING STRATEGY OPTIONS TO CUSTOMER NEEDS

Customer Needs	High Through-put	High Machine Utilization	Low Setup Time	Low Material Movement Time	Manu-facturing Flexibility	Efficient Inspection	Quick Process Modification
Low Price	V	H	M	M	N	N	N
High Feature Attractiveness	N	N	N	N	N	M	H
High Performance and Reliability	N	N	N	N	N	V	N
Rapid Delivery and Reliability	N	N	M	M	H	N	M
Broad Product Line	N	N	H	M	V	N	M
Rapid New Product Introduction	N	N	H	M	H	M	H

FIGURE 12.12: CUSTOMER NEEDS AND MANUFACTURING CAPABILITIES

bility to *customization* corresponding to the path through the jth customer need, we may express it as $u_{ij} = \text{Worst}\,(x_j, y_{ij})$ where x_j = importance of the jth customer need to customization and y_{ij} = importance of the ith manufacturing capability to the jth customer need.

For the example above, $i = 1$ for throughput, $j = 1$ for low price, and $x_1 = N$, $y_{11} = V$, so that $u_{11} = \text{Worst}\,(N,V) = N$.

In a similar way, it can be verified that $u_{12} = N$, $u_{13} = N$, $u_{14} = N$, $u_{15} = N$, and $u_{16} = N$. Hence the importance of high throughput to customization would be N. That is $u_i = \frac{\text{Best}}{j}\,(u_{ij})$. We can establish the importance of other attributes of manufacturing capability to customization as shown in Figure 12.15.

To track customization to AMT configuration, we construct paths between them that pass through manufacturing capability attributes using Figures 12.15 and 12.13. The paths linking FMS to customization are shown in Figure 12.16.

Assume that w_{ik} measures the importance of the kth AMT configuration to customization corresponding to the path passing through the ith manufacturing capability. As before, we have

$$w_{ik} = \text{Worst}\,(u_i, z_{ki})$$

Since $k = 2$ identifies the FMS, we have

Manufacturing Capabilities	AMT Configurations									
	Machining System		Materials Handling System (MHS)		Tool Change System		Inspection System		System Integration	
	Transfer Line	FMS	Manual MHS	AS/RS MHS	Manual Tool Change	Automated Tool Change	Manual Inspection	Online Inspection	CAD/ CAM	CAD/ CAM & ERP
High Throughput	V	M	L	H	L	M	N	N	N	N
High Machine Utilization	V	H	L	H	L	M	N	N	N	N
Low Setup Time	L	H	N	N	N	N	N	N	N	N
Low Material Movement Time	N	N	M	V	N	N	N	N	N	N
Manufacturing Flexibility	N	H	M	H	L	H	V	N	N	N
Efficient Inspection	N	N	N	N	N	N	L	H	N	N
Quick Process Modification	N	H	N	N	N	N	N	N	H	H

FIGURE 12.13: MANUFACTURING CAPABILITIES AND AMT CONFIGURATIONS

$$w_{12} = N, \; w_{22} = N, \; w_{32} = M, \; w_{42} = N, \; w_{52} = M, \; w_{62} = N, \; w_{72} = M.$$

To compare AMT configurations against one another, we need to establish a measure of the utility of FMS to customization. We define this as the number of manufacturing capabilities that are impacted by the AMT configuration and are at least moderately important to customization. For the FMS we see that $w_{32} = w_{52} = w_{72} = M$. Hence the utility measure of FMS to customization would be 3.

In general we let $\delta_{ik} = 1$ if w_{ik} is better than or equal to M. Letting w_k be the utility of AMT configuration k to customization, we can write

$$w_k = \sum_i \delta_{ik}$$

In the example above, with $k = 2$,

$$\delta_{32} = \delta_{52} = \delta_{72} = 1$$

Hence

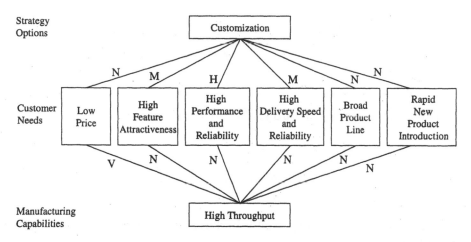

FIGURE 12.14: TRACKING CUSTOMIZATION TO THROUGHPUT

	High Throughput	High Machine Utilization	Low Setup Time	Low Material Movement Time	Manufacturing Flexibility	Efficient Inspection	Quick Process Modification
Customization	N	N	M	M	M	H	M

FIGURE 12.15: TRACKING CUSTOMIZATION TO MANUFACTURING CAPABILITY

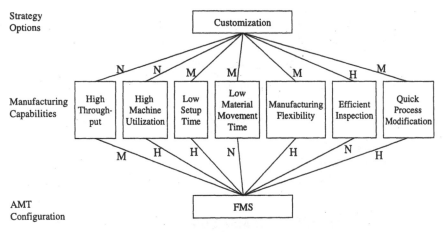

FIGURE 12.16: TRACKING CUSTOMIZATION TO FMS

	Machining		MHS		Tool Change		Inspection		Integration	
	Transfer Line	FMS	Manual MHS	AS/RS MHS	Manual Tool Change	Automated Tool Change	Manual Inspection	Online Inspection	CAD/ CAM	CAD/ CAM and ERP
Customization	0	3	2	2	0	1	1	1	1	1

FIGURE 12.17: TRACKING CUSTOMIZATION TO AMT CONFIGURATION

$$w2 = \delta_{32} + \delta_{52} + \delta_{72} = 3$$

The values of w_k ($k = 1$ to 8), established in a similar way, are shown in Figure 12.17.

To build an AMT configuration, it is clear that we need to choose one option from each of the five entities: machining, MHS, tool change, inspection, and integration software. The possible configurations that maximize the summation of w_k for all k, *for this example* would be as shown in Figure 12.18.

Observe that in Figure 12.18 there are 16 possible paths, implying that 16 AMT configurations will all have the best utility if *customization* is the way to compete. Note that manufacturing capability may be improved at one or more of the five configuration entities. Setup time is improved by FMS;

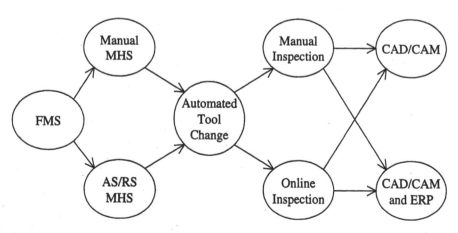

FIGURE 12.18: AMT CONFIGURATIONS

material movement time is improved by manual MHS or AS/RS, whichever is chosen. Manufacturing flexibility is improved by four of the five entities. Process modification speed is improved by two of the above entities.

12.4 MANUFACTURING CAPABILITY AND FLEXIBILITY

The QFD framework facilitates the representation of linkages between different types of capability and competitive strategy, as discussed in section 12.3. For example, the most appropriate capability for the competitive strategy of customization was found to be a flexible manufacturing system. The manner in which the linkage between flexible capability and customer satisfaction is realized, however, has not been explained. In the next few sections we examine the role of flexible capability in terms of benefits, investment, real-time control, and how such flexibility can be created through process design.

Manufacturing flexibility is a multifaceted capability. However, at a generic level, we may define flexibility very simply as a means of meeting future uncertainty. Uncertainties are caused by the diversity in the environment in which the technology must operate. Need of flexibility in different environments would also depend on the nature of the response planned. For example, the capability corresponding to taking different actions in different circumstances is not the same as preserving flexibility by postponing decisions (delaying commitment). Mandelbaum (1978) defines two types of flexibility: state flexibility and action flexibility. State flexibility is the capacity of continuing functioning effectively despite changes in environment, and action flexibility, as mentioned earlier, is the ability to switch to new actions as needed. Similarly, uncertainties can be of different types. Uncertainties in environment could be related to product volume, new products, unit price, and quality. Flexibility in capability, on the other hand, would be related to equipment, materials handling, process plans, and routing, among others. Contingencies such as equipment breakdowns, variability in task times, queuing delays, rejects, and reworks necessitate flexibility in capability to preserve customer satisfaction. Flexibility has long been recognized as a positive characteristic in manufacturing;

nearly 50 years ago, for example, Diebold (1952) conjectured about a machine that could simultaneously perform a bundle of functions. Manufacturing flexibility is usually defined for a scenario, which in turn is defined by products, processes, and procedures.

Sethi and Sethi (1990) have defined eleven types of flexibilities: three at the equipment level, five at the managerial level, and another three at a system (long-term) level. Of these, the major forms of flexibility that a manager must understand in responding to uncertainties are process flexibility, volume flexibility, and product flexibility. Process flexibility is the ability to switch to different (known) products without incurring setup cost; volume flexibility is the ability to increase or decrease production volume rapidly; and product flexibility is the ability to introduce new products quickly. The three-equipment level flexibilities (machine, materials handling, and process plan) are important, but they are driven by the flexibilities at a managerial level mentioned above.

The basic decisions in all such flexibilities are (1) determining the product portfolios, production quantities, and new products that would be most synergistic with one another, and (2) optimizing investment in flexibilities at the equipment level. We consider these two decisions together.

Capacity Mix Planning

Consider a simple case first—a single flexible facility producing products 1 and 2, as shown in Figure 12.19.

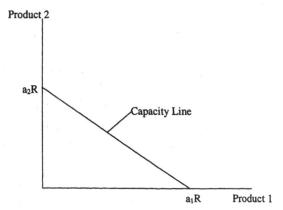

FIGURE 12.19: SINGLE FLEXIBLE FACILITY

The facility can produce a maximum of $a_1 R$ units of product 1 and $a_2 R$ units of product 2, where R is the capacity size (measured in appropriate units). It is clear that the equation of the capacity line would be

$$\frac{y_1}{a_1} + \frac{y_2}{a_2} = R \qquad (12.1)$$

where y_i denotes the maximum production quantities of the products if both are produced. Let the demand of product i be denoted as x_i with pdf of $f(x_i)$. Assuming c to be the cost of acquiring capacity per unit and p_i ($i = 1,2$) to be the unit prices of products, the profit Π for given x_i can be written as

$$\Pi(x_1, x_2, R) = p_1 x_1 + p_2 x_2 - cR, \text{ if } \frac{x_1}{a_1} + \frac{x_2}{a_2} \le R$$

Assuming $p_1 a_1 > p_2 a_2$ the expected profit would be

$$\Pi = \int_{x_1 = -\infty}^{a_1 R} \int_{x_2 = -\infty}^{a_2 (R - \frac{x_1}{a_1})} (p_1 x_1 + p_2 x_2) f(x_1) f(x_2) dx_1 dx_2 +$$

$$p_1 a_1 R \int_{x_1 = 0}^{\infty} \int_{x_2 = \alpha}^{\infty} f(x_1) f(x_2) dx_1 dx_2 - cR \qquad (12.2)$$

where

$$\alpha = \left\{ a_2 \left(R - \frac{x_1}{a_1} \right) \right\}$$

The optimization problem is to maximize equation 12.2 with respect to R.

Next, consider a more complex decision where dedicated capacities for products 1 and 2, equal to R_1 and R_2, is planned. In addition, there is a flexible facility of size R that can be used by either (or both) of the products if demand exceeds the capacity of their respective dedicated facilities. The capacity diagram is shown in Figure 12.20.

There are now seven possible demand scenarios (S_1 to S_7) that determine how much of which of the three facilities would be used. These are

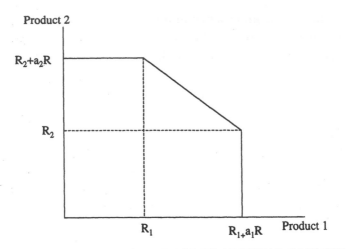

FIGURE 12.20: MIX OF DEDICATED FLEXIBLE CAPACITY

S_1: $x_1 \leq R_1$ and $x_2 \leq R_2$,

S_2: $x_1 \leq R_1$ and $R_2 < x_2 \leq R_2 + a_2R$,

S_3: $x_1 \leq R_1$ and $x_2 > R_2 + a_2R$,

S_4: $R_1 < x_1 \leq R_1 + a_1R$ and $x_2 \leq R_2$,

S_5: $x_1 > R_1 + a_1R$ and $x_2 \leq R_2$,

S_6: $x_1 > R_1$, $x_2 > R_2$ and $\frac{x_1 - R_1}{a_1} + \frac{x_2 - R_2}{a_2} < R$

S_7: $x_1 > R_1$, $x_2 > R_2$ and $\frac{x_1 - R_1}{a_1} + \frac{x_2 - R_2}{a_2} > R$

Note that the revenue function for known x_1, x_2, and sufficiently large values of R_1, R_2, and R would be

$$\Pi(x_1, x_2, R_1, R_2, R) = p_1x_1 + p_2x_2$$

Thus in scenario S_1 the expected profit would be

$$\Pi(S_1) = \int_{x_1 = -\infty}^{R_1} \int_{x_2 = -\infty}^{R_2} \{p_1x_1 + p_2x_2\} f(x_1)f(x_2)dx_1dx_2 \qquad (12.3)$$

In scenario S_2, since x_2 falls within the limits of R_2 and $R_2 + a_2R$, we let $u_2 = x_2 - R_2$. We can now write $\Pi(S_2)$ as

$$\Pi(S_2) = \int_{x_1 = -\infty}^{R_1} \left\{ \int_{u_2 = 0}^{a_2R} (p_2R_2 + p_1x_1 + p_2u_2)f(x_1)f(u_2)du_2 \right\} dx_1 \quad (12.4)$$

In scenario S_3, the number of units of product 2 sold would be $R_2 + a_2 R$. So $\Pi(S_3)$ would be

$$\Pi(S_3) = \int_{-\infty}^{R_1} \{p_2(R_2 + a_2 R) + p_1 x_1\} f(x_1) dx_1 \qquad (12.5)$$

Similar to S_2, in scenario S_4 we let $u_1 = x_1 - R_1$. $\Pi(S_4)$ would be written as

$$\Pi(S_4) = \int_{x_2=-\infty}^{R_2} \left\{ \int_{u_1=0}^{a_1 R} (p_1 R_1 + p_1 u_1 + p_2 x_2) f(u_1) f(x_2) du_1 \right\} dx_2 \quad (12.6)$$

Scenario S_5 is similar to S_3. It can be expressed as

$$\Pi(S_5) = \int_{x=-\infty}^{R_2} \{p_1(R_1 + a_1 R) + p_2 x_2\} f(x_2) dx_2 \qquad (12.7)$$

In scenario S_6, units sold would exceed R_i ($i = 1,2$), and the excess demand beyond R_i would be assigned to the flexible facility. The expression for $\Pi(S_6)$ would be

$$\Pi(S_6) = \int_{u_1=0}^{a_1 R} \int_{u_2=0}^{a_2(R-\frac{u_1}{a_1})} \{p_1(R_1 + p_2 R_2 + (p_1 u_1 + p_2 u_2)\} f(u_1) f(u_2) du_1 du_2$$

$$(12.8)$$

In scenario S_7, the flexible facility will be used to produce product 1 or 2, whichever fetches higher profit ($p_1 a_1 R$ vs. $p_2 a_2 R$). Both dedicated facilities will be completely utilized. Hence if $p_1 a_1 > p_2 a_2$, we have

$$\Pi(S_7) = \int_{x_1=R_1}^{\infty} \int_{x_2=a_2}^{\infty} \{p_1(R_1 + a_1 R_1) + p_2 R_2\} f(x_1) f(x_2) dx_1 dx_2 \quad (12.9)$$

where

$$\alpha_2 = R_2 + a_2 \left(R - \frac{x_1 - R_1}{a_1} \right)$$

Total expected profit would be

$$\Pi(R_1, R_2, R) = \sum_{i=1}^{7} \Pi(S_i) - (c_1 R_1 + c_2 R_2 - cR) \qquad (12.10)$$

Note that in scenario S_7, the flexible facility was not shared; it produced either product 1 or 2, depending on the value of $p_i a_i$. In situation where customer service rather than profit is the objective, the above rule for sharing (or assigning) the flexible facility would not be optimal. The profit maximization problem is to maximize $\Pi(R_1, R_2, R)$ with respect to R_1, R_2, and R. Chakravarty (1989) suggests a capacity rationing rule and shows the interdependence between the optimal rationing rule and investment in capacity. Assuming uniform distribution $f(x_i)$ in the interval (0 to b_i), and customer service levels of β_i, Chakravarty expresses the rationing rule as

Give priority to product 1 (in allocating flexible capacity) if

$$\frac{1-q}{a_2}(x_2 - R_2) - \frac{q}{a_1}(x_1 - R_1) \geq R(1 - 2q) \qquad (12.11)$$

where $q = AB/AC$ in Figure 12.21.

That is, if the demand corresponds to a point in region ABD, product 1 will have priority allocation of R, and if demand falls in region BCD, product 2 will be produced. He goes on to establish bounds on q as

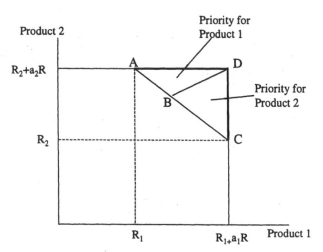

FIGURE 12.21: RATIONING RULE

$$1 - \frac{b_2(R_1 + a_1R) - b_1b_2\beta_1}{a_1a_2R^2/2} \leq q \leq \frac{b_1(R_2 + a_2R) - b_1b_2\beta_2}{a_1a_2R^2/2} \tag{12.12}$$

The capacity investment problem can be written as

$$Minimize\ c_1R_1 + c_2R_2 + cR$$

subject to

$$R_1 + a_1R \geq b_1\beta_1$$

$$R_2 + a_2R \geq b_2\beta_2$$

$$(R_1 + a_1R)(R_2 + a_2R) - \frac{R_2}{2a_1a_2} \geq b_1b_2\beta_1\beta_2$$

Chakravarty's analysis reveals that if $R_1 + a_1R = b_1\beta_1$ and $R_2 + a_2R > b_2\beta_2$, then product 1 has priority in allocation of flexible capacity. However, if $R_1 + a_1R$ also exceeds $b_1\beta_1$, then q will lie within the bounds specified in equation 12.12.

Fine and Freund (1990) have studied a capacity acquisition problem assuming that everything produced can be sold. They use a quadratic profit function to incorporate the variation of profit per unit with production quantity. They evaluate this model for different demand scenarios with discrete probabilities. Andreou (1990), in a different approach from the finance literature, shows how flexibility can be conceived of as real options.

Chakravarty (1994) has broadened the dimensions of decisions in three different ways. He models cost of flexibility in terms of system elements that contribute to flexibility. For example, in the context of an automated assembly system, he uses investment in holders, grippers, assembly heads, and workers—all contributing to flexibility in different ways—as different decision variables. Thus the cost of the flexible facility cR will be split into c^1 (holders), c^2 (grippers), c^3 (assembly heads), and c^4 (workers). He also accounts for the finite life of an assembly cell by incorporating the time dimension. Finally, the requirements for holders, grippers, and assembly heads are expressed in terms of components of products and, indirectly, commonality among components. The decision variables are technology configurations to be used. For example, configurations can be expressed as (transfer line, manual assembly), (transfer line, robotic assembly), (robotic assembly), and (robotic assembly, transfer line, and manual assembly). Corresponding to each technology configuration would be a partition of the product set indicating which products are assigned to which technology cell. The investment in manufacturing cells is formulated as a dynamic program with time periods as stages and two-state variables in each stage: technology configuration and product partition. He solves the model for a scenario with 5 products, 6 part types per product, 3 cell types, 6 types of robotic grippers, 15 types of indexing heads, and 3 types of fixtures. The indexing heads and robot grippers possess flexibility in terms of parts they could assemble.

Eppen, Martin, and Schrage (1989) have applied a formulation similar to that in Chakravarty 1994 to capacity planning at General Motors. They, however, simplify the problem by assuming that plant configuration is independent of the mix of products fabricated in the plant. Thus any product can be manufactured in any plant under any configuration. This also allows them to transfer a proportion of unsatisfied demand for a product to other products. The cost of changing from one plant configuration to another is assumed fixed and known. They solve the model for 20 products, 7 plants, 30 possible configurations, 5 time periods in the planning horizon, and 3 demand scenarios in every period. Since there is a nonzero probability of losing money, they computed the downward risk as

$$f(\Pi, z) = \begin{cases} z - \Pi, \text{ if } \Pi \le z \\ 0, \text{ if } \Pi > z \end{cases} \tag{12.13}$$

where z is the target profit obtained from their model and Π the actual profit. Since Π is a random variable with pdf $\Phi(\Pi)$, the expected downward risk can be expressed as

$$E\{f(\Pi, z)\} = \int_{-\infty}^{\infty} \Phi(\Pi) f(z, \Pi) d\Pi \tag{12.14}$$

12.5 REAL-TIME FLEXIBILITY

As illustrated in Figure 12.6, production contingencies may arise in real time from several sources, such as machine breakdowns, unexpected demand increases, new product introduction, and excessive rejects. Typically, production schedules must be adjusted to respond to these contingencies. MRP-based procedures have been used to minimize the impact of contingencies. Such procedures include automatic triggering based on exception conditions and preset guidelines (Orlicky 1975) and dampening mechanisms such as time fences and quantity rules (Steel 1975). These procedures are usually nonoptimal, and therefore they make poor use of capacity and/or they require frequent reruns of the MRP software. In any case, the impact of contingency cannot be corrected in the same time period when it occurs.

In a system where manufacturing capacity is constrained, one may add short-term capacity through overtime, additional shifts, or outsourcing. Holding inventory is another way of dampening the impact of contingencies. Make-to-order is a capability suited to a quick response to changes in customers' demand. Make-to-stock capability, on the other hand, leads to cost minimization. A mixed approach where products are assembled based on actual orders (assemble-to-order) and components are made to stock has been suggested as a way of providing a measured degree of flexibility at a low cost. In such a system, if demand in period t_j for component j exceeds capacity, the excess demand may be lost, or the manufacturer may be able to adjust capacity allocation to components to delay the onset of the shortfall to period τ $(\tau > t_j)$.

This obviously assumes that additional capacity can be found (overtime, outsourcing, etc.) in periods up to t_j. It is clear that costs may be incurred for two reasons: creating additional short-term capacity in periods prior to t_j or backordering for a duration $\tau - t_j$. The manufacturer would, of course, like to maximize $\tau - t_j$. For a given t_j, this would imply increasing τ. Observe that τ can be increased by a capacity reallocation to product i from other products in periods prior to t_j.

Assume that v_{jt} is the inventory of component j and h_j is the carrying cost per unit for the period. Assume also that x_{it} is the quantity of item i commencing production in period t. The constant a_{ij} denotes the number of units of component i required to produce one unit of the parent item j. The production lead time of component j is denoted by L_j.

Next, assume that a contingency of size u_m occurs in component m in period Γ. It is clear that this contingency will result in backordering of the end item at time $t > \Gamma$. The disturbance can "travel" from component m up the MRP tree to the end item via multiple paths, which may not all be of equal length (duration). Thus, if the disturbance travels only on the minimum-length path of length g_{mE}, t will be written as

$$t_{min} = \Gamma + g_{mE}$$

Note that disturbance will travel along a path that has no slack capacity. Similarly, for the longest path of duration f_{mE} we have

$$t_{max} = \Gamma + f_{mE}$$

Clearly the manufacturer's strategy would be to shift capacity from the longer paths to the shorter paths, starting from the shortest path.

Thus the optimization problem can be written as

$$\text{Minimize} \sum_{j \in J} \sum_{t=1}^{T} h_j v_{jt} + \sum_{t=t_{min}}^{t_{max}} \delta_t (B_{Et}) \qquad (12.15)$$

where δ_t is the cost of backordering in period t and B_{Et} is the quantity of the end product backordered. The material balance constraint for the end item E would be expressed as

$$x_{i,\,t-L_i} - \sum_{E \in S_i} a_{iE}(x_{Et} - B_{Et}) = 0, \ \forall_i, \ \forall_t, \qquad (12.16)$$

For component j it would be

$$v_{j,\,t-1} + x_{j,\,t-L_j} - v_{jt} - \sum_{k \in S_j} a_{jk} x_{kt} = 0, j \neq m, \ \forall_t, \qquad (12.17)$$

And for component m in period Γ it would be

$$v_{m,\,t-1} + x_{m,\,t-L_m} - v_{mt} - \sum_{k \in S_m} a_{mk} x_{kt} = u_{m,\,\forall_+} \qquad (12.18)$$

$$\sum_j x_{j,\,t-L_j} \leq \text{capacity}, \ \forall_j, \ \forall_t$$

To solve the optimization problem defined by expressions 12.15 to 12.18, values of δ_t will have to be input exogenously, leading to a trial-and-error solution. Second, the values of t_{min} and t_{max} must also be determined exogenously (a tedious process at best). In addition, for a real-world problem the number of components would be too large for any realistic solution. Chakravarty and Balakrishnan (1998), using the so-called frame-transition approach, have devised a fast solution procedure that produces optimal solutions.

Note that although t_{max} is the latest time to which backordering can be delayed, it may not be achievable because of inability to free up capacities in earlier periods on the shorter paths from m to E. A revised t_{max} called t_L is determined that is consistent with capacity availability. The total capabilities thus freed up in each period are determined. Next, the most appropriate item whose production quantities can be reduced in period t, and which has its short path to the end product exceeding t_{min}, is identified. The current production schedule is then adjusted. Next, the latest period at which the capacity requirement is unsatisfied is identified. An attempt is then made to free up capacity in this future period. If it is not successful, t_L is increased to $t_L + 1$. This process is repeated until all capacity needs are satisfied.

Dispatching Flexibility

We have seen that because of contingencies caused by customers, the planned production quantities need adjustment in real time. In a pure made-to-order environment, the need to adjust planned quantities is even greater. We have seen how assignment of capacity can be adjusted from one set of components to another to delay the need for backordering. However, such frequent adjustments to the production schedule expose the manufacturing system to increasing levels of stress, often described as system nervousness (Steel 1975).

Typically, a flexible machine is used by more than one part type. These parts are released to the machine at appropriate times when the machine is free. The release rules, also called dispatching rules, assign priority to the parts competing for a machine in real time. Many such dispatching rules have been devised and used. These rules are widely known as the shortest processing time (SPT), expected due date (EDD), first in first out (FIFO), and last in first out (LIFO). Application of these rules generates different performance frontiers (in terms of cost and flexibility) for the manufacturing system. It therefore follows that as the production schedule is adjusted in real time, the dispatching rule used should also change for better productivity. In practice, however, a dispatching rule is chosen through trial and error or through simulation and is usually retained throughout a machine's life cycle. The reason is that the relationship between dispatching rules and short-term scheduling changes is usually not obvious.

In recent years two approaches, by Shaw, Park, and Raman (1992) and by Chakravarty (1997), have been proposed for discerning such relationships. Bhattacharyya and Koehler (1998) have reviewed several learning rules that can be used in a dispatching context. Shaw, Park, and Raman propose the use of inductive learning with pattern-directed decisions for selecting dispatching rules opportunistically. They use simulation and machine learning techniques to extract patterns, which are then related to dispatching rules in the form pattern $(a,b) \rightarrow$ dispatching rule k. More specifically, one may state such assertions in the following form: if $x > a$ and $y > b$, then dispatching rule k. The system state values such as a and b are used to classify dispatching rules. A decision-tree-oriented classification scheme, available as a software called ID3 (Quinlan 1987), is used. The cutoff values used in branching decisions are exogenous to machine learning, however, and are obtained by trial and error. A typical ID3-generated decision tree is shown in Figure 12.22. It classifies weather conditions for outdoor activity into OK and not OK (NOK) categories. The attributes used are outlook, humidity, and wind condition, as in Quinlan 1987.

The attributes for NOK, for example, are sunny outlook and high humidity, or rainy outlook and windy.

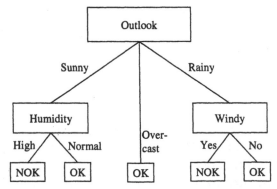

FIGURE 12.22: ID3-GENERATED DECISION TREE

Chakravarty (1997) uses a data envelopment analysis (DEA) to discern patterns from data on dispatching rules and on system state attributes (described as number of parts in the system, x_1; machining time per part, x_2; and available machines, x_3). The two performance attributes observed are throughput, y_1, and mean flow time, y_2. Consider first a simple example, shown in Figures 12.23 and 12.24. The relevant values for two recorded observations on system state are shown in Figure 12.23.

System performance (throughput and mean flow time) are shown separately for each of the two dispatching rules, SPT and EDD, in Figure 12.24.

Assume that the currently observed values of the system state attributes are

Jobs in the cell	= 20
Machining time per part	= 18
Available machines	= 21

The monotonicity property of DEA states that the values of system state attributes and performance attributes in any observation can be expressed as convex combinations (weighted sum) of previous observations. Thus, letting λ_j be the weight attached to the jth observation where $\sum_j \lambda_j = 1$, the expected system state values for current observation can be expressed as

$$E \text{ (jobs in the cell)} = 10\lambda_1 + 25\lambda_2 \qquad (12.19)$$

$$E \text{ (machining time per unit)} = 15\lambda_1 + 20\lambda_2 \qquad (12.20)$$

$$E \text{ (available machines)} = 30\lambda_1 + 10\lambda_2 \qquad (12.21)$$

Similarly, with respect to performance attributes, we may write

Observations (j)	System State Attributes		
	Number of Parts in the Cell	Matching Time per Part	Available Machines
1	10	15	30
2	25	20	10

FIGURE 12.23: SYSTEM STATE VALUES

Dispatching Rules (k)	Observation (j)	Performance Attributes	
		Throughput	Mean Flow Time
SPT (k = 1)	1	50	100
	2	200	150
EDD (k = 2)	1	80	80
	2	250	180

FIGURE 12.24: PERFORMANCE VALUES

$$E\,(\text{throughput/SPT}) \;=\; 50\lambda_1 \;+\; 200\lambda_2 \qquad (12.22)$$

$$E\,(\text{mean flow time/SPT}) \;=\; 100\lambda_1 \;+\; 150\lambda_2 \qquad (12.23)$$

$$E\,(\text{throughput/EDD}) \;=\; 80\lambda_1 \;+\; 250\lambda_2 \qquad (12.24)$$

$$E\,(\text{mean flow time/EDD}) \;=\; 80\lambda_1 \;+\; 180\lambda_2 \qquad (12.25)$$

Assume that our objective is to pick one of the two dispatching rules (SPT, EDD) that would maximize throughput, given the system state attribute values in the current observation. Let the binary variable z ($z = 0,1$) denote the dispatching rule choice decision. That is, $z = 1$ implies that SPT is chosen, and $z = 0$ implies choice of EDD.

Thus the objective function to be maximized (throughput) can be written from expressions 12.22 and 12.24 as

$$\text{Maximize } F = (50\lambda_1 + 200\lambda_2)(z) + (80\lambda_1 + 250\lambda_2)(1 - z) \quad (12.26)$$

The constraints would be expressed in terms of the values of state attributes. Note that throughput *increases* (or stays constant) as the number of available machines or the number of jobs in the cell increase. Hence from expressions 12.19 and 12.21 we have (with 20 jobs and 21 machines in the current observation)

$$10\lambda_1 + 25\lambda_2 \le 20 \qquad (12.27)$$

$$30\lambda_1 + 10\lambda_2 \le 21 \qquad (12.28)$$

Since throughput *decreases* with machining time per part, the relevant constraint from expression 12.20 would be (with machining time per part of 18 in the current observation)

$$15\lambda_1 + 20\lambda_2 \geq 18 \qquad (12.29)$$

We also have

$$\lambda_1 + \lambda_2 = 1 \qquad (12.30)$$

The optimization problem would therefore be Maximize F as in expression 12.26, subject to conditions 12.27 to 12.30. Observe that the objective function 12.26 is nonlinear, as it has terms with product of z and λ_i. A simpler approach would be to solve the problem two times, first with $z = 1$ and then with $z = 0$.

With $z = 1$, it can be verified that the optimal solution would be

$$F = 150, \lambda_1 = 1/3, \lambda_2 = 2/3$$

With $z = 0$, the optimal solution would be

$$F = 193\tfrac{1}{3}, \lambda_1 = 1/3, \lambda_2 = 2/3$$

Hence the optimal dispatching rule would be EDD, with maximum $F = 193\tfrac{1}{3}$.

Next, to minimize mean flow time, note that it increases as the number of parts in the cell or machining time per part increases. It decreases as the number of available machines increases. Hence the optimization problem would be

$$\text{Minimize } F = (100\lambda_1 + 150\lambda_2)(z) + (80\lambda_1 + 180\lambda_2)(1 - z) \qquad (12.31)$$

subject to

$$10\lambda_1 + 25\lambda_2 \geq 20 \qquad (12.32)$$

$$15\lambda_1 + 20\lambda_2 \geq 18 \qquad (12.33)$$

$$30\lambda_1 + 10\lambda_2 \leq 21 \qquad (12.34)$$

As before, solving with $z = 1$, we have

$$F = 133\tfrac{1}{3},\ \lambda_1 = 1/3,\ \lambda_2 = 2/3$$

With $z = 0$, we have

$$F = 146\tfrac{2}{3},\ \lambda_1 = 1/3,\ \lambda_2 = 2/3$$

Therefore, the optimal dispatching rule is, again, EDD.

It is clear that for this example EDD dominates SPT irrespective of system objective. In terms of generalizing the solution approach, note that the biggest problem is classification of constraints in the not-greater-than and not-less-than groups. In addition the classification scheme would have to be different for the maximization and minimization objectives. Chakravarty (1997) addresses this problem by defining two additional DEA postulates. For a performance (output) variable y_{rj}, define the index set of state attribute (input) variables P_r and N_r such that

$$P_r = \left\{ s\Big/ \frac{\partial y_{rj}}{\partial s_j} \geq 0 \right\} \tag{12.35}$$

$$N_r = \left\{ s\Big/ \frac{\partial y_{rj}}{\partial s_j} \geq 0 \right\} \tag{12.36}$$

The observation j may be a convex combination of other observed values. We first consider maximization y_{rj}.

(a) If $(x_{sj}, y_{rj}/s \in P_r) \in T$ and $(x_{si}, y_{ri}/s \in P_r) \in T$
then $\qquad x_{sj} < x_{si}$

where T is the production possibility set defined by the convex combination of observations. Note that x_{sj} may be a convex combination of the set of observed state values.

(b) If $(x_{sj}, y_{rj}/s \in N_r) \in T$ and $(x_{si}, y_{ri}/s \in N_r) \in T$
then $\qquad x_{sj} > x_{si}$

Now let us consider minimization of y_{ri}

(a) If $(x_{sj}, y_{rj} / s \in P_r) \in T$ and $(x_{si}, y_{ri} / s \in P_r) \in T$
    then $\qquad x_{sj} > x_{si}$

(b) If $(x_{sj}, y_{rj} / s \in N_r) \in T$ and $(x_{si}, y_{ri} / s \in N_r) \in T$
    then $\qquad x_{sj} < x_{si}$

For the example problem, it can be verified that for throughput maximization

$$P_1 = \{1,3\}, \ N_1 = \{2\}$$

Similarly, for mean flow time minimization

$$P_2 = \{1,2\}, \ N_2 = \{3\}$$

12.6 PRODUCT CUSTOMIZATION THROUGH PROCESS DESIGN

Letting customers have product features of their choice is one way of customization. In many such instances, the order in which various manufacturing processes are performed on the product has no impact on product quality. It may, however, impact the manufacturing cost differently. The most often cited example comes from a company called Benetton (Dapiran 1992) in the apparel industry. The two major processes involved in making sweaters are dyeing and knitting. Bleached yarns are dyed different colors and then knitted according to varying styles and sizes.

Technologically it is possible for the knitting operation to precede the dyeing operation without sacrificing quality. Lee and Tang (1998) have shown that in such cases, the sum of the variances of production quantities at manufacturing processes can exceed the variance of the total sales. They also establish that performing low-variance tasks first minimizes the total variance.

Consider a simple case of a product that requires two processes to add different features to it. Assume that process A can add a single feature, A_1, whereas process B can add two features, B_1 and B_2. The product sold to customers will therefore come in two varieties: $A_1 B_1$ and $A_1 B_2$. If process B

precedes A, the products B_1A_1 and B_2A_1 will be identical to A_1B_1 and A_1B_2, respectively. The total demand for A_1B_1 and A_1B_2 is a random variable with mean μ and variance σ^2. For a given total demand of N, demand for A_1B_1 (or A_1B_2) can be described by the binomial distribution. Assume that the demand for A_1B_1 is x_1, for A_1B_2 it is x_2, and that $x_1 + x_2 = N$ where N is arbitrary. Let $\frac{x_1}{N} = p$ and $\frac{x_2}{N} = (1 - p)$.

We consider the two possible process sequences AB and BA. For the first sequence, uncertainties in the manufacturing processes are shown in Figure 12.25.

It is clear from Figure 12.25 that for a given value of N, the total production at process A is always N. Hence the variance of process A's output is zero. Next, note that since process A is adding only a single feature, the items for production arrive at process B in a single stream. Hence, irrespective of the value of p, the production variance at process B will equal σ^2.

Next, consider the sequence BA as in Figure 12.26.

Observe that now production items arrive at the last process (process A) in two distinct streams. Hence the production variance at process A for the two streams B_1A_1 and B_2A_1 can be written as

$$\text{Variance of } B_1A_1 = p^2\sigma^2$$

$$\text{Variance of } B_2A_1 = (1 - p)^2\sigma^2$$

Hence

$$\text{Process } A \text{ variance} = \sigma^2\{p^2 + (1 - p)^2\}$$

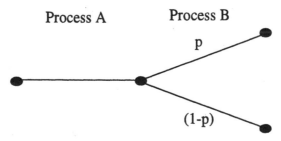

FIGURE 12.25: PROCESS *A* FOLLOWED BY PROCESS *B*

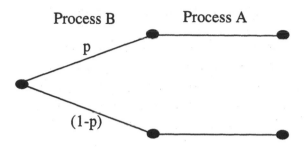

FIGURE 12.26: PROCESS *B* FOLLOWED BY PROCESS *A*

which simplifies to

$$\text{Process } A \text{ variance} = 2\sigma^2 p(1 - p) + \sigma^2 \qquad (12.37)$$

For process B, it is assumed that capacity reserved for stream $B_1 A_1$ cannot be utilized by stream $B_2 A_1$.

$$\text{Variance of } B_1 A_1 \text{ for a given } N = Np(1 - p)$$
$$\text{Variance of } B_2 A_1 \text{ for a given } N = N(1 - p)(p)$$

Hence

$$\text{Process } B \text{ variance for a given } N = 2Np(1 - p)$$

Next, note that $E(N) = \mu$. Hence

$$\text{Process } B\text{'s unconditional variance} = 2\mu p(1 - p) \qquad (12.38)$$

Thus from equations 12.37 and 12.38 we have

$$\text{Total variance of sequence } BA = 2\mu p(1 - p) + 2\sigma^2 p(1 - p) + \sigma^2$$

This simplifies to

$$\text{Total variance} = 2p(1 - p)(\mu - \sigma^2) + \sigma^2 \qquad (12.39)$$

It follows from equation 12.39 and variance of the AB sequence $(= \sigma^2)$ that

$$\text{Variance } (BA) - \text{Variance } (AB) = 2p(1 - p)(\mu - \sigma^2) \qquad (12.40)$$

It is clear from equation 12.40 that if $\mu > \sigma^2$, the *AB* sequence would be preferred to the *BA* sequence.

For the general case, assume that process *A* can add n_A features and process *B* can add n_B features. Let p_i be the probability of demand of products with feature *i* added at process *A*. Let q_j be the probability of demand of products with feature *j* added at process *B*. The total variance for sequence *AB* can be written as

$$V_{AB} = \mu \sum_{i=1}^{n_A} p_i(1 - p_i) + \sigma^2 \sum_{i=1}^{n_A} p_i^2$$

This simplifies to

$$V_{AB} = \mu - (\mu - \sigma^2) \sum_{i=1}^{n_A} p_i^2 \tag{12.41}$$

Similarly

$$V_{BA} = \mu - (\mu - \sigma^2) \sum_{i=1}^{n_B} q_i^2 \tag{12.42}$$

For sequence *AB* to be preferable,

$$V_{AB} < V_{BA}$$

That is,

$$(\mu - \sigma^2)\left(\sum_{i=1}^{n_A} p_i^2 - \sum_{i=1}^{n_B} q_i^2 \right) > 0 \tag{12.43}$$

Equation 12.43 implies that for low demand variability (i.e., σ^2 is low), sequence *AB* would be preferred over *BA* if

$$\sum_{i=1}^{n_A} p_i^2 > \sum_{i=1}^{n_B} q_i^2$$

Note that

$$\sum_{i=1}^{n_A} p_i = 1, \text{ and } \sum_{i=1}^{n_B} q_i = 1$$

With some mild assumptions, it can be shown, using equation 12.43, that process A will be scheduled before process B if

  (a) Process A has fewer available choices of features, or
  (b) The range of probability p_i of choosing feature i, is wide (note that probability pair $(\frac{1}{16}, \frac{15}{16})$ has a wider range than the pair $\frac{1}{3}, \frac{2}{3}$), or
  (c) Production variance at A is low (compared to B), and
  (d) Processing lead time at process A is long

However, the above insights hold only if $\mu > \sigma^2$.

12.7 NEW PRODUCT PHASE-IN

It is clear that product feature customization through process resequencing would work well only if the range of required features is narrow. If this is not the case, companies must introduce new products quickly, to take advantage of market opportunities. Very often such market-oriented changes cause severe disruptions in manufacturing. Disruptions may be caused for a variety of reasons, such as schedule changes, backorders, and inventory fluctuations. Though such disruptions present only short-term problems, the cost could be high. Marketing opportunities arising primarily from higher revenues, on the other hand, may last longer.

A typical problem faced by companies is described by Garvin (1991). In the early 1980s Boeing, as discussed in Chapter 1, decided to switch from three-person to two-person cockpits. Boeing had to develop separate plans for each of 30 aircraft that needed modifications. If the modifications were done as separate projects, disruption cost could have been minimized. However, problems with design integrity, space requirement, and product testing might have emerged.

Thus there is an implied manufacturing risk with the rapid introduction of new products. Billington, Lee, and Tang (1998) list several other risks associated with rollover to new products. For example, there would be a risk of not being able to develop or acquire the technology for the new products in time. There would also be a marketing risk if the company decided to drive the current inventory of old product to zero before introducing the other product. If the inventory is used up too rapidly or if the new product is delayed, a product void will be created, and the company may end up losing many customers. They recommend a dual-product rollover, where both old and new products overlap for a period before the old product is phased out, especially in environments where both market and production/product risks are high.

We study this problem from a manufacturing point of view. That is, we consider a production system driven by bill of material (BOM). Production and procurement lead times dictate that inventories of components and subassemblies are built up ahead of the completion date of the end product. Thus if it is decided to cut out the old product at time t, actions to reduce inventory levels of components must commence well before t. If the new product has components and/or subassemblies in common with the old product, then instead of depleting inventories of these components, we may want to apply them to the new product. It is also assumed that additional capacity, if needed, is outsourced. Note that we studied a similar problem in the context of contingencies in Chapter 11.

Balakrishnan and Chakravarty (1996) have studied the overlap between phase-in and phase-out of new and old products, respectively, for a market-leading firm that manufactures and sells first-generation durable products. It is assumed that competitors can clone the product, but they can make it available to customers only in the next planning horizon, which can be extended through patent protection (Purohit 1994). The total demand for old and new products would increase, as it is assumed that the new product would have innovative features that customers prefer. It is clear that the new product would partially cannibalize demand from the old product. We model these demand dynamics by letting d_t and f_t denote the committed and forecast demand of the current product before the new product is introduced. D_t and F_t denote the corresponding values after the new product is introduced. Also assume that the committed and forecast demand of the new product N are denoted by D_t^N and F_t^N.

Since the total demand increases, we have for committed demand

$$D_t + D_t^N \geq d_t$$

More specifically, the split of total demand between current and new products would be written as

$$D_t + \mu D_t^N = d_t \tag{12.44}$$

where μ $(0 \leq \mu \leq 1)$ is the elasticity of demand between D_t and D_t^N. It is clear that μ will be a function of the level of substitutability between the current and new products and of the level of innovation present in the new product.

In a similar way, for the forecast demand we would have

$$F_t + \mu_F F_t^N = f_t \tag{12.45}$$

In most cases $\mu_F < \mu$, i.e., $\mu_F = \rho\mu$ $(0 \leq \rho \leq 1)$, since customers with committed orders would be less likely to switch.

The profit function would now be written as

$$\Pi = \sum_{t \leq T} \{(s - c)S_t - \theta B_t - \beta b_t\} - \sum_{i \in I} \sum_{t \leq T} \{r_i y_{it} + h_i H_{it}\} -$$
$$\sum_{i \in J} \alpha_i H_{iT} - \sum_{i \in I} \sum_{t \leq T} c_i x_{it} \tag{12.46}$$

where S_t is the sales quantity and B_t and b_t are the backorders of forecast and committed sales. The variables y_{it}, x_{it}, and H_{it} denote quantities of i outsourced, produced, and inventoried, respectively. We drop the subscript i (from y, x and H) when we refer to the end product. Unit sales price and cost are s and c; θ and β are the cost of backordering of forecast and committed end product. The variables r_i, h_i, and α_i denote unit cost of outsourcing, carrying inventory and obsolescence.

The constraints on backorder and sales would be (for the end item)

$$B_t \geq b_t \tag{12.47}$$

$$b_t - b_{t-1} + S_t \leq D_t \tag{12.48}$$

$$S_t = D_t + F_t + B_{t-1} - B_t \tag{12.49}$$

For the new product n,

$$S_t^N = D_t^N + F_t^N + B_{t-1}^N - B_t^N \tag{12.50}$$

The inventory balance equations for the end item would be

$$S_t = R_t + x_{t-L} + y_{t-l} + H_{t-1} - H_t \qquad (12.51)$$

where R_t is the planned order receipts in period t and L is the production lead time. The outsourcing lead time of the end product is l. For the new product,

$$S_t^N = R_t^N + x_{t-L}^N + y_{t-l} + H_{t-l} - H_t \qquad (12.52)$$

For components, the inventory balance equation would be

$$\sum_{j \in \Omega_i} a_{ij} x_{it} = R_{it} + x_{i,\,t-L_i} + y_{i,\,t-l_i} + H_{t-1} - H_t \qquad (12.53)$$

where a_{ij} is the number of units of i required per unit of parent j and L_i and l_i are the production and procurement lead times of i.

The capacity constraint would be

$$\sum_{i \in I} \tau_i x_{it} \le B \qquad (12.54)$$

where τ_i is the production time per unit of item i. The other constraints would be expressions 12.44 and 12.45.

Balakrishnan and Chakravarty (1996) ran this model with different parameter settings to generate managerial insights. One of their findings was that total profit initially increased with an increase in the time before the old product would be discontinued. It reached a maximum value and then decreased. That is, there exists an optimal value for the cutout time. This supports Ho's observation (1994) that the old product should not be discontinued straightaway. They also found that in all of their experiments, except when obsolescence cost was high, the new product was introduced before the old product was cut out. That is, the overlap between the two products was statistically significant. Their other findings were that the new product should be introduced earlier if (1) the value of μ became smaller, (2) the cost of outsourcing per unit decreased, (3) the cost of backordering per unit decreased, (4) the market became more competitive, and (5) the value of ρ (= μ_F/μ) decreased.

12.8 PROACTIVE RESPONSE TO PRODUCT PROMOTIONS

Retailers use product promotions to capture short-term demand from competitors and/or to get rid of inventories. Promotions usually take the form of targeted advertising, coupons, mail-in rebates, and other such short-term drives that result in savings for customers. The sudden (short-term) surge in demand, if not coordinated well with production, can lead to shortages and/or excessive production-related costs. Production cost may increase due to disturbance in the system, discussed earlier. There is, however, a trade-off between these increases in production cost and the additional revenue accrued from increased sales. Note that if a promotion is planned simply as a reaction to accumulated inventories, it cannot be a profit-maximizing strategy. A proactive production strategy would be to produce enough in a period to cover demand generated from future planned promotions in the next few periods.

The decisions we are faced with, therefore, are (1) in which periods to set up for production, (2) how much to produce at such times, and (3) in which periods to promote the product. One of the important considerations in such decisions is the impact of promotion in period t on creating demand in periods $t + 1$ and beyond. Sogomonian and Tang (1993) use exponential decay to model demand in period r $(r > t)$. That is, demand in period r, given that the promotion is done in period t, is written as

$$D_{tr} = a \, exp(t - r), r \leq t \tag{12.55}$$

where a is a constant. With s and c as retail price and production cost per unit, respectively, the revenue would be expressed as

$$\text{Revenue in period } r = (s - c)D_{tr}$$

This additional revenue would be compared with costs, which are of three kinds: fixed cost K, incurred every time the machine is set up for production; fixed cost A, incurred every time a promotion is undertaken; and the cost of carrying inventory. We may carry more inventories to reduce the number of setups if K is high. As in Sogomonian and Tang 1993, we assume that there are no backorders, that is, the cumulative production up to any period m is not less than the total demand up to that period. The demand as in equation 12.55 is assumed to be deterministic.

The basic property we use in developing the model is that production takes place in a period if and only if inventory at the beginning of that period is zero (Wagner and Whitin 1958). We also assume that production takes place in a period only if there is a promotion in that period. Promotion of a product can, however, take place even if there is no production in that period. It is clear that in any discrete period n we have three options: promote the product, promote and produce, and do nothing. In a network model, the proxy for doing nothing would be to skip that node (period n). The remaining two options are denoted as two states, where state 1 is promote but do not produce and state 2 is both promote and produce the product.

The network model for profit maximization, over three periods, is shown in Figure 12.27. The nodes in the diagram denote decision states in any period, and the arcs denote transitions (from one period to future periods) in a network. A transition from node 2 in period n to node 2 in period $(m > n)$ implies that a quantity $Q = \sum_{r=n}^{m-1} D_r$ will be produced in period n. There will also be a promotion in period n, but no promotion or production will take place in periods $n + 1$ to $m - 1$. Secondary demands in periods $n + 1$ to $m-1$ will result from promotion in period n and would exhibit "demand decay" behavior, discussed earlier. There will be a promotion as well as production in period m. The setup cost associated with this transition is $K + A$ (does not include setup costs in period m). Inven-

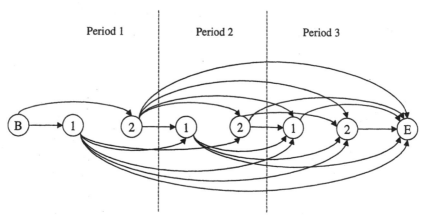

FIGURE 12.27: LONGEST-PATH MODEL OF PRODUCT PROMOTION

tory will be carried in periods $n + 1$ to $m - 1$, to satisfy secondary demands in these periods. The net profit from this transition can therefore be written as

$$R(n2,m2) = (s - c) \sum_{r=n}^{m-1} D_r - (K + A) - h \sum_{r=n+1}^{m-1} (r - n)D_r \quad (12.56)$$

Next, the net profit from node 2 in period n to node 1 in period m can be verified to have an extra promotion cost and inventory cost in period m. This can be written as

$$R(n2,m1) = (s - c) \sum_{r=n}^{m} D_r - (K + 2A) - h \sum_{r=n+1}^{m} (r - n)D_r \quad (12.57)$$

Next, consider the transition from node 1 in period n to node 2 in period m. Observe from equation 12.57 that since there is no production in period n, the primary demand from promotion in period n would have been absorbed by producing enough in some prior period j ($j = n - u$). Hence, this transition will be associated with revenue and carrying costs corresponding to only the secondary demands in periods $n + 1$ to $m - 1$. Note that carrying cost of the secondary demand in period $n + 1$ will be charged for the duration $n + 1 - (n - u)$. Hence,

$$R(n1,m2) = (s - c) \sum_{r=n+1}^{m-1} D_r - h \sum_{r=n+1}^{m-1} (r - n + u)D_r \quad (12.58)$$

With a similar observation, it is clear that

$$R(n1,m1) = (s - c) \sum_{r=n+1}^{m} D_r - A - h \sum_{r=n+1}^{m} (r - n + u)D_r \quad (12.59)$$

We can now formulate the longest-path problem as

$$f(n(k)) = \underset{m,j}{\text{Maximum}} \{R(n(k),m(j)) + f(m(j))\} \quad (12.60)$$

where $k = 1, 2$; $j = 1, 2$; $n = 1$ to N; $n < m \leq N$. Note that computation complexity is $O(N^2)$.

Consider a simple example with three periods. Let $K = 2{,}000$, $A = 400$,

$s = 40$, $c = 10$, and $h = 2$. Assume that demand due to promotion in period t and its decay in time is as follows:

t	$t-1$	$t-2$	$\ldots$	
Demand	1,000	500	140	$\ldots$
		1,000	500	$\ldots$
			1,000	$\ldots$

It is clear that the network model would be similar to the diagram in Figure 12.27. To solve the longest-path problem, we use the "backward pass" of recursion using equation 12.60, starting from period 3.

It is clear that in period 3,

$$f(3(1)) = 0$$

$$f(3(2)) = (30)(1,000) - (2,000 + 400) = 27,600$$

In period 2,

$$f(2(2)) = \text{Max} \{R(2(2),3(1)) + f(3(1)), R(2(2),3(2)) + f(3(2)), R(2(2), E)\}$$

Note that

$$R(2(2), 3(1)) = 30(2,000) - (2,000 + 400 + 400) - 2(1,000) = 55,200$$

$$R(2(2), 3(2)) = 30(1,000) - (2,000 + 400) = 27,600$$

$$R(2(2),E) = 30(1,000 + 500) - (2,000 + 400) - 2(500) = 41,600$$

Therefore

$$f(2,(2)) = \text{Max}\{55,200 + 0, 27,600 + 27,600, 41,600\}$$

$$= 55,200, \text{ via } 3(1) \text{ or } 3(2)$$

In a similar way,

$$f(2,1) = \text{Max} \{29,600 - 20,00u + 0, 0 + 27,600, 0\}$$

$$= 27,600, \text{ via } 3(2), \text{ since } u \geq 1$$

In period 1, note that realization of $f(1(1))$ is not feasible, as there are no production states prior to state $1(1)$.

$$f(1(2)) = \text{Max} \{R(1(2), 2(1)) + f(2(1)), R(1(2), 2(2)) +$$
$$f(2(2)), R(1(2), 3(1)) + f(3(1)), R(1(2), 3(2)) + f(3(2))\}$$

It can be verified that

$$f(1(2)) = \text{Max} \{55{,}200 + 27{,}600,$$
$$27{,}600 + 55{,}200, 69{,}200 + 0, 41{,}600 + 27{,}600)\}$$
$$= 82{,}800, \text{ via } 2(1) \text{ or } 2(2).$$

Therefore the maximum profit = 82,800.
This solution is not unique. The three possible solutions are

Period 1	Period 2	Period 3
Produce 2,000 and Promote	Promote	Produce 1,000 and Promote
Produce 1,000 and Promote	Produce 2,000 and Promote	Promote
Produce 1,000 and Promote	Produce 1,000 and Promote	Produce 1,000 and Promote

Consider now the solution with $K = 5{,}000$. We will have

$$f(3(1)) = 0, f(3(2)) = 24{,}600, f(2(2)) = 52{,}200$$
$$f(2(1)) = \text{Max} \{27{,}600 - 2{,}000u, 24{,}600\}$$
$$\text{For } u = 1, f(2(1)) = 25{,}600$$
$$\text{For } u = 2, f(2(1)) = 24{,}600$$

Next note that to compute $f(1(2))$, we need the value of $f(2(1))$ in one of the branches. Verify that in the context of transition from $1(2)$ to $2(1)$, $u = 1$. Therefore, $f(2,(1)) = 25{,}600$ (not 24,600).
    Hence

$$f(1(2)) = \text{Max} \{55{,}200 + 25{,}600, 24{,}600 + 52{,}200, 66{,}200 +$$
$$0, 38{,}600 + 24{,}600\}$$

That is,

$$f(1(2)) = 77{,}800$$

The above solution is unique. The decisions would be

Period 1	Period 2	Period 3
Produce 3,000 and promote	Promote	Promote

Sogomonian and Tang (1993) have suggested a much more complex procedure to solve the same problem. Cheng and Sethi (1999) suggest a solution procedure when demands are stochastic and backorders are allowed.

12.9 PRODUCTION ON DEMAND

We have seen, in the previous section, how an increase in production setup cost from \$2,000 to \$5,000 requires the company to produce 3,000 units in a single setup. When the setup cost is reduced to \$2,000, the optimal solution is to produce 1,000 units (or 2,000 units) more often. It is clear that cost of carrying inventory would increase if the number of setups decreases. Thus the trade-off between the manufacturing setup cost and the cost of carrying inventory is clear. In general, there is a three-way trade-off between cost of capacity, cost of setups, and cost of inventory. Consider, for example, a scenario where demand D_t in period t is not a constant in t. For a sufficiently high setup cost K, production for a planning horizon T will be accomplished in a single setup. A quantity $\sum_{r=1}^{T} D_r$ would be produced at the start of the planning horizon, and the cost of carrying inventory would be $\sum_{r=1}^{T} hrD_r$. At the other extreme, if the setup cost is insignificant (as in a flexible manufacturing system), it would be optimal to produce the quantity D_t in period t. This way the setup cost would be KT and inventory cost would be zero. Notice in the first case that a large manufacturing capacity would be required; its utilization can be kept at a high level, as the same facility can be used for other products in subsequent periods. In the latter case, on the other hand, the capacity size required would be smaller and would be equal to $\underset{r}{\text{Maximum}} (D_r)$. Its utilization, however, would vary tremendously from period to period. Capacity underutilization obviously constitutes a cost.

Japanese manufacturing companies have been well known for their competitiveness in the 1980s and 1990s. It has been widely documented how they achieved competitiveness through innovative manufacturing practices (Schonberger 1982). Toyota led the way through what is commonly known as just-in-time manufacturing, which seeks to attain production on demand. To attain such a system, however, the cost of setting up a machine and the variance in capacity utilization must be minimized. Toyota invested heavily in the 1970s to reduce the setup time of 800-ton presses used in forming automobile engine heads and fenders (Sugimori, Cho, and Uchikawa 1977). They also used a mixed-model assembly line in which several models of the same car could be produced simultaneously, to reduce the variance in capacity utilization. They did so by sequencing the models so as to hold the cumulative production ratio of each model at any point in time close to overall ratio of its demand to the total demand of all products.

Setup Cost Reduction

Chakravarty and Shtub (1985) discuss strategies for reducing setup cost in a multi-item, multi-stage production system. They develop closed-form solutions for the case when the cost of investment to reduce setup cost can be assumed to be a well-defined function of setup cost. For the case when this relationship can only be described as a discrete function, they develop a dynamic programming approach to determine optimal investment amount. We discuss a simplified version of their model.

First, consider a simple one-machine production scenario, where yearly product demand can be approximated by its average value D. Assume, as before, the cost of machine setup and inventory-carrying to be K per setup and h per unit per year, respectively. The total operational cost (setup cost plus inventory cost) per year can be written as

$$c(K,T) = \frac{K}{T} + \frac{T}{2}hD$$

Assuming that $K = f(x)$, where x is the amount invested in reducing setup cost, we can write

$$c(x,T) = \frac{f(x)}{T} + \frac{T}{2}hD$$

With optimal T, we can write

$$c(x) = \{2f(x)hD\}^{1/2} \qquad (12.61)$$

It is clear that the setup cost K will rapidly decrease with x initially, and then it would become asymptotic with the x axis. Hence, we use the exponential decay function for $f(x)$. That is,

$$f(x) = ae^{-bx}$$

It is clear that $c(x)$ is convex in x.

The NPV of $c(x)$ can be written as

$$NPV = \int_0^\infty c(x)e^{-\alpha a}dt = \frac{1}{\alpha}c(x)$$

Hence the function to be minimized will be written as

$$\Phi(x) = \frac{1}{\alpha}c(x) + x$$

where $\Phi(x)$ is convex in x. Note that $\Phi(x)$ will be minimized when

$$\frac{d\Phi(x)}{dx} = 0$$

That is,

$$\frac{1}{\alpha}\frac{d}{dx}c(x) + 1 = 0 \qquad (12.62)$$

From equation 12.61 observe that

$$\frac{d}{dx}c(x) = hD\frac{df(x)}{dx}/c(x)$$

Hence from equation 12.62

$$\frac{d}{dx}f(x) = \frac{-\alpha}{hD}c(x)$$

That is,
$$f'(x) = -\alpha\left\{\frac{2f(x)}{hD}\right\}^{1/2} \tag{12.63}$$

Since $f(x) = ae^{-bx}$, we have from equation 12.63

$$x = \frac{1}{b}\ln\left(\frac{ahDb^2}{2\alpha^2}\right) \tag{12.64}$$

Porteus (1985) has also analyzed a single-item, single-stage problem to arrive at a similar result. Note that Porteus implicitly assumes that $f(x) = e^{(a-x)/b}$, so that the optimal value of x is the solution of $e^{(a-x)/b} = (2\alpha^2b^2)/hD$.

Next, consider a multi-echelon system, shown in Figure 12.28.

Crowston, Wagner, and Williams (1973) have established that a solution with T_j as an integer multiple of T_{j-1} produces an optimal or near-optimal solution to the above multi-stage problem. Hence we let $T_j = m_jT$, where T is the cycle time at the most upstream stage (stage 1) and m_j is an integer. So

$$c(\tilde{x}_j, \tilde{m}_j, T, \lambda) = \sum_{j=1}^{N}\frac{f_j(x_j)}{m_jT} + \frac{T}{2}\sum m_jh_jD \tag{12.65}$$

where x_j and h_j are the investment and inventory carrying cost at stage j, respectively. B is the budget for setup cost reduction. $\tilde{x}$, and $\tilde{m}_j$ are the respective vectors.

Considering the Lagrangian function

$$L(\tilde{x}_j, \tilde{m}_j, T, \lambda) = c(x_j, m_j, T) - \lambda\left(\sum_{j=1}^{N}x_j - B\right)$$

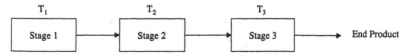

FIGURE 12.28: MULTI-ECHELON PRODUCTION SYSTEM

we can easily establish that

$$\frac{d}{dx_j} f_j(x_j) = \gamma m_j, \text{ where } \gamma = \lambda T, \text{ for all } j \tag{12.66}$$

For optimal T, using equation 12.65 we have

$$c(\tilde{x}_j, \tilde{m}_j) = \left[\left\{ 2 \sum_{j=1}^{N} \frac{f_j(x_j)}{m_j} \right\} \sum_{j=1}^{N} \{ m_j h_j D \} \right]^{1/2}$$

Note, as before, $c(\tilde{x}_j, \tilde{m}_j)$ is convex in x_j if $f_j(x_j)$ is an exponentially decaying function.

Next consider the Lagrangian function

$$L(\tilde{x}_j, \tilde{m}_j, \lambda) = \frac{1}{\alpha} c(\tilde{x}_j, \tilde{m}_j) - \lambda \left(\sum_{j=1}^{N} x_j - B \right) \tag{12.67}$$

Equating $\frac{\partial L}{\partial x_j} = 0$, we have

$$\frac{\partial}{\partial x_j} c(\tilde{x}_j, \tilde{m}_j) = \alpha \lambda$$

This simplifies to

$$DH f_j'(x_j) / m_j = \alpha \lambda c(\tilde{x}_j, \tilde{m}_j) \tag{12.68}$$

where

$$H = \sum_{i=1}^{N} m_i h_i$$

Letting $f_j(x_j) = a_j e^{-b_j x_j}$ we have, from equation 12.66,

$$\frac{f_j'(x_j)}{m_j} = -b_j a_j e^{-b_j x_j} = \gamma$$

That is,

$$a_j e^{-b_j x_j} = -\gamma/b_j \qquad (12.69)$$

Equation 12.68 can now be simplified to

$$\gamma = -2D(\alpha\lambda)^2 / \left(H \sum_{i=1}^{N} b_i \right)$$

Hence, from equation 12.69 we have

$$e^{-b_j x_j} = \left(\frac{2D\alpha^2\lambda^2}{bH} \right)\left(\frac{1}{a_j b_j} \right), \text{ where } b = \sum_{i=1}^{N} b_i$$

Thus

$$x_j = -\frac{1}{b_j} \ln(\lambda^2) + \frac{1}{b_j} \ln\left(\frac{bH}{2D\alpha^2} \right) + \frac{1}{b_j} \ln(a_j b_j) \qquad (12.70)$$

Next, since $\sum_{i=1}^{N} x_i = B$, we have from equation 12.70

$$B = -\ln(\lambda^2) \sum_{i=1}^{N} \frac{1}{b_i} + \ln\left(\frac{bH}{2D\alpha^2} \right) \sum_{i=1}^{N} \frac{1}{b_i} + \sum_{i=1}^{N} \frac{\ln(a_i b_i)}{b_i} \qquad (12.71)$$

Eliminating λ^2 from equations 12.70 and 12.71, we can express x_j as

$$x_j = \theta_j + \mu$$

where $\theta_j = \frac{\ln(a_j b_j)}{b_j}$

$$\mu = \left(B - \sum_{i=1}^{N} \theta_i \right) \Big/ \left\{ b_j \sum_{i=1}^{N} (1/b_i) \right\}$$

The somewhat surprising implication of the above result is that investment x_j in stage j, is independent of the magnitude of demand, carrying cost, and integer multipliers of stage cycle times. However, if budget is not a constraining factor, the above factors do impact the value of x_j.

Capacity Utilization Variance

Just-in-time production systems have used mixed-model assembly lines to even out capacity utilization from period to period (Schonberger 1982). A family of similar products (models) is run down the line in a predetermined order. Consider a family of n models with demand D_i, $i = 1$ to n, in a period of length t. The total demand $D_T = \sum_{i=1}^{n} D_i$. Next, assume that the period t is divided into D_T production stages, so that in stage r exactly one unit of a product (one of the n models) is produced. We use the binary variable x_{ir} to model the decision in stage r. That is, $x_{ir} = 1$ if model i is produced in stage r; it is zero otherwise. Let y_{ir} denote the total quantity of i produced in stages 1 to r. It is clear that

$$\sum_{i=1}^{n} y_{ir} = r$$

$$y_{ir} = \sum_{k=1}^{r} x_{ik}$$

Therefore, the average number of units of i produced in a production stage would equal y_{ir}/r. So if $y_{ir} = \frac{D_i r}{D_T}$ for any r, customers of model i will be fully satisfied. Since it may not be possible to satisfy the above criterion, we would like to minimize the variance $(y_{ir} - \frac{D_i r}{D_T})^2$ over all models and stages. Thus a possible objective would be to

$$\text{Minimize} \sum_{r=1}^{D_T} \sum_{i=1}^{n} \left(y_{ir} - \frac{D_i r}{D_T} \right)^2$$

Miltenberg (1989) has developed an efficient procedure that approximates $\frac{r D_i}{D_T}$ by closest integer values ($= y_{ir}$).

As an example, consider $D_1 = 5$, $D_2 = 4$, and $D_3 = 2$. Therefore $D_T = 11$. There would be 11 production stages. In stage 1 ($r = 1$), the values of

$\frac{rD_i}{D_T}$ $(= y_{ir})$ would be $(\frac{5}{11}, \frac{4}{11}, \frac{2}{11})$. Since $\sum_{i=1}^{3} y_{ir} = 1$, one of the $\frac{D_i}{D_T}$ values would be rounded up to 1. Note that $\frac{D_1}{D_T} = \frac{5}{11}$ is closest to 1. Hence we set $y_{11} = 1$ and $y_{21} = y_{31} = 0$. For $r = 2$, $\frac{rD_i}{D_T}$ values are $(\frac{10}{11}, \frac{8}{11}, \frac{4}{11})$. It can be verified that $y_{12} = 1$,

$y_{22} = 1$, and $y_{32} = 0$, so that $\sum_{i=1}^{3} y_{i2} = 2$. In a similar way, we can estimate values of y_{ir} for $r = 1$ to 9 as shown below. The entry z_r in the last column is defined as, $z_r = i$ if $x_{ir} = 1$.

Stage (r)	$\frac{rD_1}{D_T}$	$\frac{rD_2}{D_T}$	$\frac{rD_3}{D_T}$	y_{1r}	y_{2r}	y_{3r}	z_r
1	5/11	4/11	2/11	1	0	0	1
2	10/11	8/11	4/11	1	1	0	2
3	15/11	12/11	6/11	1	1	1	3
4	20/11	16/11	8/11	2	1	1	1
5	25/11	20/11	10/11	2	2	1	2
6	30/11	24/11	12/11	3	2	1	1
7	35/11	28/11	14/11	3	3	1	2
8	40/11	32/11	16/11	4	3	1	1
9	45/11	36/11	18/11	4	3	2	3
10	50/11	40/11	20/11	4	4	2	2
11	55/11	44/11	22/11	5	4	2	1

Note that in stage $r = 10$, the initial estimates of y_{ir} would be $(5,4,2)$, with

$\sum_{i=1}^{3} y_{i,10} = 11$. However, since $\sum_{i=1}^{3} y_{i,10}$ must equal 10, $y_{1,10}$ is reduced to 4, as $5 - \frac{50}{11} > 2 - \frac{20}{11}$. The value of z_r identifies the model produced in stage r. The sequence of products scheduled by stage would thus be 1, 2, 3, 1, 2, 1, 2, 1, 3, 2, 1. It is clear that this mixed-model schedule does not have a repeating pattern. The sequence obviously would change in different periods based on the values of D_i.

REFERENCES

Andreou, S. (1990), "A Capital Budgeting Model for Product Mix Flexibility," *Journal of Manufacturing and Operations Management*, vol. 3, pp. 5–23.

Balakrishnan, N., and A. Chakravarty (1996), "Managing Engineering Change: Market Opportunities and Manufacturing Costs," *Production and Operations Management*, vol. 5, no. 4, pp. 335–356.

Bhattacharya, S., and G. Koehler (1998), "Learning by Objectives for Adaptive Shop Floor Scheduling," *Decision Sciences*, vol. 29, no. 2, pp. 347–375.

Billington, C., H. Lee, and C. Tang (1998), "Successful Strategies for Product Rollovers," *Sloan Management Review*, spring, pp. 23–30.

Boyer, K., G. Leong, P. Ward, and L. Krajewski (1997), "Unlocking the Potential of Advanced Manufacturing Technologies," *Journal of Operations Management*, vol. 15, pp. 331–347.

Chakravarty, A. (1989), "Analysis of Rationing for a Mix of Manufacturing Facilities," *The International Journal of Flexible Manufacturing Systems*, vol. 2, pp. 43–62.

Chakravarty, A. (1994), "Assembly Capacity Mix Planning with Product Flexible Technology and Fixtures," *IIE Transactions*, vol. 26, no. 4, pp. 19–35.

Chakravarty, A. (1997), "A Model for Switching Dispatching Rules in Real Time in a Flexible Manufacturing Cell," *Production and Operations Management*, vol. 6, no. 1, pp. 398–418.

Chakravarty, A., and N. Balakrishnan (1998), "Reacting in Real Time to Production Contingencies in a Capacitated Flexible Cell," *European Journal of Operational Research*, vol. 110, pp. 1–19.

Chakravarty, A., and S. Ghose (1993), "Tracking Product-Process Interactions: A Research Paradigm," *Production and Operations Management*, vol. 2, no. 2, pp. 72–93.

Chakravarty, A., and A. Shtub (1985), "New Technology Investments in Multistage Production Systems," *Decision Sciences*, vol. 16, pp. 248–264.

Cheng, F., and S. Sethi (1999), "A Periodic Review Inventory Model with Demand Influenced by Promotion Decisions," *Management Science*, vol. 45, no. 11, pp. 1510–1523.

Clark, K. (1996), "Competing Through Manufacturing and the New Manufacturing Paradigm: Is Manufacturing Strategy Passé?" *Production and Operations Management*, vol. 5, no. 1, pp. 42–58.

Crowston, W., M. Wagner, and J. Williams (1973), "Economic Lot Size Determination in Multi-Stage Assembly Systems," *Management Science*, vol. 19, pp. 517–527.

Dapiran, P. (1992), "Benetton—Global Logistics in Action," *Asian Pacific International Journal of Business Logistics*, vol. 5, pp. 7–11.

Diebold, J. (1952), *Automation: The Advent of the Automated Factory*, Van Nostrand, New York.

Eppen, G., R. Martin, and L. Schrage (1989), "A Scenario Approach to Capacity Planning," *Operations Research*, vol. 37, no. 4, pp. 517–527.

Fine, C., and R. Freund (1990), "Optimal Investment in Product Flexible Manufacturing Capacity," *Management Science*, vol. 36, no. 4, pp. 449–466.

Garvin, D. (1991), "The Boeing 767: From Concept to Production," Harvard Business School Case 9-688-040, Boston, Mass.

Hayes, R., S. Wheelwright, and K. Clark (1988), *Dynamic Manufacturing*, The Free Press, New York.

Ho, C. (1994), "Evaluating the Impact of Frequent Engineering Changes on MRP System Performance," *International Journal of Production Research*, vol. 32, no. 7, pp. 619–641.

Jaikumar, R. (1986), "Postindustrial Manufacturing," *Harvard Business Review*, November–December, pp. 69–76.

Krajewski, L., and L. Ritzman (1992), *Operations Management: Strategy and Analysis*, 3rd ed., Addison-Wesley, Reading, Mass.

Lee, H., and C. Tang (1998), "Variability Reduction Through Operations Reversal," *Management Science*, vol. 44, no. 2, pp. 162–172.

Lefly, F., and J. Sarkis (1997), "Short-Termism and the Appraisal of AMT Capital Projects in the U.S. and U.K.," *International Journal of Production Research*, vol. 35, pp. 341–368.

Mandelbaum, M. (1978), "Flexibility in Decision Making: An Exploration and Unification," Ph.D. thesis, Dept. of Industrial Engineering, University of Toronto.

Miltenberg, J. (1989), "Level Schedules for Mixed Model Assembly Lines in Just-in-Time Production Systems," *Management Science*, vol. 35, no. 2, pp. 192–207.

Naik, B., and A. Chakravarty (1992), "Strategic Acquisition of New Manufacturing Technology: A Review and Research Framework," *International Journal of Production Research*, vol. 30, no. 7, pp. 1575–1601.

Orlicky, J. (1975), *Material Requirement Planning*, McGraw-Hill, New York.

Porter, M. (1980), *Competitive Strategy*, The Free Press, New York.

Porteus, E. (1985), "Investing in Reduced Setups in the EOQ Model," *Management Science*, vol. 31, no. 8, pp. 998–1010.

Purohit, D. (1994), "What Should You Do When Your Competitors Send in the Clones," *Marketing Science*, vol. 13, no. 4, pp. 392–411.

Quinlan, J. (1987), "Simplifying Decision Trees," *International Journal of Man-Machine Studies*, vol. 27, pp. 221–234.

Saaty, T. (1980), *The Analytical Hierarchy Process*, McGraw-Hill, New York.

Schemener, R. (1977), "Searle Medical Investment Group (A) Teaching Note," Harvard Business School Teaching Note 5-677-242, Boston, Mass.

Schonberger, R. (1982), *Japanese Manufacturing Techniques—Hidden Lessons in Simplicity*, The Free Press, New York.

Sethi, A., and S. Sethi (1990), "Flexibility in Manufacturing: A Survey," *International Journal of Flexible Manufacturing Systems*, vol. 2, pp. 289–328.

Shaw, M., S. Park, and N. Raman (1992), "Intelligent Scheduling with Machine Learning Capabilities: The Induction of Scheduling Knowledge," *IIE Transactions*, vol. 24, no. 2, pp. 156–168.

Sogomonian, A., and C. Tang (1993), "A Modeling Framework for Coordinating Promotion and Product Decisions Within a Firm," *Management Science*, vol. 39, no. 2, pp. 191–203.

Steel, D. (1975), "The Nervous MRP Systems: How to Do Battle," *Production and Inventory Management*, vol. 16, no. 4, pp. 83–89.

Sugimori, Y., K. Cho, and S. Uchikawa (1977), "Toyota Production System and

Kanban System: Materialization of Just-in-Time and Respect-for-Human System," *International Journal of Production Research,* vol. 15, pp. 533–564.

Teece, D., and G. Pisano (1994), "The Dynamic Capabilities of Firms: An Introduction," *Industrial and Corporate Change,* vol. 3, no. 3, pp. 39–63.

Wagner, H., and T. Whitin (1958), "Dynamic Version of the Economic Lot Size Model," *Management Science,* vol. 5, pp. 89–96.

Warbalickis, R. (1988), "Justification of FMS with the Analytic Hierarchy Process," *Journal of Manufacturing Systems* , vol. 7, no. 3, pp. 175–182.

Wernerfelt, B. (1984), "A Resource Based View of the Firm," *Strategic Management Journal,* vol. 5, pp. 171–180.

SUBJECT INDEX

Action flexibility, 469
Activity based costing, 15
Adaptive customization, 132
Advanced manufacturing technology, 460
Advertising budget, 456
After-sale service, 456
Aggregate design benchmark, 205
Allocation rules, 437
AMT configurations, 461, 465
Analytic hierarchy process, 461
Arbitrage, 408
Architecture of supply chain, 308
Artificial intelligence, 15, 134, 186
AS/RS, 469
As-is modeling, 382
Assemble-to-order, 477
Assemby heads, 475
Attribute bundling, 119, 121
Attribute preference, 244
Automated storage and retrieval, 461
Automated tool change, 468
Autonomous transfer mode (ATM), 342, 362
Axiom of functional independence, 149

Back-end processes, 366
Batching policy, 279
Beer distribution game, 315
Better business bureau, 393
Bidding scheme, 99
Bill of material, 490
Binary choice problem, 116
BOM structure, 375, 388
Buffer inventory, 372
Built-to-order, 91
Bull whip effect, 315, 316, 317, 318, 339
Bundle based negotiations, 394
Business models, 308, 312
Business process architecture, 382
Business process reeingeneering (BPR), 3
Business-to-business e-commerce, 433
Buy-back policies, 437

CAD, 388
CAD/CAM, 461, 468
Capacity, 408

Capacity allocation, 402
Capacity mix planning, 470
Capacity planning model, 287
Capacity utilization variance, 504
Cash-to-cash cycle, 366, 369
Centralized coordination, 24
Channel assembly, 361
Channel assembly model, 364
Choice model, 127
Choice-based market analysis, 115
Clock speed, 315, 320, 325, 333
Collaborative customization, 132
Collaborative planning, 318
Commercial chain, 3
Commonality, 208
Competetive strategy, 461, 462
Competitive advantage, 3, 27, 29, 48, 51, 241
Competitive positioning model, 104
Complementors, 241
Component interdependence, 137
Concept generation, 243
Conceptual design, 137
Concurrency, 197
Conformance quality, 9
Conjoint analysis, 121–127, 129, 142, 220
Conjunctive utility model, 120
Consignment inventory, 427
Contract manufacturers, 334
Contract structuring, 402
Contractual threshold, 437
Convex combinations, 481, 484
Convex cost function, 444
Coordination, 23, 26,195, 337
Coordination architecture, 50, 51
Coordination in a supply chain, 417
Corollary, 150
Cosmetic customization, 132
Cost linkage, 45
Cost of compromise, 6
Cost of coordination, 7
Cost of design, 162
Cost of inflexibility, 7
Cost of manufacturing, 162
Cost of ownership, 308, 314–15
Coupons, 493
Covisint, 361
Cross-functional team, 28
Culturally-correct QFD, 193
Customer commitment-process, 91
Customer economics, 240, 241

Customer interface, 188
Customer needs, 462
Customer targeting, 240
Cycle inventory, 372
Cycle time, 370

Data envelopment analysis, 15, 20, 205, 481
Data flow diagram, 18
Data-networks-systems (DNS), 340
Data warehouse, 192
DEA, 481
Decentralized coordination, 24, 402
Decentralized system, 417, 432
Decision rights, 437
Decision-making sequence, 66
De-coupled system, 149
DEFENDER model, 103
Delayed differentiation, 373
Delivery lead time, 372, 411
Delivery time, 52
Delta architecture, 387
Demand accuracy, 370
Demand chains, 303
Demand distribution, 443
Demand lead time scenarios, 412
Demand management, 11
Demand scenarios, 435
Demand variance, 422
Demand volatility, 315, 316
Design attributes, 173, 185
Design axioms, 145
Design complexity, 455
Design engineering, 136
Design evolution, 280, 281, 282
Design factory, 287, 291
Design for installation, 341
Design for manufacturing, 341
Design incompatibility, 272, 275
Design integrity, 489
Design matrix, 147
Design process, 136
Design review, 244, 287, 291, 293–95
Design rules, 145
Design sequence, 158
Design structure matrix (DSM), 158
Design variants, 142
Designing tolerance, 150
Deterministic models, 127
Development cost, 244, 254
Development cycle, 244

Development cycle time, 255
Development time, 244, 264
Digital business community, 375, 381
Digital signature, 381
Disc operating system (DOS), 321
Discounted cash flow, 460–61
Discrete choice model, 116
Disjunctive utiliy model, 120
Dispatching flexibility, 479
Distributied coordination, 48
Distribution, 303
Distribution of demand, 423
Domain bids, 56
Domain knowledge, 66
Domain transparency, 49, 50
Domain view, 3, 20, 22, 28
Downstream operation, 271, 277
Dynamic capabilities, 452
Dynamic enterprise, 27
Dynamic heuristic, 130
Dynamic replenishment, 318, 433
Dynamic work distribution, 313

e-Business, 306
Echelon holding costs, 446
Echelon stock, 372
e-Commerce, 381, 437
Economies of scale, 337, 452
EDD, 482, 483, 484
EDI, 339, 370, 386
Elasticity of demand, 491
Electronic proximity, 307, 312
Encryption, 381
Engineering analysis, 243
Engineering change, 162, 272, 278
Engineering solution, 141, 142
Enterprise resource planning (ERP), 68, 91, 339, 468
Exchange curve, 327, 331
Expected due date, 479
Extended enterprise, 320, 368
Extended QFD, 196
Extranet, 312

Fill rates, 370
Financial evaluation, 464
Financial service, 303
Financing costs, 407
First in first out, 479
Flexible manufacturing, 12
FMS, 461, 465, 468
Forcast error, 422
Forecasting, 318
Form design, 137, 145
Frame-transition, 479
Freight forwarding, 303
Front-end agreement, 380

Front-end processes, 366
Fully distributed coordination, 25
Function coordination, 138–39
Function design, 137

Generic product, 205
Greedy interchange heuristic, 128
Grippers, 475
Groupware, 370

Handoff, 305
Holders, 475
Horizontal view, 3
Hotelling's positioning model, 103–109
House of quality, 172
HTML/XML, 381
Hurdle rate, 460
Hybrid Architecture, 313, 314

ID3, 480
ID3-generated decision tree, 480
IDEF$_0$, 16, 18
If-needed characterization, 93
Incentive contract, 327
Incentive programs, 56, 60, 61
Industrial dynamics, 315
Industrial engineering, 163
Information flow, 314
In-house assembly, 362, 365
Innovation, 22, 198, 240, 245, 250, 252
Innovation linkage, 37
Integral architecture, 308, 313, 314
Integral product, 311
Integral product design, 306
Integral supply chain, 307, 308, 311, 333
Integral supply chain, 311, 437
Intelligent agent, 26, 390
Intensive communication, 273
Inter enterprise synergy, 379
Inter-domain linkages, 36, 49
Interface cards, 362
Interface set, 45
Interfunctional communication, 195
Interlinked intra-domain decision, 77
Intermediate node, 83
Intra-domain linkage, 47
Intranet, 339
Inventory balance, 455
Inventory control, 385
Inventory holding, 12, 13
Inventory replenishment, 71, 75
Inventory sharing, 433, 434
Inventory turn, 370
Investment cost, 323
Invoice routing, 385

Is-a characterization, 92
ISM, 411

JIT delivery, 375
Joint venture, 405
Just in time scheduling, 12, 14, 16, 68
Just-in-time (JIT), 4, 6, 307, 453, 499

Knowledge based outsourcing, 326
Knowledge linkage, 69
Knowledge modularization, 85
Knowledge sharing, 214
Kuhn & Tucker consition, 291

Lagrangean function, 440
Langrange multiplier, 153
Last in first out, 479
Lead time, 437
Linear compensatory value model, 118
Linkage, 23, 170
LINMAP model, 103, 109–112
Logistic distribution, 116
Logistics service, 303
Logistics management, 385
Logit model, 116, 117, 221
Longest path problem, 495
Longest-path formulation, 235

Made-to-order, 133
Mail-in-rebates, 493
Make-to-order, 477
Make-to-stock, 477
Management control, 308, 311, 315
Managing variety, 240
Manual assembly, 475
Manual inspection, 468
Manual MHS, 468
Manufacturer's gain, 420
Manufacturer's inventory accumula-
tion, 419
Manufacturer's total cost, 447
Manufacturing capabilities, 462
Manufacturing cycle, 417
Manufacturing domain, 4, 33–35, 29, 46, 66, 67
Manufacturing engineering, 162
Manufacturing flexibility, 469
Manufacturing lead time, 455, 456
Manufacturing-marketing interface, 417
Manufacturing operations, 419
Manufacturing synergy, 308, 315
Manufacturing technology, 410
Manugfacturing cost of variety, 235
Mapping, 170, 186–92, 200
Mapping process, 188
Market development, 380

Market imperfection, 408
Market knowledge, 326
Market positioning, 452
Marketing domain, 4, 33–35, 29, 46, 68
Markovian model, 292
Mass customization, 132–134
Mass production, 10
Master production scheduling (MPS), 68
Master scheduling, 162
Material flow, 314
Material requirement planning (MRP), 14, 68, 162
Materials Acquisition, 303
Materials handling, 162
Mathematical modeling of QFD, 176
Mean flow time, 482
Mental model, 66
Merge in transit, 366
Meta system, 22
Minimum purchase commitments, 437
Minimum purchase quantity contract, 437
Mix of dedicated and flexible capacity, 472
Mixed-integer optimization model, 239
Mobile agents, 393
Modular architecture, 306
Modular product, 311
Modular product design, 9, 10
Modular structure, 308
Modular supply chain, 307, 308, 311, 366, 402, 430, 436, 437, 451
Monopoly, 107
Monotonicity, 481
MRP, 402
MRP based procedures, 477
MRP tree, 477
Multiechelon inventories, 402
Multi-echelon system, 501
Multi-function coordination, 194
Multi-functional team, 186, 194, 198
Multinomial logit model, 119
Multiple-chain strategy, 460
Multi-tier pricing contract, 441

Net present value (NPV), 405
Net profit after tax, 414
Neural net, 15, 20, 186
Node linking, 74, 77
Nominal modification, 279
Nonstationary demand, 317
Noncompensatory utility model, 120

Object oriented representation, 93
Octopus' structure, 389

OEM, 305, 314, 334, 378
Okumalink, 430
Online inspection, 468
Operating profit, 414
Operational effectiveness, 241
Optimal batching interval, 319
Optimal inventory level, 424
Optimal review strategy, 293
Order batching, 318
Order engineering, 337
Order execution, 337
Order interval, 362
Order management, 16, 337, 385
Order-to-delivery cycle, 49, 362
Organizational proximity, 307
Outsourcing, 11, 322, 477
Overhead absorption, 408
Overlapping, 255, 272, 273, 281, 287
Overtime, 477
Ownership and coordination, 333

Partner interface, 381
Parts deployment, 171
Pattern recognition model, 185
Pay-back period, 461
Pentium, 343
Performance frontier, 452
Performance improvement, 251
Performance parameters, 173
Performance quality, 9
Pilot production, 244, 247, 252, 255, 256
Plant location, 410
Platform architecture, 219
Platform composition, 228
Platform design, 228
Platform dimensions, 217
Platform-based system, 133
Point of indifference, 103
Point-of-sale data, 370
Posterior distribution, 423
Postulate, 67, 69, 81, 94
Preferred region, 114
Price discount, 428
Price incentive, 420
Pricing, 437
Principal agent paradigm, 60
Probability density function, 265
Probit model, 116
Process costs, 370
Process design, 244
Process innovation, 37
Process knowledge, 326
Process modeling, 16
Process plan complexity, 455
Process planning, 171
Process technology, 38
Process transparency, 305

Process view, 15, 20, 28
Product architecture, 306, 311
Product attribute, 136
Product attribute space, 100, 114
Product breadth, 219, 220, 229
Product customization, 485
Product definition, 33, 246, 252, 255, 256
Product demand, 456
Product depth, 219
Product design, 8, 244, 380, 252
Product development, 380
Product differentiation, 452
Product economics, 240, 241, 410
Product evaluation, 245
Product family, 13
Product feature, 168–69
Product flow, 385
Product innovation, 37
Product introduction, 337
Product knowledge, 326
Product launch, 245, 255, 266
Product line design, 132
Product mix, 410
Product performance, 154, 244, 259, 262, 263
Product planning, 171, 196
Product platform, 208, 244
Product positioning, 100
Product promotions, 493
Product quality, 456
Product realization, 245
Product scheduling, 14
Product technology, 37
Product variety, 6, 8, 203, 219
Product variety linkage, 39, 55
Production capacity schedule, 456
Production cost, 323
Production management, 385
Production planning, 170
Production quantity, 457
Production quantity schedule, 456
Production supervision, 163
Product-process, 92
Product-process interactions, 455
Profit/Welfare maximization, 127–129
Profit function, 256, 406
Project coordination, 245
Prototyping, 246, 272
Purchasing, 162, 385

QFD, 469
Quality, 43, 100, 437
Quality assurance, 9
Quality control, 162
Quality function deployment, 177

Quality linkage, 43
Quantitative modeling, 402
Quantity flexibility, 437, 444
Quick response linkage, 41

Ramp-up, 245, 247, 252, 255
Rate of transshipment, 431
Rationing rule, 474
Real-time customization, 133
Realtime flexibility, 476
Relationship matrix, 186
Receiving, 162
Relational database, 92
Relationship chain construction, 82
Relationship chain management, 460
Replenishment, 318
Resource allocation, 245, 289
Resource sharing, 214
Retail operations, 418
Retailer's gain, 420
Retailers total cost, 447
Retailing, 303
Return map, 255
Return on Investment, 334
Robotic assembly, 476
Robust design, 154
ROI, 461
Rosetta business community, 384
RosettaNet, 381

Sales promotion, 456
SAP, 50, 88
Scope of variety, 201
Scoring techniques, 461
S-curve, 264
Sequencing rules, 68
Sequential chain, 318
Setup cost reduction, 499
Shipping, 162
Shipping modes, 410
Shipping quantity, 410
Shortage cost, 423
Shortest processing time, 479

Simultaneity ratio, 273
Single flexible facility, 470
Single period model, 423
Single-chain strategy, 460
Social knowledge, 23, 61
Spacial coordination, 312
Spirometer, 176
SPT, 481,482,484
Star(web) architecture, 386
State flexibility, 469
Steady state probability, 430
Strategic evaluation, 462
Subcontract, 405
Supplier assembly, 361, 363
Supplier capability development, 326
Supplier capability updating, 313
Supplier evaluation, 402
Supply chain, 3, 13
Supply chain accounting, 445
Supply chain configuration, 403
Supply chain demand, 411
Supply chain knowledge, 326
Supply chain models, 402
Supply chain platform, 410
Supply chain strategies, 340
Supply chain with uncertainities, 409
Supply contract, 436
Supply webs, 303
Support process, 92
System economics, 240, 241
System state values, 481

Tacit knowledge, 23, 61, 93, 307
Target costing, 217
Targeted advertising, 493
Tariffs, 403, 406
Task interdependence, 274
Technology driven products, 230
Technology performance, 252
Temporal coordination, 312
Testing, 244
Throughput, 14, 482
Time-to-market, 255

Total delay, 280
Total system cost, 447
Total value chain, 303
TQM, 453
Traditional (linear) supply chain, 314
Transaction web, 318
Transaction web architecture, 312
Transaction web supply chain architecture, 313
Transfer line, 475
Transparency, 93, 94, 199
Transparent customization, 132
Transportation cost, 406

Uncoupled system, 149
Upstream design, 273
Utility, 102, 103
Utility model, 116

Value based resale, 303
Value chain, 3, 304
Value-added resellers (VARs), 362, 373
Variable cost, 162
Variance of demand, 315
Variety, 201
VE, 386
Vendor managed inventory (VMI), 339, 376
Vendor support, 162
Virtual enterprise, 384
Virtual enterprise architecture, 312, 313
Volume commitment, 428

Warehousing, 303
Warranty budget, 456
"What-if" analysis, 292
Wide area networks (WAN), 362
Wind condition, 480
Windows, 322, 343
WIP inventory, 458

XML, 393

AUTHOR INDEX

Adler, P., 287, 291
Agrawal, N., 437
Ahmadi, R., 287
Alavi, M., 22
Ali, A., 264, 271
Alles, M., 23
Anand, K., 23, 24, 61
Anderson, E., 201
Anderson, J., 429, 433
Andreou, S., 475
Angiolillo, P., 4
Anupindi, R., 433, 434
Arersa, N., 68, 120
Axsater, S., 430

Baiman, S., 23
Balakrishnan, N., 20, 49–50, 186, 187, 189, 203, 228, 479
Baldwin, C., 202, 241
Baldwin, Y., 375
Banerjee, A., 23, 319, 422
Bartlett, C., 23
Barua, A., 20
Bassok, Y., 433, 434
Baum, J., 119, 236, 239
Beitz, W., 140, 141, 142, 145
Belhe, U., 184
Beltman, J., 120
Ben-Akiva, M., 116
Berger, J., 4, 422, 423, 427, 428
Bhattacharya, S., 245, 480
Billington, C., 489
Black, T., 158
Blackburn, J., 272
Blackmon, D., 14
Blattberg, R., 68
Boyer, K., 460
Bu-Huliga, M., 93
Byrne, J., 21

Cachon, G., 448
Chakravarty, A., 10, 11, 13, 15, 19, 20, 23, 41, 48, 49, 66, 77, 86, 87, 93, 203, 205, 207, 228, 236, 239, 295, 319, 372, 408, 416, 418, 419, 422, 430, 433, 438, 440, 444, 455, 461, 475, 479, 480, 481, 490, 492, 499
Chang, T., 35
Chang, Y., 41
Chard, A., 22
Chen and Associates, 89, 92
Cheng, F., 498

Chhajed, D., 239
Chikan, A., 74
Cho, H., 15
Cho, K., 499
Christensen, C., 41
Clark, K., 20, 41,162, 254, 272, 303, 375, 451
Clausing, D., 11
Clausing, D., 172, 196
Cohen, M., 262, 410
Crowston, W., 501
Cusumano, M., 272

Dada, M., 23, 319
Dapiran, P., 485
Datar, S., 23
Davenport, T., 3, 9, 14, 16, 17, 22, 68, 88
DeGroote, X., 220
Diebold, J., 470
Dincer, 408, 409
Dobson, G., 68, 129, 239

Eisenhardt, K., 272
Eliashberg, J., 107, 262
Elmarahy, H., 153
Elmarahy, W., 153
Eppen, G, 442, 476
Eppinger, S., 41, 160, 210, 276, 280, 284

Fan, M., 394
Farlow, D., 222
Feldman, S., 392
Fine, C., 158, 306, 320, 322, 325, 475
Fisher, M., 14, 235
Flaherty, T., 440
Fornell, C., 201
Forrester, J., 315
Freund, R., 475
Fujimoto, T., 272

Garvin, D., 7, 9, 43, 489
Gebala, D., 277
Gensch, D., 68, 120
Ghiaseddin, N., 74
Ghose, S., 19, 20, 48, 66, 77, 186, 187, 189, 203, 205, 207
Ghoshal, S., 23
Gilmore, J., 132
Goldberg, J., 225
Goyal, S., 13
Grahovac, J., 430, 433
Green, P., 68, 128

Greis,, N., 312, 337
Griffin, A., 191, 195
Gupta, S., 235
Gurbaxani, V., 23
Guttman, R., 392

Ha, A., 293
Hales, R., 193
Hamel, G., 11
Hauser, J., 11, 112, 171, 172, 174, 188, 191, 195
Hausman, W., 8
Hax, A., 201, 202, 239
Hayes, R., 3, 11, 240, 303, 451
Henderson, J., 406
Hendriks, H., 154
Hendrix, E., 154
Heyman, D., 317
Hill, C., 23, 28
Ho, C., 262, 492
Ho, T., 201
Hodder, J., 408, 409
Hoedemaker, G., 272
Hoffer, J., 92
Holsapple, C., 75, 77, 79
Hout, T., 41
Hsu, C., 18–19

Imai, K., 272
Ives, B., 23
Iyer, A., 422, 427, 428

Jaikumar, R., 11, 451, 460
Jain, H., 93
Jakiela, M., 162
Jarvenppa, S., 91
Jennings, N., 23, 26
Jensen, M., 23
Jones, 370
Jones, G., 23, 28

Kalish, S., 68, 129, 239
Kalwani, M., 264, 271
Kamien, M., 264
Kasarda, J., 312, 337
Kasbah, 393
Katz, M., 436
Kekre, S., 205
Koehler, G., 480
Kohli, R., 130
Konda, S., 191
Kotler, P., 35, 41, 66, 103, 243
Kovenock, D., 264, 271
Krajewski, L., 457

Krieger, A., 128
Krishnamurti, R., 130
Krishnan, V., 41, 219, 230, 231, 235, 245, 280, 284
Kumar, M., 392
Kusiak, A., 184

Lal, R., 319
Learmunth, G., 23
Lee, C., 20
Lee, H., 316, 318, 319, 410, 418, 437, 445, 485, 489
Lefly, F., 461
Lerman, R., 116
Levy, S., 191
Lillien, G., 41, 66,103, 122
Liu, J., 93
Loch, C., 41, 278
Loury, G., 264
Louviere, J., 220
Lovejoy, W., 444

Maes, P., 392
Magretta, J., 366
Mahajan, V., 245
Malone, T., 4
Mandelbaum, A., 469
Mandelson, H., 23, 24, 61
Manrai, A., 107
Martin, G., 23, 319, 372, 418, 422, 442, 476
Mason, T., 4
Matta, K., 74
McBride, R., 128
McFadden, F., 92
McGuire, T., 69
Mecking, C., 154
Meckling, W., 23
Miltenberg, J., 13, 504
Mishina, K., 12
Mohanan, J., 422
Monarch, I., 191
Monden, Y., 12
Montgomery, D., 8
Moore, S., 68, 120
Moore, W., 220
Moukas, A., 392
Murakoshi, T., 42, 133
Murthy, K., 66, 103

Nahmias, S., 437
Naik, B., 11, 15, 461
Narus, J., 429, 433
Nash, J., 23
Nazareth, D., 93
Nelson, S., 68
Nguen, V., 287
Nonaka, I., 272

Orlicky, J., 477

Padmanabhan, P., 316, 318
Page, A., 123, 220
Pahl, G., 141, 142, 145, 146
Park, S., 15, 480
Parker, G., 326, 328
Paterson, R., 429
Pearson, S., 162
Petzinger, T., 26
Pine, J., 132
Pisano, G., 9, 240, 251, 452
Porter, M., 3, 6, 36, 452
Porteus, E., 23, 60, 61, 293, 501
Prahalad, C., 11
Purohit, D., 490
Pyke, D., 429

Quandt, R., 406
Quinlan, J., 480

Raman, N., 15, 239, 480
Ramdass, K., 235
Rangaswamy, A., 122
Reich, Y., 192
Reinganum, J., 264
Riggers, B., 389
Ritzman, L., 457
Robertson, D., 211
Rockart, J., 4
Rosenbaum, H., 123, 220
Rosenblatt, M., 319, 418
Rust, R., 201

Saaty, T., 461
Sabbagh, K., 272
Sachs, E., 158
Sarkis, J., 461
Sartorius, D., 162
Scheer, A., 89
Schemener, R., 460
Schmidt, G., 222
Schonberger, R., 499, 504
Schrage, L., 442, 476
Schwartz, N., 264
Schwerer, E., 287
Selby, R., 272
Sethi, A., 470, 498
Shapiro, B., 8
Shaw, M., 15, 396, 480
Sherbrooke, C., 430
Shocker, A., 109
Shtub, A., 295, 499
Shugan, S., 112
Sikora, R., 396
Silver, E., 429
Simon, H., 406
Singh, R., 219, 230
Singhal, J., 142
Singhal, K., 142
Sinha, D., 74, 77
Skinner, R., 422
Smith, R., 160

Sobel, M., 317
Sogomonian, A., 493, 498
Srikant, K., 23, 319
Srinivasan, V., 68, 109, 205
Staelin, R., 69, 319
Stalk, G., 41
Stallert, J., 394
Steel, D., 477
Sterman, J., 315
Steward, D., 158
Subramaniam, E., 191
Sueyoshi, T., 41
Sugimori, Y., 499
Suh, N., 149, 150, 221
Sullivan, R., 41
Sycara, K., 26

Tabriz, B., 272
Takeuchi, H., 272
Tang, C., 201, 485, 489, 493, 498
Teece, D., 452
Tenenbaum, J., 375
Terwiesch, C., 41, 272, 277
Thompke, S., 249
Tirupati, D., 219, 230
Topkis, D., 20
Tsay, A., 222, 437, 444
Tuomi, I., 91
Tyndall, G., 305

Uchikawa, S., 499
Ulrich, K., 163, 211, 235

Verma, R., 220
Von Hippel, E., 138

Wagner, H., 494, 501
Wang, R., 287
Warbalickis, R., 461
Wassenhove, L., 272
Wasserman, G., 174
Weeks, A., 142
Wernerfelt, B., 452
Whang, S., 23, 60, 61, 316, 318, 437, 445
Wheelright, S., 3, 11, 20, 41, 162, 272, 303, 451
Whinston, A., 20, 75, 77, 79, 394
Whitin, T., 494
Whitney, D., 41, 280, 284, 322, 325
Wilde, D., 201, 202, 239
Williams, J., 501
Womak, 370
Wooldridge, M., 26
Wu, Z., 153
Wysk, R., 15, 35

Zangemeister, C., 145
Zeng, D., 26
Zhu, J., 225
Zipkin, P., 448
Zufryden, F., 128